Tissue Repair

Xiaoming Li
Editor

Tissue Repair

Reinforced Scaffolds

Springer

Editor
Xiaoming Li
Key Laboratory for Biomechanics and Mechanobiology of Ministry of Education, School of Biological Science and Medical Engineering
Beihang University
Beijing, China

ISBN 978-981-10-3553-1 ISBN 978-981-10-3554-8 (eBook)
DOI 10.1007/978-981-10-3554-8

Library of Congress Control Number: 2017943345

Printed on acid-free paper

This Springer imprint is published by Springer Nature
The registered company is Springer Nature Singapore Pte Ltd.
The registered company address is: 152 Beach Road, #21-01/04 Gateway East, Singapore 189721, Singapore

Contents

About the Editor

Dr. Xiaoming Li (1977) is a professor at the School of Biological Science and Medical Engineering, Beihang University, Beijing, China. Currently serving as editor in chief of the journal *Biomaterials Advances*, he has also been an invited guest editor for several special issues published in SCI journals, including *BioMed Research International*, *Journal of Nanomaterials*, *International Journal of Polymer Science*, etc. Further, he has authored more than 60 SCI articles in the fields of biomaterials, tissue engineering, and regenerative medicine.

Chapter 1
Overview of Scaffold Reinforcement for Tissue Repair

Xuenan Gu, Fan Wang, and Xiaoming Li

Abstract Nowadays, tissue loss or defect is a major health problem that directly affects the life quality and length of human being. The traditional way to solve the problem is using tissue grafts. However, the source of autografts is limited and this method ineluctably brings additional pains to the patients. Allografts usually causes immunologic rejection and the spread of diseases. Fortunately, tissue engineering scaffolds can overcome these shortages and present a promising method for tissue repair. To achieve desired repair efficacy, the scaffolds need possess appropriate physicochemical performances, among which satisfactory mechanical properties are fundamental. Therefore, to reinforce tissue engineering scaffolds has always been a hot research point. In particular, there are accumulated great achievements in the studies of reinforcing the scaffolds by fibers or tubes during last 20 years. This book will show you the basic knowledge and research progress of the scaffolds reinforced by fibers or tubes, which consists of 8 chapters, concerning comprehensive related issues, such as the mechanism of the tissue engineering scaffolds reinforced by fibers or tubes, the selection or fabrication of novel fibers or tubes to reinforce scaffolds, processing techniques to improve adhesive strength between the matrix and the fibers or tubes, how to get the homogeneous structure and composition throughout the reinforced scaffolds, effects of fibers or tubes on the biocompatibility of the scaffolds, effects of fibers or tubes on the biodegradability of the scaffolds, interactions between fiber or tube reinforced scaffolds and cell/tissue, instrumentation and technology to analyze the interactions between fibers or tubes and matrix, special design and fabrication of novel scaffolds reinforced by fibers or tubes for hard tissue repair, special design and fabrication of novel scaffolds reinforced by fibers or tubes for soft tissue repair, etc. This book will provide special knowledge of materials for the persons with biomedical background, and special biomedical knowledge for the persons with the background of materials, which will hopefully become a valuable informative read for the researchers and students of biomedical engineering major.

Keywords Scaffolds • Tissue repair • Reinforce • Fibers • Tubes

---

X. Gu • F. Wang • X. Li (✉)
Key Laboratory for Biomechanics and Mechanobiology of Ministry of Education, School of Biological Science and Medical Engineering, Beihang University, Beijing 100191, China
e-mail: x.m.li@hotmail.com

X. Li (ed.), *Tissue Repair*, DOI 10.1007/978-981-10-3554-8_1

1.1 Background of Tissue Engineering Scaffolds

Tissue loss or defect, resulting from traumatic or nontraumatic destruction and failing to heal spontaneously, has been a major health problem that directly affects the life quality and length of human being in modern society. A large fraction of the nation's health care costs are attributable to tissue loss or defect, and approximately 8 million surgical procedures are performed annually in the United States to treat these disorders. Over 10 million new cases of various tissue defects are annually reported. It is pressed for the need for the development of functional therapeutic strategies. Current treatment of organ failure or tissue loss involves tissue grafts or surgical reconstruction or mechanical devices such as kidney dialyzers [1]. These therapies have revolutionized medical practice but have disadvantages. The source of autografts is extremely limited and this method ineluctably brings additional pains to the patients. They have also been treated with the help of tissue or organ procured from the donors. But these allografts usually causes immunologic rejection and the spread of disease. And the source of appropriate donor organs is also limited. More than 70,000 patients are currently awaiting using tissue graft, but fewer than 11,000 tissue graft are available annually [2]. Furthermore, the mismatch of supply and demand has been wider and wider. In these circumstances, some additional functional devices are used to replace some functions of the injured tissues or organs. However, these implants have a limited service life. Therefore, a revolutionary strategy has long expected to achieve permanent tissue repair rather than transient tissue replacement [3–5]. Fortunately, tissue engineering has been developed as a promising method to obtain the goal [6–9].

To date, it has been shown that nearly every tissue and organ in the human body has been attempted to engineer in vitro. Work has been proceeding worldwide for many years in the soft tissue engineering and hard tissue engineering, such as cartilage [10], bone [11], hair [12], nerves [13], tendons [14], ligaments [15], fascia, skin [16], fibrous tissues, blood vessels [17] and even heart muscle [18] and valves [19]. The highest levels of success have been achieved in the areas of skin, bladder, airway and bone, where tissue-engineered constructs have been successfully used in patients [5, 20]. Generally, tissue engineering seeks to fabricate functional implants, which can induce tissue regeneration [21]. The implants are artificial synthesized materials, called scaffolds or the complexes of cells and scaffolds. A major advantage of this approach is that the desired tissues can be regenerated to precisely repair the tissue defect or loss. Therefore, tissue engineering can hopefully overcome the shortages of tissue grafts, surgical reconstructions and mechanical devices for tissue repair, and be only one approach that can meet the current huge demand quantity of tissue repair.

At present, tissue engineering has been developed into two types. One is *in vitro* tissue engineering, which is that a functionalized complex of cells and scaffold, cultured together under appropriate conditions *in vitro* that lead to desired new tissue formation, is implanted to repair tissue defect or loss [5, 22]. The other is *in vivo* tissue engineering, which is that only a high-performanced scaffold is implanted, and that the ample tissue regeneration needed cells can be successfully recruited by the functionalized scaffolds *in vivo*, which are further induced into the desired tissues.

One obvious advantage of *in vitro* tissue engineering is that the cells can be freely selected and cultured on the scaffolds according to the tissue regeneration requirements without the restriction of the complicated biological environment. And, their behaviors, such as attachment, proliferation, differentiation and biomineralization, etc., can be controlled by the techniques *in vitro*. Furthermore, the prepared complexes of cells and scaffolds can be timely tracked, observed, analyzed and studied, and it is possible to further improve their functions by additional techniques *in vitro* if necessary. Normally, three key factors should be considered to achieve the satisfactory tissue repair efficency for the *in vitro* tissue engineering: (i) the accurate selection of the appropriate cells, contributing to the formation of the desired tissue and normally harvested from the patients or donors, (ii) the successful preparation of the effective scaffolds, serving as a mimic of extracellular matrix, which can not only offer enough mechanical support for the embedded cells, but also create a three-dimensional superior growth environment for the cells to contribute to the desired tissue formation, and (iii) reasonable control of the interactions between the cells and the scaffolds to achieve functionalized cell-scaffold complexes that can direct the course of the desired tissue regeneration *in vivo* by control of various interactions with components of living systems, which finally results in tissue repair. Of course, the disadvantages of *in vitro* tissue engineering are also obvious. Because the cell-scaffold complexes are developed *in vitro*, it is still doubtable that they can fully accomplish the expected mission after implanted into human body, where they have to undergo a variety of unprecedented impacts *in vivo*, such as human normal body fluid environment, metabolism, blood circulation, biomechanical effects, interactions together with extremely abundant types of proteins and enzymes, etc. For example, it has been proved that biomechanical stimulation is an important factor for cell functions, and can significantly affect the structures and functions of new formed tissues.

In vivo tissue engineering can overcome these disadvantages of *in vitro* tissue engineering because only scaffolds are implanted. However, it has a higher requirement for the performances of scaffolds than *in vivo* tissue engineering. Above all, no matter which type of tissue engineering, scaffolds lie at their hearts and play crucial role for their success. Therefore, scaffolds have been regarded as a hot and fashionable research area in this field [23–26].

1.2 The Functions of an Ideal Scaffold

In general, appropriate tissue engineering scaffolds should be able to guide cells to grow, provide enough structural support, assist the transport of essential nutrients and waste products, and facilitate the formation of functional tissues and organs [27]. Firstly, the scaffolds should provide a satisfactory cell-adhesion substrate because the majority of mammalian cell types are anchorage-dependent and will die if no cell-adhesion substrate is available. Secondly, the scaffolds should be designed as a three-dimensional environment with appropriate structure and functions, where the tissue repair needed cells can grow well with high loading efficiency and cellular functions can be properly regulated. Thirdly, the scaffold can provide enough and stable structural support for the embedded cells to be active and for the new tissue to regenerate inside and grow into the required shape and size without the need for further treatment. The mechanical support can effectively resist against various forces *in vivo* such that the predefined three-dimensional structure of the scaffolds is maintained during tissue development. Fourthly, the scaffolds should possess special structure that could provide interconnected three-dimensional space and channels for essential nutrients to reach the cells and tissues and bring them indispensible living energy. Meanwhile, the structure can allow for the discharge of the metabolic waste products of the cells and tissues inside the scaffolds. Furthermore, based on their high performances of every aspect, the scaffolds should facilitate the formation of desired tissues. Especially for *in vivo* tissue engineering scaffolds, they can not only spontaneously recruit or attract the needed cells to attach well on them, but also act as a template to guide or support the organization of the cells, promote their proliferation, and induce them into the desired tissues directly or by adsorbing or interacting with appropriate bioactive substances in the human body (e.g. cell adhesion peptides, growth factors, etc.) [28].

1.3 The Properties Needed for an Ideal Scaffold

As mentioned above, satisfactory tissue engineering scaffolds should be able to replicate the biological functions and structural support of the native extracellular matrix. Firstly, the clinical success of the scaffolds is largely dependent on their material composition. One of the major concerns in scaffolds is the selection of appropriate raw materials. Various materials, such as organics, inorganics and metals, etc., have been tried to prepare scaffolds for many years. Actually, the selection of scaffold materials depends on the tissue type that they need to repair. For example, because calcium phosphate ceramics exhibit high degree of bioactivity, osteoconductivity and wear resistance, they have been used to prepare hard tissue engineering scaffolds for decades.

However, although scientist have tried to evaluate the specific abilities of various single type of materials in order to search a perfect environment for the regeneration of tissues, most of the results are not satisfactory. In many cases, the scaffolds made of single type of material can hardly meet fully the requirements for the desired tissue repair. Therefore, biocomposite scaffolds have been more and more developed because they can combine all the advantages of every component whilst minimizing respective disadvantages. Hence, a composite usually has better properties than each component. On the other hand, based on the bionics principle, the tissue engineering scaffolds should be prepared as biocomposites because most of natural tissues are composed of more than one component, such as bone, cartilage, skin, nerve and blood vessel, etc. Especially, since many scaffolds made of single type of material have insufficient mechanical properties, other materials with high mechanical properties are normally needed to reinforce them.

Besides the composition, the second deciding factor for the performance of a scaffold is its structure. The architecture and microstructure of the scaffolds define the ultimate formation productiveness, shape, structure and properties of the regenerated tissue and organs. An ideal scaffold should possess porous structure, which had better contain both macropores and small micropores. The macropores facilitate the penetration and ingrowth of cells, blood vessels and tissues [23]. The pore size may vary depending on the type of tissue engineering. The pore size should be normally at least 100 μm in diameter for cell survivability and successful desired tissue regeneration [29]. In the case of bone engineering, the appropriate pore sizes are in the range of 200–350 μm [30]. The porosity should be normally at least 80%. However, because the increase of the porosity unavoidably leads to the decrease of mechanical properties, the porosity should be decided according to the requirements of the regenerated tissue and the special implantation site. Furthermore, the pores should be interconnected to provide possibility for the necessary nutrient substances and oxygen to diffuse and arrive at the sites of the cells or tissues inside the scaffolds, and for the their metabolites and the scaffold biodegradation products to discharge out. Meanwhile, the micropores of smaller than 10 μm can concentrate more proteins that will improve further the bioactivity of the scaffolds and thereby promote the functions of the attached cells [31]. Most importantly, the structure of the scaffolds should maintain its frame integrality during the whole period *in vivo.* to ensure the desired activities of the cells and satisfactory performances of the tissues. Finally, the external shape of the construct must be considered, particularly if the scaffold is customized individually [32].

As for its overall performance, an ideal scaffold for tissue engineering should possess all the qualities of a native extracellular matrix (ECM) and should function in the same way as that of ECM under physiological conditions. However, there is no clear and recognized guidance yet, according to which, the so-called ideal scaffold is defined. To date, numerous scaffolds produced from many types of biomaterials have been used to regenerate different tissues and organs, including bone, cartilage, blood vessels, nerves, skin, liver, etc. On this base, an ideal tissue scaffold should meet several performances. Since many scaffold-based tissue engineering approaches are still experimental, it is not yet clear and hard to say how to accurately design an so-called ideal scaffold. Additionally, each tissue type requires specific properties of the scaffolds. However, regardless of the tissue type, scaffold design should follow a set of minimum requirements [5, 32].

(1) Biocompatibility

Biocompatibility is the basic requirement for the tissue engineering scaffolds. The most distinct difference between the tissue engineering scaffolds and other functional materials is that they must possess satisfactory biocompatibility, which can ensure that the tissue defect or loss is without causing any inflammation. At the time of scaffold material selection, its biocompatibility should be primarily considered. If the biologically incompatible materials were implanted into the human body, they would definitely bring about inflammatory or foreign-body responses that eventually lead to rejection and/or necrosis. On the other hand, the scaffold must elicit a negligible immune reaction or clinically detected foreign body reactions to prevent the rejection of the body. Moreover, each material, which the scaffolds are made of, should support the appropriate regulation of cell behaviors, such as adhesion, proliferation, migration, differentiation, biomineralization, etc., directly based on its specific physicochemical propertie or by multiple stirring interactions with the bioactive factors *in vivo* [33] so that the new desired functional tissue can form. In most cases, the tissue engineering scaffolds are made of several components to get optimized biocompatibility by utilizing the coordination between the components. For example, Li et al. [34] prepared poly-L-lactic acid (PLLA)/chitin scaffolds, in which there was a good coordination between PLLA and chitin. The alkaline degradation products of the chitin could neutralize the acidity caused by PLLA degradation, thus providing a better environment for the growth of cells and tissues than PLLA or chitin alone.

(2) Biodegradability

Biodegradation is a physiological process, during which materials are gradually excreted from the body by dissolution, enzymatic hydrolysis, cell phagocytosis, etc. The tissue engineering scaffolds are not intended as permanent constructs. They should gradually biodegrade to allow the cells to produce their own extracellular matrix and eventually achieve the ingrowth of new tissues or the formation of new organs. The new tissues grow into the implantation site at the same time of material biodegradation. To achieve this goal, the scaffold degradation profile should be designed. At the time of scaffold material selection, the consideration of the coordination of its biodegradability and other components' should be put emphasis on. Ideally, the degradation rate of the scaffold should match the new tissue formation rate so as to provide a smooth transition of the load transfer from the scaffold to the new tissue [35]. When the scaffolds are biodegraded completely, the new desired tissues fully replace the scaffolds and fill the implantation site. Additionally, biodegradation products should be removed from the body via metabolic pathways at an adequate rate that can keep the concentration of these degradation products in the tissues at a tolerable level [36]. Furthermore, the biodegradation rate of the scaffolds should be controllable to match the different tissue growth rate. For example, the degradation behavior of the scaffolds should vary based on the applications, such as 9 months or more for scaffolds in spinal fusion, 3–6 months for scaffolds in cranio-maxillofacial applications [29]. Moreover, tissue regeneration rate varies with the age and sex of the patients. To achieve controllable biodegradation rate, researchers usually develop biocomposites as tissue engineering scaffolds, which contain the components with faster biodegradation rate and the components with slower biodegradation rate. For example, Wu et al. [37] prepared porous scaffolds of zein/poly(epsilon-caprolactone) biocomposites with satisfactory porosity and well interconnected network for bone tissue engineering. Their results indicated that the degradation rate could be tailored by adjusting the amount of the zein in the scaffolds. Another sample is that Mukundan et al. prepared Nanofibrous composite scaffolds of poly(ester amides) with tunable physicochemical and degradation properties by controlling the different additive amount of polycaprolactone.

(3) Mechanical Properties

It has been unanimously recognized that enough mechanical properties lay the foundation of the successful implant surgery of tissue engineering scaffolds. Firstly, the scaffolds must provide sufficient mechanical strength and stiffness to allow surgical handling during implantation initially, and later being consistent with the host tissue. The implanted tissue engineering scaffolds must have sufficient mechanical integrity to function for the new tissues from the time of implantation to the completion of the remodeling process [28, 38]. Secondly, only if they possess proper mechanical properties, can the scaffolds keep their shape and characters after being embedded in the body. Thirdly, since the scaffolds, as artificial three-dimensional frames, must provide enough structural support for cellular adhesion, migration, proliferation, differentiation, and the desired tissue regeneration, they should possess and maintain satisfactory mechanical properties. Moreover, in most instances, mechanical properties of the tissue engineering scaffolds have been found to influence their other properties, such as biocompatibility, biodegradability, etc. It has been fully shown that lack of mechanical properties could restrict the use of tissue engineering scaffolds to great extent [39]. On the other hand, as a dynamic and hierarchically organized composite, native ECM not only provides mechanical support for embedded cells but also provides mechanical stimulations to regulate various cellular behaviors and tissue regenerations. So, ideal scaffolds should possess adequate mechanical properties from both view of practical applications and biomimetic sense.

Therefore, at the time of scaffold material selection, it is essential to make sure that this material can contribute to the satisfactory mechanical properties of the whole scaffold by itself or the appropriate interactions with other components. Furthermore, all the materials in the scaffolds should coordinate appropriately to make sure that the mechanical properties of the scaffolds can match those of the growing tissues during the gradual biodegradation process. In this case, a further challenge is that the growing rate and mechanical properties of the new tissues vary with the age and sex of the patients. Since it has been highly recognized that the scaffolds, which are made of a single material, hardly meet all the requirements of tissue engineering according to the results of an enormous amount of research in this field during last several decades, more and more studies have been focused on the development of biocomposites as the tissue engineering scaffolds. Especially, in many cases some materials with high mechanical properties are used to reinforce matrix materials so that the mechanical properties of the biocomposites can meet the requirement of the tissue engineering scaffolds. Currently, more and more attention has been paid to the study of reinforcing materials. The traditional reinforcing materials include particles and fibers or tubes.

1.4 Particle Reinforced Composites (PRCs)

Currently, particle reinforced composites have been widely used for industrial applications. Particulate systems have the dimensions that are approximately equal in all directions and thus they remain predominantly isotropic in their physical behaviors. According to the particle size, they fall into two categories: large particle reinforced and small particle reinforced composites. For the large particle reinforcement, the sizes of the particles are between 1 and 50 μm in diameter. The interactions between the particles and matrix cannot be brought about at the atomic or molecular level. The particulate phase is generally hard and stiff. The reinforcement mechanism in the large particle reinforced composites is that the matrix can transfer some of the loaded stress to the particles and the reinforcing particles tend to restrain the movement of the matrix phase in the vicinity of each of them. The degree of reinforcement depends on the bonding interactions between the particles and matrix. In the case of small particle reinforced composites, the sizes of the particles are between 10 and 100 nm in diameter.

The reinforcement mechanism in the small particle reinforced composites is that the dispersed small particles can hinder or impede the motion of dislocations and thus the plastic deformation is restricted [40]. In one situation, with the increase of the applied load, the dislocations inside the matrix are impeded by the small particles and finally form dislocation loops. In the other situation, with the increase of the applied load, the dislocations cut the small particles. In both cases, it is obvious that the particles can significantly decrease the imposed destructive energy on the matrixes, thereby increasing their mechanical properties.

The matrixes of particle reinforced composites, which are use as tissue engineering scaffolds, are usually polymers. And the reinforcing particles are normally polymers, ceramics or bioglasses. For example, 3D porous silk scaffolds have good biocompatibility and biodegradability as tissue engineered scaffolds. However, their mechanical properties are still too weak for musculoskeletal tissue repair. The study of Kim et al. [41] showed that the strength of the silk scaffolds remained weaker than that of natural bone (wet compressive modulus < 200 kPa and yield strength <40 kPa). Rajkhowa et al. [42] used crystalline silk particles to reinforce the silk scaffolds. With the addition of 200% particle reinforcement, the specific compressive modulus and the yield strength of the reinforced biocomposites increased about 40 times, compared to the matrix. The excellent reinforcement achievement was based on not only the satisfactory interfacial compatibility between the particles and the matrix, but also the high interfacial cohesion between them because of the partial solubility of the particles in the matrix. Furthermore, some drug encapsulated polymeric microparticles could not only be used to reinforce tissue engineering scaffolds but also perform as growth factor or drug carriers for their controlled delivery *in vivo* [43]. Besides polymer particles, some bioactive glass and bioceramic ones have also been used to reinforce polymer matrix to prepare tissue engineering scaffolds because of their high stiffness. Meanwhile, in many cases, based on their satisfactory biocompatibility, bioactivity, osteoconductivity and even osteoinductivity, they can also improve the biofunctions of the polymer matrixes, especially for bone repair. For example, the study of Niu et al. [44] showed that the biocomposites of PLLA and mesoporous bioglass (m-BG) particles possessed significantly higher compressive strength and compressive modulus than pure PLLA scaffolds. Moreover, it was indicated that the incorporation of the m-BG particles into the PLLA scaffolds significantly improved their water absorption, degradability and bioactivities, which was dependent on the content of the particles. Furthermore, their data demonstrated that the addition of the particles could enhance the attachment, proliferation and ALP activity of MC3T3-E1 cells on the scaffolds *in vitro*, and osteogenesis and bone repair ability of the scaffolds *in vivo*. Another study example is about the use of bioceramic particles to reinforce polymer matrixes. Lou et al. [45] prepared PLLA scaffolds reinforced by beta-tricalcium phosphate (β-TCP) particles. Their study results showed that the particles not only significantly enhanced the mechanical properties of the scaffolds, but also improve the proliferation, penetration, and ECM deposition of the osteoblasts cultured on the scaffolds.

1.5 Fiber or Tube Reinforced Composites

Fibers or tubes are another kind of reinforcing materials, which have long been used in the repair or replacement of tissues ranging from their use as sutures to more complicated applications. Although particle reinforced composites with lower cost have been widely used in many industrial fields, in recent years there have been great achievements in fiber or tube reinforced composites, especially for being used as tissue engineering scaffolds [46, 47]. Furthermore, the addition of fibers or tubes is also useful for maintaining the original dimensions of the scaffolds. Gentleman et al. [48, 49] have characterized the mechanical performances of collagen fibers created from type I collagen. It seems that the addition of the fibers in a collagen gel significantly improved maintenance of the gel dimensions [49, 50]. On the other hand, technologically, the most important composites are exactly those, in which the dispersed phase is in the form of a fiber or tube. Fiber or tube reinforcement, as a controllable variable, can be manipulated when engineering the mechanical and physical properties of biomaterials to suit the applications of repairing specific tissues because the requirements for the mechanical properties of hard and soft tissue engineering scaffolds are really different. The reinforcing fibers or tubes can take on various configurations, such as woven, knitted, braided, continuous randomly aligned, staple randomly aligned, and continuous uniaxially aligned ones, etc., which have significant effects on the morphology, spatial orientation and mechanical properties of the scaffolds. Therefore, the structures of fibers or tubes reinforced composites represent a wide range of morphological and geometric possibilities that can be tailored for each specific tissue engineering applications. Technically, the continuous fiber systems normally have a preferred orientation, which can provide the ideal reinforcing effects [51]. A distinctive characteristic features of fibers or tubes reinforced composites is that they can be designed unrestrictedly by the selection of various reinforcements, and thus their mechanical, physical or biochemical properties can be tailored in a large range to satisfy the demands of the different host tissues. In particular, the mechanical properties of the scaffolds reinforced by fibers or tubes will largely depend on many factors, such as fiber or tube length, volume fraction and aspect ratio of the fibers or tubes, the organization of fibers or tubes, fiber-fiber or tube-tube contacts, impregnation of the fibers or tubes with the matrix, the interactions between the matrix and the fibers or tubes, the polymerization of the shrinkage of the matrix, and the individual properties of the fibers or tubes and the matrix, etc. [20, 52]. To date, it has been shown that many kinds of fibers and tubes can effectively reinforce scaffolds. These reinforced scaffolds have been largely applied to repair hard and soft tissues. They include ceramic fibers/tubes, bioglass fibers/tubes, carbon fibers/tubes, polymer fibers/tubes and so on [53] the diameter of which is normally smaller than 10 μm.

The main reinforcement mechanism in the fiber or tube reinforced composites is that the dispersed fibers or tubes can hinder or impede the motion of cracks in the matrix, which are caused by the applied load. In one situation, with the increase of the applied load, the restrain of the crack motion lead to the extraction of the fibers or tubes from the matrixes. In the other situation, with the increase of the applied load, the restrain of the crack motion lead to the deflection of the crack, which will be deflected again by another fiber or tube. In both cases, it is obvious that the fibers or tubes can significantly decrease the imposed destructive energy on the matrixes, thereby increasing their mechanical properties. For example, continuous fiber reinforced polymer scaffolds can offer significantly improvement in stiffness and strength over the parent matrix.

For the polymeric matrix, short fibers are the most commonly used reinforcements to enhance the mechanical properties while minimizing the addition of the reinforcements. Polylactic acid (PLA), a non-toxic, biocompatible, and biodegradable polymer, is currently widely used to fabricate tissue engineering scaffolds. However, the mechanical properties of pure PLA scaffold are much lower than those of natural bone. Furthermore, along with the degradation of the scaffold, the overall strength decreases too fast. So, it is very difficult to repair large defect using pure PLA scaffold. On the other hand, chitin is a unique cationic polysaccharide with satisfactory biocompatibility. It has been found that chitin fibers could effectively reinforce PLA scaffold, and that the reinforced scaffold could repair larger bone defect than pure PLA scaffold [34, 54]. In another recent study reported by Slivka et al. [55], the poly(glycolic acid) (PGA) fibers reinforced poly(lactic-co-glycolic acid) (PLGA) scaffolds had increased compressive modulus from 10 to 50 MPa, by the increasing the amount of the reinforcing fibers from 0 to 20 wt %. The resulted strength of the reinforced scaffolds was approximately fivefold higher than that of the cartilage, which might be beneficial for tissues to bear high loads.

Bioceramics are another kind of matrixes, which are often reinforced by fibers or tubes. Continuous fiber reinforced ceramic scaffolds can offer significantly enhanced toughness without compromising the high stiffness and wear resistance of the parent ceramic matrix. Calcium phosphate ceramics are commonly used materials for the restoration of bone defects with excellent biocompatibility and bioactivity. However, brittleness and low flexural/tensile strength so far limit their application to non-load bearing areas. It has been shown that reinforcement of calcium phosphate cements with collagen fibers can substantially improve its toughness and has been one major strategy to overcome the present mechanical limitations of calcium phosphate cements. Fiber reinforced calcium phosphate cements thus bear the potential to facilitate the use of degradable bone substitutes in load bearing applications [39]. The calcium phosphate cements (CPC) was approved by FDA for repairing craniofacial defects. However, the CPC applications were limited to the non-stress-bearing repair because of their brittle and weak nature. The resorbable fibers have been incorporated to provide the early-strength and then to create macropores after fiber dissolution. The bending strength was improved from 2.7 to 17.7 MPa with 45 vol % polyglactin fibers incorporation in the CPC. The strengthening effect can be further increased to 40.5 MPa by the simultaneously incorporation of chitosan lactate into the matrix [56, 57]. Zhao et al. [58] reported a significant increased fatigue resistance of CPC with the addition of 15% chitosan and 20% polyglactin fibers. The maximum stress of CPC composite scaffolds was 10 MPa at 2×10^6 cycles, compared to 5 MPa of CPC control. The mean stress-to-failure was 9 MPa for the reinforced composites, while 5.8 MPa for the CPC control. Besides by themselves, fibers can also reinforce tissue engineering scaffolds by the cooperation with particles. The synergic addition of PLA fibers and tricalcium phosphate (TCP) particle reinforcements in PCL scaffolds resulted in a high porosity (80%) structure and the elastic modulus was increased from about 0.4 MPa up to about 2.2 MPa [59]. In many cases, some fibers can improve the biocompatibility of the matrix at the time of serving as reinforcing materials. Guarino et al. [60] found that their prepared PLA fiber reinforced polycaprolactone (PCL) scaffolds could support higher activities of the human osteoblasts and human marrow stromal cells than pure PCL scaffold.

Similar to fibers, tubes are another kinds of reinforcing materials. Among all of the reinforcing materials, the nanostructured ones occupy significant position. Especially, nanotubes are a type of one-dimensional macromolecules with large surface area, and normally exhibit similar strengthening mechanism as nanofiber reinforced composites. For instance, carbon nanotubes (CNTs) are graphitic sheets grown as hollow cylinders, the diameter of which are at nanometer scale, and the length of which are at micrometer scale. They exist as either rolled-up single sheet of graphene, commonly called single-walled carbon nanotubes (SWNTs), or concentric rolled-up multiple sheets of graphene, commonly called multi-walled carbon nanotube (MWNTs). CNTs have exceptionally high strength and stiffness and thus are a kind of appropriate candidates to reinforce tissue engineering scaffolds. In addition, they possess superior electrical conductivities, which may aid in directing cellular functions, since the electrical stimulation has been shown to induce the cell differentiation, proliferation and dendritic extension, etc. Therefore, there is an increasing amount of publications investigating the fabrication and properties of the scaffolds reinforced by CNTs for tissue engineering [61, 62]. When CNTs are used to fabricate tissue engineering scaffolds, the major concern may be their compatibility

Table 1.1 In vitro mechanical performance of the different Hap-CNT composites [69]

Samples	Fracture toughness (MPa $m^{1/2}$)	Change in fracture toughness (%)	Flexural strength (MPa)	% change in flexural strength	Impact strength (J/m^2)	Change in impact strength (%)
HAp	0.81	–	11.23	–	47.25	–
HAp + 1% CNTs	1.88	132.10	26.28	134.02	133.1	181.69
HAp + 2% CNTs	1.17	44.44	29.34	161.26	122.95	160.21
HAp + 1% f-CNTs	1.23	51.85	22.58	101.07	78.26	65.53
HAp + 2% f-CNTs	1.7	109.88	21.1	87.89	144.96	206.79

with cells and surrounding tissue [63, 64]. A great number of investigations have concentrated in the *in vitro* and *in vivo* biocompatibility of CNTs, which showed that CNTs had no toxicity and could support well cell adhesion, spreading and proliferation at low concentrations, and had high ability to absorb proteins [63, 65, 66], which guaranteed the satisfactory biocompatibility of CNT reinforced scaffolds. With the adsorption of the proteins and recombinant human bone morphogenetic protein-2 (rhBMP-2) in advance, MWNTs exhibited increased alkaline phosphatase (ALP)/DNA and protein/DNA and induced ectopic bone formation in the dorsal musculature of ddy mice [66]. However, most CNTs reinforced scaffolds contain up to 5 wt % CNTs because the hydrophobic CNTs tend to form agglomerates at high concentrations. Shin et al. [67] prepared the gelatin methacrylate (GelMA) hybrid reinforced by CNTs and a dose-dependent increase in mechanical properties was observed with increasing amount of the CNTs. The tensile modulus increased about 3 times and the compressive modulus increased about 2 times with the addition of 0.5 mg/ml CNTs in GelMA hybrids. The CNT reinforced scaffolds supported well the adhesion, spreading and proliferation of NIT-3 T3 cells and human mesenchymal stem cells. Mikael et al. found that the compressive strength and modulus of PLGA/MWNTs scaffolds were significantly increased (35 MPa, 510.99 MPa) compared to the pure PLGA scaffolds (19 MPa and 166.38 MPa) by adding only 3% MWNTs. Furthermore, their *in vitro* and *in vivo* study indicated that the PLGA scaffolds reinforced by MWNTs had satisfactory cellular and tissue compatibility [68]. Besides the random dispersion of CNTs into the matrixes, it may be possible to achieve high electrical conductivity and mechanical strength at their low concentrations by aligning CNTs to form a connective pathway throughout the structure and get efficient charge transfer. This kind of scaffolds may benefit the regeneration of neural tissues by promoting neurite outgrowth, orientation and branching, etc. [61]. Mukherjee et al. [69] used CNTs to reinforce hydroxyapatite (HAp) for bone regeneration. Their results showed that the mechanical properties, including fracture toughness, flexural strength and impact strength, have been greatly improved by the addition of CNTs and functionalised CNTs (f-CNTs) (Table 1.1). Moreover, it was observed that the bone defects with the CNTs reinforced scaffolds implanted contained greater extent of golden yellow zone than the defects with pure HAp scaffolds implanted (Fig. 1.1), indicating that the reinforced scaffolds promoted greater extent of new bone formation than the matrix scaffolds.

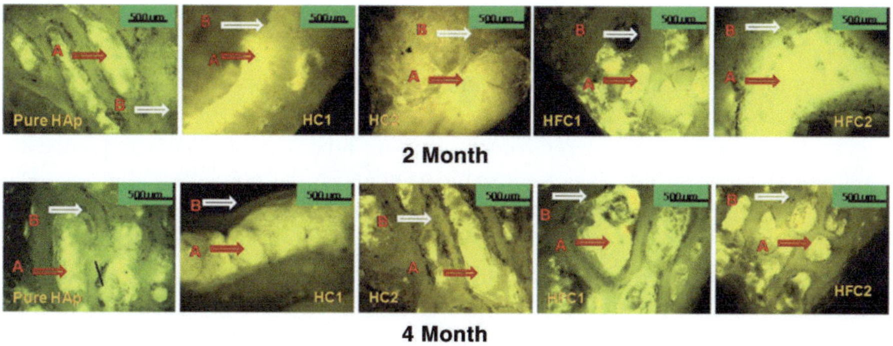

Fig. 1.1 Fluorochrome labeling images of Hap; HAp + 1% CNTs (HC1); HAp + 2% CNTs (HC2); HAp + 1% f-CNTs (HFC1); HAp + 2% f-CNTs (HFC2) at 2 and 4 month post-operatively showing the *golden yellow* new bone (NB) and the deep sea *green* old bone (OB) zones (Adapted with permission from Ref. [69]. Copyright 2016 Elsevier Ltd)

Titania (TiO_2) nanotubes are another kind of satisfactory candidates to reinforce tissue engineering scaffolds. TiO_2 is known as one kind of suitable biomaterials based on its excellent biocompatibility, especially for orthopedic applications, and has been widely used as coating for bone implants. Byrne et al. prepared polystyrene composites reinforced by TiO_2 nanotubes, the Young's modulus and tensile strength of which increased 18% and 30% respectively relative to the pristine polymer at extremely low nanotube loading levels (between 0 and 1 wt%) [70]. Another study shows that TiO_2 nanotubes could reinforce acrylic cement resin [71]. A small amount of nanotube content (1 wt%) was shown to be able to provide a significant increase in fracture toughness (73%), flexural strength (42%) and flexural modulus (56%). One of the main reasons why the titania nanotubes can reinforce so significantly is that they can achieve satisfactory interactions with the matrix. Moreover, in many cases, TiO_2 nanotubes can improve the biocompatibility and bioactivity of the matrix besides serving as reinforcing materials. To date, a great number of studies have demonstrated the TiO_2 nanotubes exhibit positive effects biocompatibility and bioactivity, such as accelerated apatite formation, enhanced osteoblast adhesion, proliferation and differentiation, etc. [72, 73]. The *in vivo* studies demonstrated that the TiO2 nanotubular surfaces can significantly improve the bone bonding strength [74].

1.6 The Structure and Main Content of This Book

Above all, it has been well recognized that the use of fibers or tubes to reinforce desirable tissue engineering scaffolds is a very efficient means to further improve their mechanical properties so as to meet different needs of clinical applications. Since the fiber or tube reinforced scaffolds have been playing more and more important roles and gained substantial interest in the field of tissue engineering and regenerative medicine, we would like to launch this book. As for this kind of reinforced scaffolds, it has been well recognized that several important issues should be paid high attention to.

Firstly, the addition of the reinforcing fibers or tubes can obviously influence the structure of the scaffolds. It is impossible to enhance the porosity and mechanical properties of the scaffolds simultaneously. The addition of fibers or tubes normally decreases the porosity [26]. However, the satisfactory mechanical properties, porosity and interconnected porous structures are all required to provide necessary support and space for the desired cell growthin and new tissue regeneration [75]. Therefore, how to balance the two contradictory aspects needs to be concretely and correctly considered according to the practical application requirements.

Secondly, since the reinforced biomaterials are composed of two or more materials or phases that are different in composition, structure, and properties, etc., defining a continuous matrix phase and at least one reinforcing phase, the reinforcing fibers or tubes need to be homogeneously dispersed in the matrix so as to achieve satisfactory structures and properties. Therefore, how to utilize techniques and methods to achieve homogeneous composition and structure throughout the fiber or tube reinforced scaffolds as much as possible is another crucial point to consider comprehensively.

Thirdly, since the mechanical properties of the scaffolds reinforced by fibers or tubes are largely dependent on the bonding strength between the reinforcements and matrixes, it is a constant research hotspot to analyze the interactions between fibres or tubes and matrix with advanced instruments and technology, and enhance or improve further the interaction force between them by selecting or fabricating appropriate fibers or tubes and utilizing effective processing techniques. For example, Li et al. [24] crosslinked chitin fibers with nano-hydroxyapatite/collagen/PLLA (nHACP) by dicyclohexylcarbodimide (DCC) to enhance the bonding strength between the fibers and matrix (Fig. 1.2), thereby improving the reinforcing efficiency of the fibers. Their results showed that the chitin fibers with the crosslink treatment could enhance the compressive strength of the scaffold about three times while those without the crosslink treatment could only enhance the compressive strength of the scaffold about once. Moreover, it was shown that the crosslink was effective to improve the structural homogeneity of the scaffolds. In this case, on the other hand, the optimization of the matrix has been paid close attention to because fibers or tubes can only play its due role on the base of the matrix.

Fourthly, besides having the above physical and structure effects, the addition of the reinforcing materials actually influences the biocompatibility of the scaffolds. It has been showed that the mechanical properties of the scaffolds actually affect the cellular functions. For example, it has been shown that the substrate stiffnesss significantly promoted the osteogenic differentiation of the cultured rat bone marrow and adipose tissue derived mesenchymal stem cells (Fig. 1.3) [76]. Moreover, it is obvious that as components in the reinforced scaffolds, the fibers or tubes themselves also have important effects on the biocompatibility and bioactivity of the scaffold. Therefore, during the design and fabrication of the scaffolds reinforced by

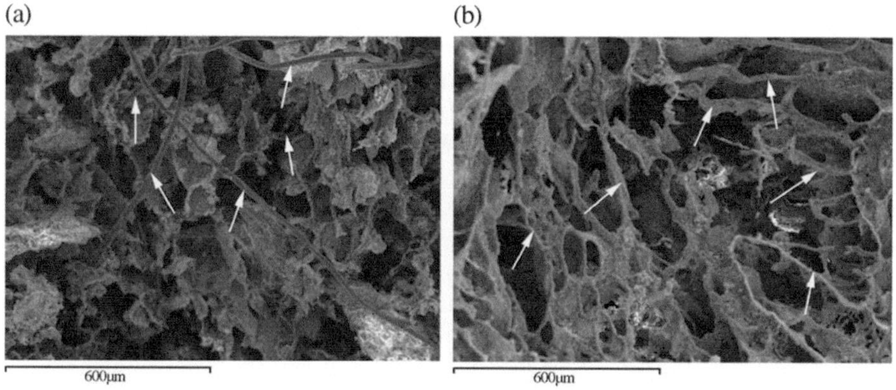

Fig. 1.2 Scanning electron microscopy of the fiber reinforced scaffolds without (**a**) and with (**b**) the crosslink treatment. The *white arrow* points to the fibers, indicating that the crosslink treatment has made the fibers bind to the matrix more tightly (Adapted with permission from Ref. [26]. Copyright 2005 Elsevier Ltd)

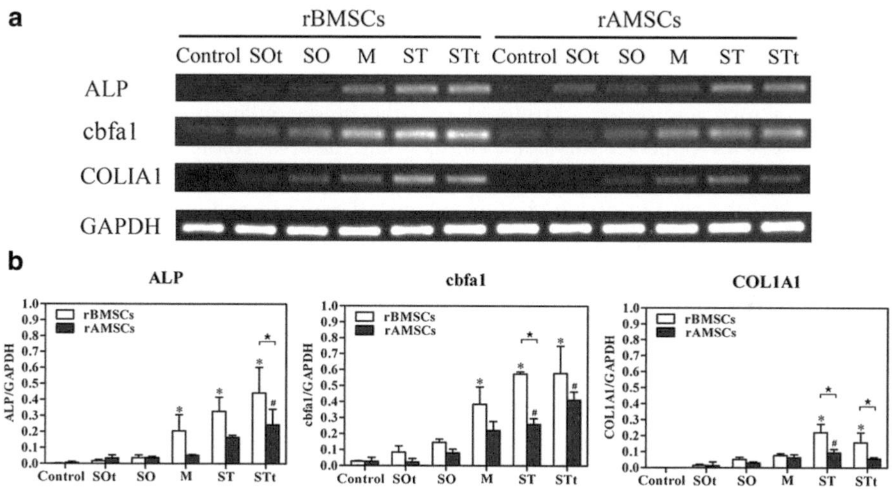

Fig. 1.3 Results of the osteoblast-related gene expressions of rBMSCs and rAMSCs cultured on the different substrates for 7 days. Rat bone marrow derived mesenchymal stem cells (rBMSCs) and rat adipose tissue derived mesenchymal stem cells (rAMSCs) were divided, respectively, into six groups: the control, SOt (*softest*), SO (*soft*), M (*middle*), ST (*stiff*), and STt (*stiffest*) group. (**a**) Representative pictures of RT-PCR product bands. (**b**) Image analysis of (**a**). The expressions of osteoblast-related genes was detected in the M, ST and STt groups, especially in the STt group, which suggest the substrate stiffness was helpful for the osteogenic differentiation of rBMSCs and rAMSCs. rBMSCs expressed more osteoblast-related markers than rAMSCs cultured on the substrates with the same stiffness, which suggest that to start the osteogenic differentiation of rBMSCs needs the substrates to have higher stiffness than to start that of rAMSCs. Results are shown as the mean ± SD values (n=3). * # $p < 0.05$, compared with the control group; ★$p < 0.05$, rBMSCs group compared with rAMSCs (Adapted with permission from Ref. [76] Copyright 2013 Wiley Periodicals, Inc.)

fibers or tubes, the efforts to maintain or even improve further the biocompatibility and bioactivity of the scaffold, meanwhile enhancing their mechanical properties, are also indispensable. And to investigate into the interactions between fibre or tube reinforced scaffolds and cell or tissues is necessary. Mryam et al. studied the biocompatibility and bioactivity of fiber reinforced hydrogel scaffolds for heart valve tissue engineering. Part of the results showed that the cells initially assumed round shape and then began to spread over time. By day 21, the cells had spread fully in the fiber reinforced scaffold The cell number on the fiber reinforced scaffold was significantly more than that on the matrix. Moreover, the cells on the reinforced scaffold also spread and distribute better than those on the matrix [77]. The *in vitro* study of Li et al. showed that their prepared collagen based scaffold reinforced by chitin fiber could support better the attachment and proliferation of human bone marrow mesenchymal stem cells better than the matrix (Fig. 1.4), the main reason of which might be that the alkalescent degradation product of the chitin fibers neutralized the acid degradation product of PLLA in the matrix, thereby creating a more advantageous microenvironment for the cells to grow [23, 78, 79]. And their further *in vivo* study showed that the fiber reinforced scaffolds could achieve a significantly larger bone defect repair than the matrix scaffolds [80].

Fifthly, it is another agreed view that the behaviors of the fibers or tubes during the biodegradation of the whole reinforced scaffolds are also crucial for their final functions. To maintain or reduce the decrease rate of the mechanical properties of the fiber or tube reinforced scaffolds during the biodegradation has been another important research point in this field. For example, Li et al. studied the degradation *in vitro* of the porous nano-hydroxyapatite/collagen/PLLA scaffold reinforced by chitin fibers, the results of which showed that the reinforced scaffolds with the crosslink between the fibers and matrix could keep better compressive strength during degradation than those without the crosslink treatment (Fig. 1.5), the main reason of which is that the formed chemical bond by the crosslink can regenerated again after being destroyed during the degradation. Furthermore, along with the degradation, PLLA in the matrix is degraded into smaller moleculars, thereby providing more functional groups to restructure the chemical bonds with the fibers [26].

Finally, since requirements for soft tissue engineering scaffolds and those for hard tissue engineering scaffolds are certainly different, special design and fabrication of the novel fiber or tube reinforced scaffolds for each kind of tissue engineering have been always paid close attention to.

Therefore, this book include other 7 chapters besides the first one as follows, in which the above important issues are included:

Chapter 2: The potential matrixes and reinforcing materials for the preparation of the scaffolds reinforced by fibers or tubes for tissue repair will be introduced and discussed in this chapter. Also the general design and preparation methods of this kind of reinforced scaffolds will be included.

Chapter 3: The mechanical properties of the scaffolds reinforced by fibers or tubes for tissue repair will be discussed in this chapter, mainly to elaborate the reinforcement mechanisms of the fibers or tubes in the scaffold, present the efforts to get

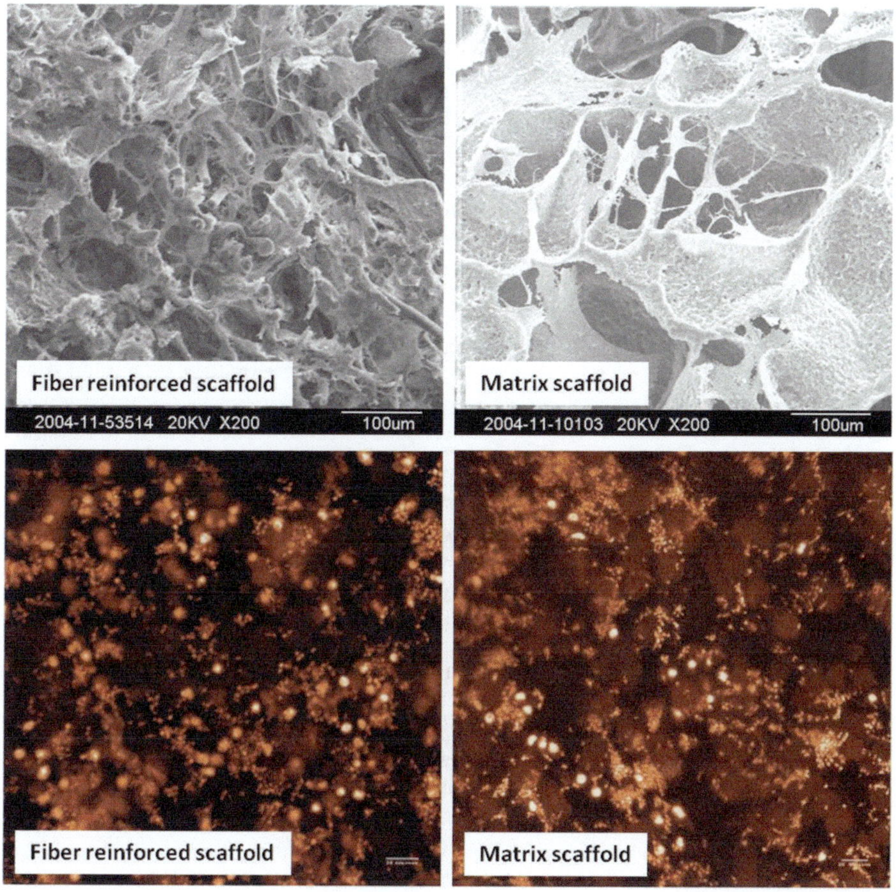

Fig. 1.4 Scanning electron microscopy (*upper two*) and laser scanning confocal microscopy (*lower two*) of human bone marrow-derived mesenchymal stem cells cultured on the scaffolds, showing that the cells attached and proliferated better on the fiber reinforced scaffold than on the matrix scaffold (Adapted with permission from Ref. [23]. Copyright 2005 Wiley Periodicals, Inc.)

the homogeneous structure and composition throughout the reinforced scaffolds, and introduce the processing techniques to improve adhesive strength between the matrix and the fibers or tubes, etc.

Chapter 4: The biodegradability of the scaffolds reinforced by fibers or tubes for tissue repair will be discussed in this chapter, mainly to expatiate how the addition of fibers or tubes affects the degradation rate, structure and mechanical property change of the scaffolds during the degradation *in vitro* and *in vivo*, and how the interactions between the matrix and the fibers or tubes affect the biodegradability of the scaffolds.

Chapter 5: The biocompatibility of the scaffolds reinforced by fibers or tubes for tissue repair will be discussed in this chapter, mainly to set forth how the addition

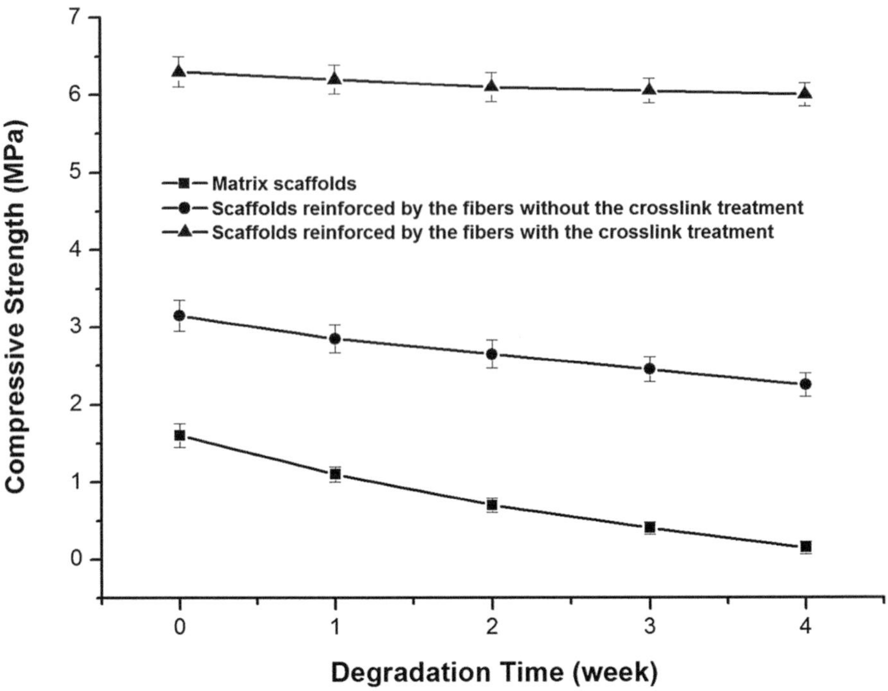

Fig. 1.5 Compressive strength of the scaffolds during the degradation *in vitro*, showing that the fiber reinforced scaffolds with crosslink treatment could keep better the mechanical property than those without crosslink treatment and matrix scaffolds (Adapted with permission from Ref. [26]. Copyright 2005 Elsevier Ltd)

of fibers or tubes affects the interactions between the scaffolds with biomacromolecules, cells and tissues *in vitro* and *in vivo*, and how the interactions between the matrix and the fibers or tubes affect those performances of the scaffolds.

Chapter 6: The potential tissues and their properties will be introduced in this chapter, mainly to discuss about all the potential tissues and their structures and properties, which need to be repaired with the scaffolds reinforced by fibers or tubes.

Chapter 7: Scaffolds reinforced by fibers or tubes for hard tissue repair will be introduced and discussed in this chapter, mainly to state special design and fabrication of novel scaffolds reinforced by fibers or tubes for hard tissue repair, and how the addition of fibers or tubes affects the special functions *in vitro* and *in vivo* to repair hard tissues.

Chapter 8: Scaffolds reinforced by fibers or tubes for soft tissue repair will be introduced and discussed in this chapter, mainly to state special design and fabrication of novel scaffolds reinforcing by fibers or tubes for soft tissue repair, and how the addition of fibers or tubes affects the special functions *in vitro* and *in vivo* to repair soft tissues.

All in all, this book presents a review of the current understanding of the scaffolds reinforced by fibers or tubes for tissue repair, mainly including the selection of matrix and reinforcing materials, the reinforcement mechanism, properties affected by the addition of fibers or tubes, special design and fabrication of this kind of reinforced scaffolds for special tissue engineering application, etc., which provides special knowledge of materials for the persons with biomedical background, and special biomedical knowledge for the persons with the background of materials, which will hopefully become a valuable informative read for the researchers and students of biomedical engineering major. It gives a comprehensive and accurate summary of the recent related research progress besides basic knowledge and provides a valuable reference for the future development of this field. Moreover, it may provide a general guide or the design and fabrication of biomaterials and regenerative medicine.

Acknowledgements The authors acknowledge the financial supports from the National Natural Science Foundation of China (No. 31370959), Beijing Nova Programme Interdisciplinary Cooperation Project (No. xxjc201616), Fok Ying Tung Education Foundation (No. 141039), Key Laboratory of Advanced Materials of Ministry of Education of China (Tsinghua University), International Joint Research Center of Aerospace Biotechnology and Medical Engineering, Ministry of Science and Technology of China, and the 111 Project (No. B13003).

References

1. Vacanti JP, Langer R. Tissue engineering: the design and fabrication of living replacement devices for surgical reconstruction and transplantation. Lancet. 1999;354(Suppl 1):S32–4.
2. Guo TJ, Daily OP, Davies DB. The UNOS OPTN waiting list: 1988 to 1994. United Network for Organ Sharing. Organ Procurement and Transplantation Network. Clin Transpl. 1994, 1994, 86:69.
3. Jones JR. Scaffolds for tissue engineering. In: Hench LL, Jones JR, editors. Biomaterials, artificial organs and tissue engineering. Cambridge: Woodhead Publishing Limited; 2005.
4. Hollister SJ. Porous scaffold design for tissue engineering. Nat Mater. 2005;4(7):518–24.
5. O'Brien FJ. Biomaterials & scaffolds for tissue engineering. Mater Today. 2011;14(3):88–95.
6. Bettinger CJ, Kulig KM, Vacanti JP, Langer R, Borenstein JT. Nanofabricated collagen-inspired synthetic elastomers for primary rat hepatocyte culture. Tissue Eng A. 2009;15(6):1321–9.
7. Fuchs JR, Nasseri BA, Vacanti JP. Tissue engineering: a 21st century solution to surgical reconstruction. Ann Thorac Surg. 2001;72(2):577–91.
8. Niklason LE, Langer R. Prospects for organ and tissue replacement. JAMA. 2001;285(5):573–6.
9. Stock UA, Vacanti JP. Tissue engineering: current state and prospects. Annu Rev Med. 2001;52(1):443–51.
10. Iwasa J, Engebretsen L, Shima Y, Ochi M. Clinical application of scaffolds for cartilage tissue engineering. Knee Surg Sports Traumatol Arthrosc. 2009;17(6):561–77.
11. Chen G, Lv Y. Immobilization and application of electrospun nanofiber Scaffold-based growth factor in bone tissue engineering. Curr Pharm Des. 2015;21(15):1967–78.
12. Itoh M, Hiraoka Y, Kataoka K, Huh NH, Tabata Y, Okochi H. Novel collagen sponge reinforced with polyglycolic acid fiber produces robust, normal hair in murine hair reconstitution model. Tissue Eng. 2004;10(6):818–24.

13. Tang ZP, Chen XY. Current application of scaffold materials for nerve tissue engineering. J Clin Rehabil Tissue Eng Res. 2008;12(1):189–92.
14. Funakoshi T, Majima T, Iwasaki N, Suenaga N, Sawaguchi N, Shimode K, Minami A, Harada K, Nishimura S. Application of tissue engineering techniques for rotator cuff regeneration using a chitosan-based hyaluronan hybrid fiber scaffold. Am J Sports Med. 2005;33(8):1193–201.
15. Funakoshi T, Majima T, Iwasaki N, Yamane S, Masuko T, Minami A, Harada K, Tamura H, Tokura S, Nishimura SI. Novel chitosan-based hyaluronan hybrid polymer fibers as a scaffold in ligament tissue engineering. J Biomed Mater Res A. 2005;74(3):338–46.
16. Wu RD. Application of synthetic and natural scaffold materials in skin tissue engineering. J Clin Rehabil Tissue Eng Res. 2009;13:2317–20.
17. Scherner M, Reutter S, Klemm D, Sterner-Kock A, Guschlbauer M, Richter T, Langebartels G, Madershahian N, Wahlers T, Wippermann J. In vivo application of tissue-engineered blood vessels of bacterial cellulose as small arterial substitutes: proof of concept? J Surg Res. 2014;189(2):340–7.
18. Blan NR, Birla RK. Design and fabrication of heart muscle using scaffold-based tissue engineering. J Biomed Mater Res A. 2008;86(86):195–208.
19. Fallahiarezoudar E, Ahmadipourroudposht M, Idris A, Yusof NM. A review of: application of synthetic scaffold in tissue engineering heart valves. Mater Sci Eng C. 2015;48:556–65.
20. Li X, Yang Y, Fan Y, Feng Q, Cui FZ, Watari F. Biocomposites reinforced by fibers or tubes as scaffolds for tissue engineering or regenerative medicine. J Biomed Mater Res A. 2014;102(5):1580–94.
21. Bell E. Tissue engineering in perspective. In: Principles of tissue engineering. 2nd ed. San Diego: Academic Press; 2000. p. xxxv–xxli.
22. Murugan R, Ramakrishna S. Design strategies of tissue engineering scaffolds with controlled fiber orientation. Tissue Eng. 2007;13(8):1845–66.
23. Li X, Feng Q, Wang W, Cui F. Chemical characteristics and cytocompatibility of collagen-based scaffold reinforced by chitin fibers for bone tissue engineering. J Biomed Mater Res B Appl Biomater. 2006;77B(2):219–26.
24. Li X, Feng Q, Jiao Y, Cui F. Collagen-based scaffolds reinforced by chitosan fibres for bone tissue engineering. Polym Int. 2005;54(54):1034–40.
25. Li X, Wang L, Fan Y, Feng Q, Cui FZ, Watari F. Nanostructured scaffolds for bone tissue engineering. J Biomed Mater Res A. 2013;101A(8):2424–35.
26. Li X, Feng Q, Cui F. In vitro degradation of porous nano-hydroxyapatite/collagen/PLLA scaffold reinforced by chitin fibres. Mater Sci Eng C. 2006;26(4):716–20.
27. Sitharaman B, Shi X, Walboomers XF, Liao H, Cuijpers V, Wilson LJ, Mikos AG, Jansen JA. In vivo biocompatibility of ultra-short single-walled carbon nanotube/biodegradable polymer nanocomposites for bone tissue engineering. Bone. 2008;43(2):362–70.
28. Kim BS, Mooney DJ. Development of biocompatible synthetic extracellular matrices for tissue engineering. Trends Biotechnol. 1998;16(5):224–30.
29. Rouwkema J, Rivron NC, Blitterswijk CAV. Vascularization in tissue engineering. Trends Biotechnol. 2008;26(8):434–41.
30. Murphy CM, Haugh MG, O'Brien FJ. The effect of mean pore size on cell attachment, proliferation and migration in collagen-glycosaminoglycan scaffolds for bone tissue engineering. Biomaterials. 2010;31(3):461–6.
31. Li X, Blitterswijk CAV, Feng Q, Cui F, Watari F. The effect of calcium phosphate microstructure on bone-related cells in vitro. Biomaterials. 2008;29(23):3306–16.
32. Hutmacher DW, Sittinger M, Risbud MV. Scaffold-based tissue engineering: rationale for computer-aided design and solid free-form fabrication systems. Trends Biotechnol. 2004;22(7):354–62.
33. Pegg DE. Principles of tissue engineering. London: Academic Press; 2013.
34. Li X, Feng Q. Porous poly-L-lactic acid scaffold reinforced by chitin fibers. Polym Bull. 2005;54(1–2):47–55.

35. Cheung HY, Lau KT, Lu TP, Hui D. A critical review on polymer-based bio-engineered materials for scaffold development. Compos Part B. 2007;38(3):291–300.
36. Bergsma JE, Rozema FR, Bos RRM, Boering G, Bruijn WCD, Pennings AJ. In vivo degradation and biocompatibility study of in vitro pre-degraded as-polymerized polylactide particles. Biomaterials. 1995;16(4):267–74.
37. Wu F, Wei J, Liu C, O'Neill B, Ngothai Y. Fabrication and properties of porous scaffold of zein/PCL biocomposite for bone tissue engineering. Compos Part B Eng. 2012;43(5):2192–7.
38. Hutmacher DW. Scaffolds in tissue engineering bone and cartilage. Biomaterials. 2000;21(24):2529–43.
39. Li X, Dunne N, Li X, Aifantis KE. Scaffolds reinforced by fibers or tubes for tissue repair. Biomed Res Int. 2014;2014(2014):309–22.
40. Callister W. Materials science and engineering: an introduction. Mater Des. 1991;12(1):59.
41. Kim HJ, Kim UJ, Leisk GG, Bayan C, Georgakoudi I, Kaplan DL. Bone regeneration on macroporous aqueous-derived silk 3-D Scaffolds. Macromol Biosci. 2007;7(5):643–55.
42. Rajkhowa R, Gil ES, Kluge J, Numata K, Wang L, Wang X, Kaplan DL. Reinforcing silk Scaffolds with silk particles. Macromol Biosci. 2010;10(6):599–611.
43. Wenk E, Meinel AJ, Wildy S, Merkle HP, Meinel L. Microporous silk fibroin scaffolds embedding PLGA microparticles for controlled growth factor delivery in tissue engineering. Biomaterials. 2009;30(13):2571–81.
44. Niu Y, Guo L, Liu J, Shen H, Su J, An X, Yu B, Wei J, Shin JW, Guo H. Bioactive and degradable scaffolds of the mesoporous bioglass and poly(L-lactide) composite for bone tissue regeneration. J Mater Chem B. 2015;3(15):2962–70.
45. Lou T, Wang X, Song G, Gu Z, Yang Z. Fabrication of PLLA/β-TCP nanocomposite scaffolds with hierarchical porosity for bone tissue engineering. Int J Biol Macromol. 2014;69:464–70.
46. Moutos FT, Freed LE, Guilak F. A biomimetic three-dimensional woven composite scaffold for functional tissue engineering of cartilage. Nat Mater. 2007;6(2):162–7.
47. Moutos FT, Guilak F. Composite scaffolds for cartilage tissue engineering. Biorheology. 2008;45(3–4):501–12.
48. Gentleman E, Lay AN, Dickerson DA, Nauman EA, Livesay GA, Dee KC. Mechanical characterization of collagen fibers and scaffolds for tissue engineering. Biomaterials. 2003;24(21):3805–13.
49. Gentleman E, Nauman EA, Dee KC, Livesay GA. Short collagen fibers provide control of contraction and permeability in fibroblast-seeded collagen gels. Tissue Eng. 2004;10(3–4):421–7.
50. Gentleman E, Nauman EA, Livesay GA, Dee KC. Collagen composite biomaterials resist contraction while allowing development of Adipocytic soft tissue in vitro. Tissue Eng. 2006;12(6):1639–49.
51. Campbell FC. Structural composite materials. In: Abstracts of IUMRS 11th International Conference in Asia (IUMRS-ICA2010); 2010.
52. Mccullen SD, Haslauer CM, Loboa EG. Fiber-reinforced scaffolds for tissue engineering and regenerative medicine: use of traditional textile substrates to nanofibrous arrays. J Mater Chem. 2010;20(40):8776–88.
53. Li X, Cui R, Liu W, Sun L, Yu B, Fan Y, Feng Q, Cui F, Watari F. The use of nanoscaled fibers or tubes to improve biocompatibility and bioactivity of biomedical materials. J Nanomater. 2013;2013(23):4361–9.
54. Jiao Y, Liu Z, Zhou C. Fabrication and characterization of PLLA–chitosan hybrid scaffolds with improved cell compatibility. J Biomed Mater Res A. 2007;80(4):820–5.
55. Slivka MA, Leatherbury NC, Kieswetter K, Niederauer GG. Porous, resorbable, fiber-reinforced scaffolds tailored for articular cartilage repair. Tissue Eng. 2001;7(6):767–80.
56. Krüger R, Groll J. Fiber reinforced calcium phosphate cements – on the way to degradable load bearing bone substitutes? Biomaterials. 2012;33(25):5887–900.
57. Xu HHK, Burguera EF, Carey LE. Strong, macroporous, and in situ -setting calcium phosphate cement-layered structures. Biomaterials. 2007;28(26):3786–96.

58. Zhao L, Burguera EF, Xu HHK, Amin N, Ryou H, Arola DD. Fatigue and human umbilical cord stem cell seeding characteristics of calcium phosphate–chitosan–biodegradable fiber scaffolds. Biomaterials. 2010;31(5):840–7.
59. Guarino V, Ambrosio L. The synergic effect of polylactide fiber and calcium phosphate particle reinforcement in poly ε -caprolactone-based composite scaffolds. Acta Biomater. 2008;4(6):1778–87.
60. Guarino V, Causa F, Taddei P, Foggia MD, Ciapetti G, Martini D, Fagnano C, Baldini N, Ambrosio L. Polylactic acid fibre-reinforced polycaprolactone scaffolds for bone tissue engineering. Biomaterials. 2008;29(27):3662–70.
61. Edwards SL, Werkmeister JA, Ramshaw JA. Carbon nanotubes in scaffolds for tissue engineering. Expert Rev Med Devices. 2014;6(5):499–505.
62. Harrison BS, Atala A. Carbon nanotube applications for tissue engineering. Biomaterials. 2007;28(2):344–53.
63. Jia G, Wang H, Yan L, Wang X, Pei R, Yan T, Zhao Y, Guo X. Cytotoxicity of carbon nanomaterials: single-wall nanotube, multi-wall nanotube, and fullerene. Environ Sci Technol. 2005;39(5):1378–83.
64. Tian F, Cui D, Schwarz H, Estrada GG, Kobayashi H. Cytotoxicity of single-wall carbon nanotubes on human fibroblasts. Toxicol in Vitro. 2006;20(7):1202–12.
65. Walker VG, Li Z, Hulderman T, Schwegler-Berry D, Kashon ML, Simeonova PP. Potential in vitro effects of carbon nanotubes on human aortic endothelial cells. Toxicol Appl Pharmacol. 2009;236(3):319–28.
66. Li X, Liu H, Niu X, Yu B, Fan Y, Feng Q, Cui FZ, Watari F. The use of carbon nanotubes to induce osteogenic differentiation of human adipose-derived MSCs invitro and ectopic bone formation in vivo. Biomaterials. 2012;33(19):4818–27.
67. Shin SR. Carbon nanotube reinforced hybrid microgels as scaffold materials for cell encapsulation. ACS Nano. 2012;6(1):362–72.
68. Mikael PE, Amini AR, Basu J, Josefina A-JM, Laurencin CT, Sanders MM, Barry CC, Nukavarapu SP. Functionalized carbon nanotube reinforced scaffolds for bone regenerative engineering: fabrication, in vitro and in vivo evaluation. Biomed Mater. 2014;9(9):306.
69. Mukherjee S, Nandi SK, Kundu B, Chanda A, Sen S, Das PK. Enhanced bone regeneration with carbon nanotube reinforced hydroxyapatite in animal model. J Mech Behav Biomed Mater. 2016;60:243–55.
70. Byrne MT, Mccarthy JE, Bent M, Blake R, Gun'Ko YK, Horvath E, Konya Z, Kukovecz A, Kiricsi I, Coleman JN. Chemical functionalisation of titania nanotubes and their utilisation for the fabrication of reinforced polystyrene composites. J Mater Chem. 2007;17(22):2351–8.
71. Khaled SM, Miron RJ, Hamilton DW, Charpentier PA, Rizkalla AS. Reinforcement of resin based cement with titania nanotubes. Dent Mater. 2010;26(2):169–78.
72. Minagar S, Li Y, Berndt CC, Wen C. The influence of titania–zirconia–zirconium titanate nanotube characteristics on osteoblast cell adhesion. Acta Biomater. 2014;12:281–9.
73. Tan AW, Pingguan-Murphy B, Ahmad R, Akbar SA. Review of titania nanotubes: fabrication and cellular response. Ceram Int. 2012;38(6):4421–35.
74. Bjursten LM, Rasmusson L, Oh S, Smith GC, Brammer KS, Jin S. Titanium dioxide nanotubes enhance bone bonding in vivo. J Biomed Mater Res A. 2009;92(3):1218–24.
75. Reis RL, Román JS. Biodegradable systems in tissue engineering and regenerative medicine. Boca Raton: CRC Press; 2005.
76. Li X, Huang Y, Zheng L, Liu H, Niu X, Huang J, Zhao F, Fan Y. Effect of substrate stiffness on the functions of rat bone marrow and adipose tissue derived mesenchymal stem cells in vitro. J Biomed Mater Res A. 2014;102(4):1092–101.
77. Eslami M, Vrana NE, Zorlutuna P, Sant S, Jung S, Masoumi N, Khavari-Nejad RA, Javadi G, Khademhosseini A. Fiber-reinforced hydrogel scaffolds for heart valve tissue engineering. J Biomater Appl. 2014;29(3):399–410.

78. Li X, Liu X, Dong W, Feng Q, Cui F, Uo M, Akasaka T, Watari F. In vitro evaluation of porous poly(L-lactic acid) scaffold reinforced by chitin fibers. J Biomed Mater Res B Appl Biomater. 2009;90(2):503–9.
79. Li X, Wang Z, Zhao T, Yu B, Fan Y, Feng Q, Cui FZ, Watari F. A novel method to in vitro evaluate biocompatibility of nanoscaled scaffolds. J Biomed Mater Res A. 2016;104(9):2177–25.
80. Li X, Feng Q, Liu X, Dong W, Cui F. Collagen-based implants reinforced by chitin fibres in a goat shank bone defect model. Biomaterials. 2006;27(9):1917–23.

Chapter 2
The Potential Matrix and Reinforcement Materials for the Preparation of the Scaffolds Reinforced by Fibers or Tubes for Tissue Repair

Wei Wang, Lei Mei, Fan Wang, Baoqing Pei, and Xiaoming Li

Abstract In recent years, tissue engineering has shown great promise in both traumatic and non-traumatic tissue repair and regeneration. Abundant experiments have been conducted on the three key factors in tissue engineering, which are commonly considered to be seeding-cell, biomaterials and the development of newly-formed tissue. It is widely accepted that tissue engineering is defined as the use of biological, chemical and engineering methods to repair, restore and regenerate living tissues by using biomaterials, cells and growth factors. The scaffold is considered to be one of the vital components, which emulates the extracellular matrix (ECM) for inoculating cells and simultaneously provides templates for new tissue growth. It is highly expected that the scaffold has ideal mechanical properties for various applications in different tissues. Recently, a reasonable scaffold design has developed an effective way to enhance the performance of matrix scaffolds. Numerous studies have demonstrated that the mechanical properties of the scaffold can be greatly enhanced by the addition of nanoscale fibers, tubes or other additives. In this chapter, we mainly demonstrate potential materials for the matrix and reinforcements and the required manufacturing techniques used for the reinforced scaffold.

Keywords Scaffolds • Reinforcement • Matrix • Fabrication techniques

2.1 Introduction

Tissue engineering is a promising solution to overcome the limitations of traditional therapeutic strategy. That is the diseased or injured tissues and organs, which are repaired and replaced by using synthetic functional scaffolding material cultured with appropriate cells, harvested from patient or donor, and then re-implanted in the

W. Wang • L. Mei • F. Wang • B. Pei • X. Li (✉)
Key Laboratory for Biomechanics and Mechanobiology of Ministry of Education, School of Biological Science and Medical Engineering, Beihang University, Beijing 100191, China
e-mail: x.m.li@hotmail.com

X. Li (ed.), *Tissue Repair*, DOI 10.1007/978-981-10-3554-8_2

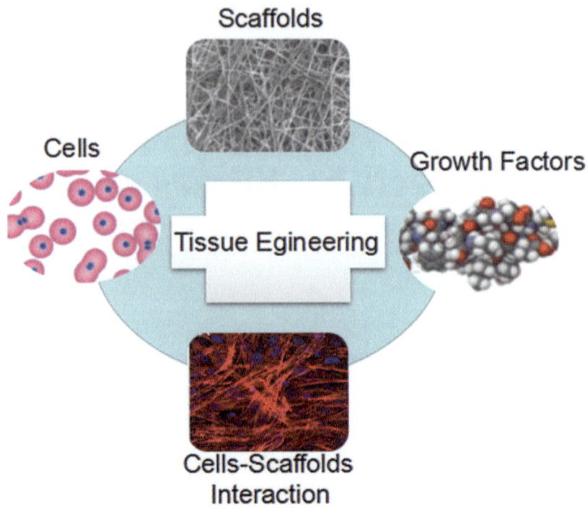

Fig. 2.1 Schematic diagram showing key components of scaffold-based tissue engineering

patient's body [1, 2]. The core of the method is the manufacture and use of scaffolds to provide a sufficient alternative framework for ex vivo cell expansion and maturation, the growth of natural tissues, and restoration of the original tissue quality according to the tissue's biochemical composition, morphology and function. In this regard, scaffolds for regenerative medicine play an important role in accommodating and directing cell colonization, migration, growth, and differentiation into specific tissues. The basic idea is to combine the specific cells with a scaffolding material under appropriate conditions for the proper growth of cells and subsequent tissues, as shown in Fig. 2.1. As one more key determinant factor for the success of scaffolds, the interaction of the cell-matrix (scaffold) directs the growth of new tissue [3]. In terms of the biomaterials, the cell-matrix interactions are governed by the surface properties of the implant material, and their imperative interactions have been found within a few nanometer of the surface of the implant [4]. Therefore, the starting material and manufacturing methodology are the most fundamental key points, which determine whether the scaffold is a suitable supporting base for successful cell transplantation therapy.

The native extracellular matrix (ECM) in the body provides a complex and dynamic nano-featured environment of pores, ridges and fibers that exhibit tissue-specific structure and properties for cells growth in the human body [5, 6]. It has been suggested that the characteristics of fibers in the native ECM play an important role in the cell orientation and phenotypic expression. Thus, the scaffold structure that has the ability to facilitate the cell-matrix interactions is critical not only to cell growth behavior but also a key prerequisite for the success of engineering physiologically functional tissues [7]. To achieve this, the use of fibers and tubes as an underlying skeleton or reinforcing agents with or without three-dimensional (3D) matrix provides an effective and workable way to produce composite scaffolds with one or more materials that can better mimic the ECM components and structures. Recent studies have shown that reinforced composite scaffolds not only have 3D porous structures, but also exhibit suitable mechanical properties, high cellularity and better mimicry of the natural tissue organization and composition [1]. Reinforcing scaffolds with fibers or tubes has been suggested to be an effective means of developing engineering materials to provide adequate topographical architecture and mechanical properties for enhancing cells adsorption and selectively altering the tissue orientation of tissue regeneration.

As we all known, it is important not only to select a kind of fiber/tube or the matrix but also to develop a satisfactory processing technique so as to get the homogeneous structure and composition throughout the reinforced scaffolds and obtain good adhesive strength between the matrix and the fibers or tubes [8, 9]. The processing technique should also be able to orient the fibers or tubes in the main trajectories and then lead to an expected final shaping of the reinforced scaffold. Therefore, the manufacturing techniques play a role that can't be ignored in the development of engineering materials for tissue engineering.

In this chapter, we state the potential matrix and reinforcement materials for the preparation of the scaffolds reinforced by fibers or tubes for tissue repair, as well as their relevant characteristics and properties in conjunction with current research presentations. Furthermore, we also critically review the typical techniques that commonly used to form fiber- or tube- reinforced scaffolds. At the end, concluding remarks and future trends of materials for the fibers or tubes reinforced scaffolds in tissue engineering are included.

2.2 Substrate Material for Fiber/Tube Reinforced Scaffolds

One of the major concerns in developing novel fiber- or tube- reinforced scaffolds is the selection of appropriate substrate biomaterials. For the material properties and load bearing requirements of the native tissue are important design considerations determining the success of the reinforced scaffolds. Implanted scaffolds as spaceholders to mimic the native EMC of damaged or diseased tissues or organs should possess some important characteristics for tissue engineering applications [10], as shown in Table 2.1. The basic characteristics of tissue engineering scaffold listed below are key factors to be considered during the design and manufacturing process.

At present, different types of scaffolds are compiled with their associated properties and applications. Typically, four individual groups of biomaterials, metals, ceramics, and polymers, have been proven with experimental data on their viability for biomedical materials, whose characteristics and applications are summarized in Table 2.2. For each scaffold material has different associated properties and mechanism, the choice of materials depends on the target tissue. For example, ceramics exhibit a high degree of biological activity and abrasion resistance, and have been used for bone repair for decades. In this respect, ceramics are widely used for processing scaffolds for hard tissue engineering. While, in most cases, there is a growing need for improved biomaterials to balance microstructural and mechanical

Table 2.1 Basic characteristics of scaffolds for tissue engineering applications

Properties	Remarks scaffold	Material
Bioresorbable	Foreign material and bulk degradation products should eliminate from body by natural pathways	Natural and synthetic bioresorbable polymer (collagen, PLA, hyaluronic acid, etc.)
Controlled porosity with interconnected pores	Tailor-made cellular adhesion, growth, extracellular matrix secretion, angiogenesis, nutrition, and oxygen transport without compromising mechanical strength	Polymeric hydrogels (PLG, PTFE, PLGA, PLCL) and composite materials (nano HA-collagen)
Biodegradable	Breakdown products of macromolecular degradation should not be toxic or immunogenic	Polymeric, composites, and ceramic scaffolds (nano TCP, nano HA, silica)
Controlled pore structure	To provide greater diffusivity and higher diffusion coefficient for waste removal and nutrient transport	Polymeric hydrogels and nano- fibrous structures (PLL/PLGA, gelain microspheres, PCL)
High surface area-to-volume ratio	For increased cell density, cell adhesion, proliferation, migration, and differentiation	Nanofibrous hydrogels, nano-structured components, and micro- fluidic scaffolds
Mechanical stability	Mechanical properties should match the replaced natural tissue to withstand in vivo stimuli	Metallic and alloy based, ceramics and composite scaffolds (nanostructured titanium and its alloys, nano alumina)
Three-dimensional templating	To assist cellular ingrowth and provide a natural three-dimensional in vivo micro-environment	Nanofibrous and polymeric hydro-gels, metallic, ceramics, and composites
Mimic natural ECM	Properties and structures should be matched with natural ECM components to coordinate with biological cues	Polymeric and composite hydrogels and nanofibrous scaffolds
Cellular compatibility	Scaffold surfaces must show cellular compatibility and should not repel cells	Polymeric and composite scaffolds (PET, PEGT/PBT)
Vascular support	To support angiogenesis and healthy regeneration	Polymeric, ceramic, and composite hydrogels and nanofibrous scaffolds
Biocompatible	Should not provoke any rejection, inflammation, immune response, etc.	Polymeric, ceramic, metallic, and composite materials
Surface modifiable	Scaffold surfaces should allow chemical or biomolecular functionalization to increase cell-material interactions	Metallic, composites, ceramic, and polymer materials (PEOT/PBT, PEG)
Non-toxic	Should not evoke toxicity to tissues	Natural polymeric scaffolds and composites
Non-immunogenic	Immunogenic response to tissue must not be evoked	Natural polymeric, composite, and ceramic scaffolds

(continued)

Table 2.1 (continued)

Properties	Remarks scaffold	Material
Non-corrosive	Should not become corroded at physiological pH and body temperature	Polymeric, ceramic, and composite scaffolds
Sterilizable	Surfaces must be receptive to sterilization processes to avoid contamination	Metallic, ceramic, composite, and polymeric scaffolds
Degradability rate matching with regrowth rate	For gradual transfer of loadbearing and support functions to newly growing tissues	Ceramic, polymeric, and composite scaffolds (PLL/HA, PDLLA, PCL)
High water content	Helps in generating hydrated in vivo environment	Polymeric and composite hydrogels

Table 2.2 A broad classification of substrate scaffold materials

Scaffolding materials	Advantages	Disadvantages	Applications	Examples
Metals	Strong, tough, ductile	Bioinert, difficult to make	Bone and dental implants, etc.	Tialloy
Ceramics	Bioactive, bioresorbable, high resistance to wear	Brittle, low toughness, not resilient	Low-weight-bearing bone implants, dental restoration, tissue scaffolds, bone drug delivery, etc.	Nano HA Nano tri-calcium phosphate
Polymers	Biodegradable, flexible, low density, resilient, surface modifiable, chemical functional groups	Low stiffness, Local acid	Tissue scaffolds, drug delivery, sutures, skin augmentation, blood vessels, heart valves, etc.	Collagen PLLA PLGA
Composites	Strong, design flexibility	Varying properties	Tissue scaffolds, drug delivery, etc.	HA/collagen

design standards, which facilitates the research on new composite materials, as mentioned below in Table 2.2.

2.2.1 Metals

Metal is typically a hard and shiny material (an element, compound or alloy) with good electrical and thermal conductivity. They usually have the great plasticity to be industrially processed into ideal models for different applications. In recent decades, metals have shown promising applications in tissue engineering because of their outstanding mechanical properties, especially in bone regeneration. It has been shown that various metals are suitable for biomedical design from operating instruments to engineered scaffolds. Here we prefer to put more emphasis on porous metal materials, which meet the oxygen and nutrition demand in tissue engineering. In recent years, researchers have been looking into stainless steel, co-base alloy, memory alloy, electron etc., and the devolvement of porous metals has been setting the stage for the concept of osteointegration of metallic implants [11]. For example, Nover et al. [12] proved that porous titanium supported robust cartilage growth, making it a viable bone-like base material for the fabrication of engineered osteochondral tissue constructs, a clinically relevant alternative to allografts. Porous metal materials can be classified as independent-hole type materials and continuous-hole type materials. The independent-hole material has the features, such as lower density, better rigidity, higher intensity, excellent vibration absorption and sound absorption. Apart from these properties, the continuous-hole material owns the advantage of being permeable and ventilated as well.

However, the traditional metallic biomaterials have the tendency of corrosion and stress shielding in human body environment and are not the best candidate for bone engineered scaffolds. It is reported that fiberglass-reinforced fibers containing glass fibers and matrix metals have high tensile strength and elasticity while accelerating the rate of degradation [13]. Recently, magnesium alloys have been recognized as a new class of biodegradable biomaterials with interesting orthopedic engineering applications, as increased bone growth has been reported repeatedly [14–16]. The Fig. 2.2 shows the entangled 3D porous structures of magnesium scaffolds with different pores [17], and the magnesium scaffolds with different architectures also have been investigated. Witte et al. [18, 19] studied the possibility of an open-porous magnesium scaffolds used as temporary bone replacements. The enhanced formation of unmineralized ECM and the enhanced mineral apposition rate adjacent of the degraded magnesium scaffolds were accompanied by an increased osteoclastic bone surface, which resulted in higher bone mass and a tendency to a more mature trabecular bone structure around the magnesium scaffolds. It indicated that even fast degrading magnesium scaffolds showed a good biocompatibility and reacted in vivo with an appropriate inflammatory host response. That is to say, magnesium was a promising matrix selection. Zhang et al. [20] prepared the porous magnesium scaffolds using the fiber deposition hot pressing (FDHP) technology, magnesium as the matrix material. This kind of fiber-reinforced magnesium scaffolds, having the 3D interconnected microstructures with the porosity ranging from 33 to 54%, showed comparable elastic modulus and compressive strength to those of cancellous bone. Cheng et al. [21] successfully prepared two open porous magnesium scaffolds using the titanium nanotube as the space holder. The porosity and pore size could be easily, precisely and individually controlled, and the mechanical properties also could be regulated within the range of human cancellous bone by altering the orientation of pores without sacrificing the requisite porous structure. The animal tests demonstrated that the scaffold exhibited acceptable inflammatory responses and could be fully degraded *in vivo*. In addition, the scaffolds with larger pore size promoted early vascularization and up-regulate

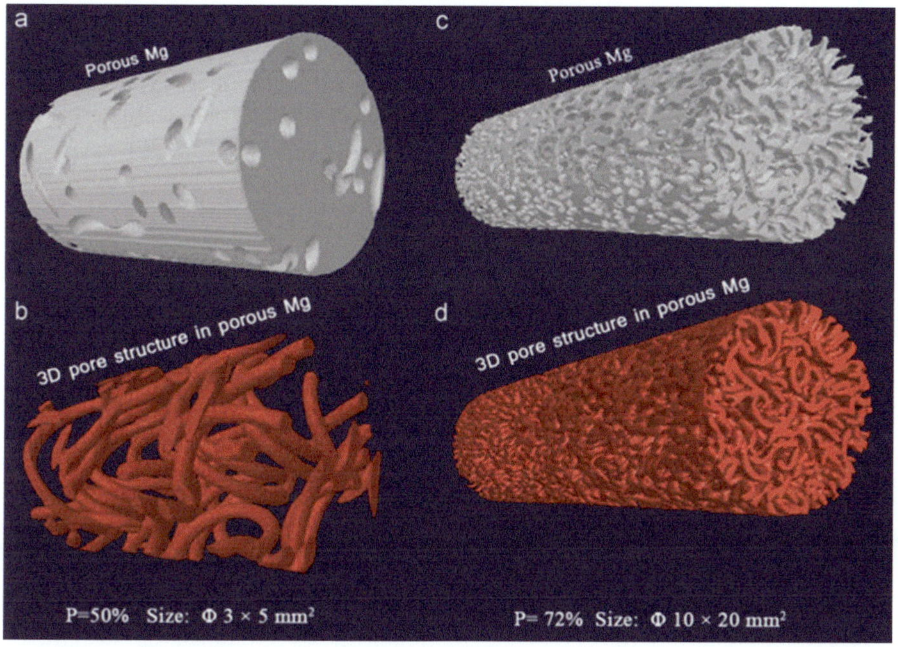

Fig. 2.2 Tomography images of entangled architectured porous Mg (**a** and **c**) and the 3D spatial structures of the pores (entangled channels) in Mg (**b** and **d**). [17] (Adapted with permission from Ref. [17]. Copyright 2016 Elsevier Ltd)

collagen type 1 and OPN expression, leading to higher bone mass and more mature bone formation.

However, magnesium alloys have not get the FDA approval to date, which places a big obstacle in the path of magnesium scaffolds development for bone engineering applications. The researchers tend to use the materials that already have the FDA approval to guarantee the safety of scaffolds in the clinic. The researchers are unwilling to risk using new materials considering the time and financial cost. So the nanofibers or nanotubes are great choices to make up for this shortcoming.

2.2.2 Ceramics

Ceramics are usually made up of two or more elements, containing metallic and non-metallic elements, and they are typically crystalline solid, prepared by applying heat and pressure to a powder. Based on X-ray diffraction characteristics, ceramics can be divided into crystalline ceramics, which have long-range order, and glassy ceramics, which in turn possess only short-range order. In general, ceramic materials have the high melting temperature, elastic modulus, compressive strength, and wear resistant, but low thermal conductivity and density. All of these properties

Table 2.3 Four different types of ceramic materials and usages in biomedical applications

Type	Characteristics	Examples	Applications
Nonporous, nearly inert ceramics	Nonporous, inert, strong and stiff	Alumina and Zirconia	Orthopedics (bone screws and plates); dental restorations
		Pyrolytic carbon	Coating material; blood contacting applications; joint components
Porous ceramics	Porous, inert, lower strong, coating for metallic implants	Calcium hydroxy-apatite (HA)	Primary structural component of bone; bone fillers; coating
Bioactive ceramics	Nonporous, bioactive, establish bonds with bone tissue	Silicate glass Soda-lime glass Borosilicate glass	Vertebral prostheses; vertebral spacers; Iliac crest prostheses; bone grafting
Resorbable ceramics	Resorbable, porous/ nonporous, degrade with time	Calcium phosphate β-Tricalcium phosphate (TCP)	Repair material for bone damaged; space filling material for bone loss Orthopedic applications

make ceramics be prospecting biomaterials in biomedical applications. In recent years, ceramics have developed into an important subset of biomaterials because of their great biocompatibility in the human body, which varies from inert ceramic oxides to the other extreme of resorbable materials. Hence bioceramics are particularly used for dental and orthopedic tissue replacements. For example, joint substitutes are commonly coated with bioceramic materials to reduce wear and inflammatory response. Other medical instruments like pacemakers, kidney dialysis machines and respirators also have a big demand on surgical cermets. Ceramics have broad applications in many aspects from medical or research intentions, and Table 2.3 above shows the four different types of ceramic materials and general usages in biomedical applications.

For example, Calcium phosphate (CaP), as a major constituent of bone, has been extensively studied. And the representative CaPs, hydroxyapatite (HA), tricalcium phosphate (TCP) and the mixture of HA and TCP, known as biphasic calcium phosphate (BCP), have been successfully used in the clinic as the bone powder filler and coatings on the metal surface to improve the implant integration with bone. It is also one of the matrix materials for the fiber- or tube- reinforced scaffolds. Both HA and TCP have similar chemical composition, structural and elastic modulus to the mineral phase of native bone. From the biocompatibility perspective, they are excellent candidates as bone tissue engineering scaffolds (see Fig. 2.3). Teixeira et al. [22] prepared HA and BCP scaffolds with 70% interconnected porosity (68% pores are 400 μm and 2% are 0.7 μm in diameter), which successfully supported new bone formation in immune deficient male mice and were more prominent in the case of HA scaffolds. Woodard et al. [23] observed the lamellar and woven bone formation in a multi-scale porous HA scaffolds containing macroporosity (250–350 μm) and microporosity (2–8 μm). High compressive strength of 30.2 ± 6.0 MPa was reported

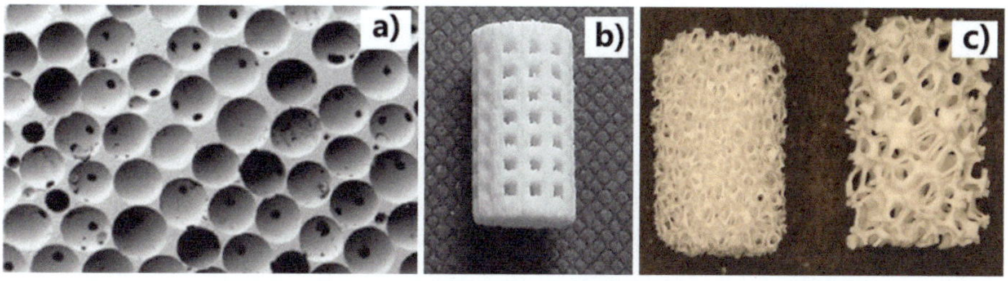

Fig. 2.3 (**a**) β-TCP scaffolds prepared using polymethylmethacrylate balls with 500–600 μm in diameter [27]; (**b**) 3-D printing of calcium sulfate cylindrical scaffolds with the sample length 12 mm [28]; (**c**) commercially available HA cylindrical scaffolds with different porosity produced by sponge replica method [29] (Adapted with permission from Refs. [27–29]. Copyright 2009, 2015, 2015 Elsevier Ltd)

with 70% porosity and the pore size at 50–600 μm. The macroporous HA scaffolds also allowed the cell adhesion, proliferation, and more importantly, internal migration, which will benefit the HA scaffolds to form a secure bonding with bone tissue [24]. HA and TCP scaffolds with internal controlled architectures have been prepared by direct 3D printing [25, 26]. Combined with microwave sintering, the maximum compressive strength was improved to 10.95 ± 1.28 MPa for TCP scaffolds with 42% porosity and the pore size at 400 μm. The TCP scaffolds with >60% porosity increased osteoblast activity and facilitated the new bone formation in the pores [25].

The calcium phosphate cement (CPC) is approved by FDA for repairing craniofacial defects. However, the CPC applications are limited to the non-stress-bearing bone because of their brittle and weak nature, which is characterized by a sudden fracture without any significant preceding plastic deformation [30]. To overcome the brittle fracture behavior of conventional CPCA, a large number of materials have been used to reinforce CPCs. The reinforcements include technical reinforcements (E-glass, carbon fibers/nano-tubes, SiC whiskers or polypropylene, aramid and polyamide fibers), calcium silicates (wollastonite whiskers), phosphates (hydroxyapatite whiskers) or carbonates (aragonite whiskers), and degradable polymers (chitosan, PLGA and PCL) [31]. In all the composite approaches, the CPC reinforced by fibers is one of the most successful techniques, particularly using long continuous fibers [32]. Contributions to the macroscopic behavior come from strength and stiffness of both fiber and cementitious matrix. Fiber reinforced CPCs are an emerging class of biomedical materials to overcome the brittle fracture behavior of conventional CPC [33, 34]. The application of fiber reinforcements in medical applications is not feasible now, for there are many important questions need to be solved.

However, ceramics show limited clinical applications for tissue engineering because of the brittle characteristic and difficulty of shaping for implantation [35]. Janas et al. [36] provided a hard tissue scaffold comprising a resorbable ceramic fiber. The scaffold was formed by first creating unfired (green) bioresorbable ceramic fibers via the viscous suspension spinning process (VSSP) and successfully used the ceramic fibers to reinforce ceramic tissue scaffolds with sacrificing the intrinsic mechanical properties. Similarly, CaPs are commonly used materials for the restoration of bone defects with excellent biocompatibility and bioactivity. Even through, brittleness and low flexural/tensile strength so far limit their application to non-load bearing areas. Researchers now use collagen fibers to enhance the CPCs, which greatly improve their strength and toughness and have been a major strategy to overcome the current mechanical limitations of CPCs. Despite the fact that there is not so much research on the fiber or tube reinforced tissue scaffolds. In the near future, the fiber or tube reinforcement will become a hot point on ceramic tissue scaffolds.

2.2.3 Polymers

Biodegradable synthetic polymers offer a number of advantages for developing scaffolds in tissue engineering [35, 37, 38]. The vital advantages include their great mechanical properties and degradation kinetics, which can be tailored by adjusting different ratios of monomers, concentrations and reaction conditions during the fabrication process. They can also be fabricated into various shapes with different surface topography and porosity to suit desired applications. Hence, biodegradable polymers are a popular choice as materials for tissue engineered scaffolds, and plenty of experiments and researches have been done on this field [39–41]. They are also promising matrix materials for the fiber or tube reinforced tissue scaffolds. Basically, there are two types of polymeric substrate materials, natural polymers and synthetic polymers, as shown in Table 2.3. Natural polymers, such as collagen, various proteoglycans, alginate-based substrates and chitosan, have all been used in the production of scaffolds for tissue engineering [41–43]. They have high degree of biocompatibility that can promote cell adhesion and proliferation. Moreover, natural polymers can be degraded gradually, and the adhered donor cells can produce their own ECM to replace the scaffolds finally. Malafaya et al. [44] found a chitosan scaffold produced by a particle aggregation method that allowed the ingrowth of the connective tissues and promoted the neo-vascularization even in the early stages of implantation. Additionally, chitosan films and collagen sponges are also the commercial product as the wound dressing. Chen et al. [45] reported the composite membranes, using type I collagen and chitosan fabricated by electrospinning technique, which exhibited better wound healing rate than the commercial collagen sponge. Although natural polymers have known biological activity, they are difficult to process into scaffolds and the lack of quantity limits their use. Furthermore, the natural polymers may also stimulate an immune response, which leads the concerns over antigenicity and delivery of diseases for allograft [46, 47].

A great number of synthetic polymers have been used as scaffold materials, including PLA, PGA, PLGA and PCL. These synthetic polymers have already passed FDA regulation and have been successfully used as degradable sutures for many years. And thus scaffolds made from these materials can provide a quick route to a commercial and clinical product [36]. The PLGA scaffolds have been used to repair defects in different tissues, such as bone, liver, nerve, skin and blood vessels. PLGA scaffolds loaded with neuron stem cells (NSC) permitted the NSCs to differentiate toward neurons, establish connections and exhibit synaptic activities [48]. Successful cartilage constructions were observed after 8 weeks in a joint cavity of a sheep in the PLGA scaffolds loaded with autologous mesenchymal stem cells [49]. Good shaping properties and controllable mechanical and degradation characteristics by altering chemical compositions and distributions of polymers are the additional advantages. Kim et al. [50] found that the PCL scaffolds with oriented nanofiber webs showed a mechanically anisotropic behavior and a higher hydrophilic property in comparison with the randomly distributed fibrous mats.

Additionally, most synthetic polymers are degraded by the hydrolysis process, and the degradation rate is not affected by the enzyme. Unlike the natural polymers, the degradation of the synthetic polymers scaffolds may not vary from patients to patients. Their major drawback is that the degradation products are not as compatible as those of the natural polymers. For example, CO_2, the degradation product of PLLA and PLGA, lowers the local pH value and induces cell and tissue necrosis [51].

There are, however, some problems with using polymer scaffolds for bone regeneration. Although the mechanical properties of polymers can be matched with collage, their elastic modulus is much lower than that of bone, which limits their use in load-bearing anatomic sites undergoing compressive forces. Secondly, polymers exhibit fast strength degeneration during the degradation process, especially under load, which is too rapid for bone regeneration. On the base of these shortcomings, fiber- or tube- reinforced tissue scaffolds become a popular study point. For example, Jiao et al. [52] found that the mechanical properties of pure PLA scaffolds were much lower than those of natural bone, while chitosan was a unique cationic polysaccharide with satisfactory biocompatibility. By using chitosan fibers to reinforce PLA scaffold, the reinforced PLA scaffold could be used to repair larger bone defect. This kind of examples goes on and on. From years' research reports, polymers have large scale use in fiber or tube reinforced tissue scaffolds, as shown in Table 2.4.

Table 2.4 Examples of polymer scaffolds employed in tissue engineering

Natural polymers	Synthetic polymers
Collagen (type I, II, III, IV)	Polylactic acid
Collagen-glycosaminoglycan copolymer	Polyglycolic acid
Fibrin	Poly-dl-lacic-co-glycolic acid
Poly(hydroxybutylate), PHB	Polytrimethylene carbonate
Sodium alginate	Polycaprolactone
Chitin and chitosan	Poly(ortho esters)

2.2.4 Composites

The scaffold is a temporary supporting structure for growing cells and tissues, whose role is to mimic the native ECM [53]. It is widely known that there are various molecules in ECM, and they make up the whole complicated and transient extracellular environment. In the design phase of the scaffold in tissue engineering, a great many of things should be taken into consideration, including the fact that the scaffold under research should meet the relevant organic application, as well as simulate the function of ECM as possible as they can be achieve. In recent decades, scientists have been striving hard to carry out a large amount of research and experiments on biocomposites for developing a combined new material that consisting of at least two different materials, which are usually among metals, ceramics and polymers, as we mentioned before [54–56].

The ECM regulates a cellular dynamic behavior and intercellular communication; in the meantime, it also functions as a warehouse storing for all kinds of growth factors. Thus, the engineered scaffold should be biocompatible and bioactive to meet all the demands when replacing the abnormal tissue in the human body. In addition, other features are required for special tissue replacement or repair. For example, bone tissue engineering requires highly strength and rigidity; vascular tissue engineering requires appropriate mechanical properties, like long-time shear resistance and compatibility with blood ingredient. Certain types of bioactive ceramics (e.g., TCP and HA) and bioactive glasses (e.g., 45S5 Bioglass®) react with physiologic fluids to form firm bonds with hard (and in some cases soft) tissue. However, these bioactive materials are relatively stiff, fragile and difficult to form complex shapes. In contrast, synthetic bioresorbable polymers are easy to fabricate into desirable structures, yet they are too weak to meet the requirements of surgery and the *in vivo* physiologic environments [57]. Hence, biocomposites can, in some ways, help them obtain well-combined performance from mixed materials, which can achieve multi-aspects goals at the same time.

Bioactive glass containing biodegradable polymers is a commonly used composite material for scaffolds. Niiranen et al. [58] prepared a self-reinforced PL/DLLA (70:30) matrix composite containing bioactive glass (BG 13–93). They found that the addition of bioactive glass could modify the degradation kinetics of the scaffold both *in vivo* and *in vitro*, as well as improve the material morphology, dimensional stability and apatite formation on its surface. It is obvious that the bioabsorbable polymers, such as PLA, PGA, PCL, polyethylene oxide (PEO), and poly(3-hydroxybutytrate) (PHB), containing other osteoconductive fillers show great potential as bone-repairing materials in hard tissue engineering. And CaPs including HA, TCP and CPCs also play an important role in the development of scaffolds for bone tissue engineering [59]. Miao et al. [60] produced porous CaP ceramics with interconnected macropores (>200 μm), micropores (5 μm) and high porosities, and the open micropores were infiltrated with PLGA to form an interpenetrating bioactive ceramic-biodegradable polymer composite structure. In this manufacturing method, the compressive moduli of the material was significantly improved, and the bioactivity of the composite scaffolds was confirmed by the apatite layer formation in the simulated body fluid. In some cases, the addition of a polymer phase into the ceramic material might have extra functions, such as the biodegradable polymer that can act as a carrier for some special biomolecules, growth factors, antibiotics and other signal molecules, thereby increasing the capability of tissue engineering constructs [59]. Furthermore, the addition of carbon nanotubes and fibers to the composite coating will enhance the mechanical properties and induce nano-topographical surface features, which will enhance better cell attachment and survival.

The metal-ceramic composite is also used as bone grafts in clinical trials. For example, Marsich et al. [61] have been carried out experiments on adding silver nanoparticles into alginate/HA ceramic composite scaffold, and they found the new bone grafts have been endowed with antibacterial properties owing to the presence of silver nanoparticles. The results of in vitro test and the release profile of silver nanoparticles indicated the prospective characteristics of these biocompatible antimicrobial scaffolds for the future tissue engineering applications. The ceramic material can perform excellent scaffolds when adding with metal infiltration, because the stable chemical structure of ceramics ensures their highly biocompatibility all the time. On the other hand, the metal is modified by the bond with ceramics, which can show a decrease in the corrosion rate and an increase in mechanical properties. Gu et al. [62] adopted the molten metal infiltration technique to fabricate the MgCa-HA/TCP, which exhibited the inferior mechanical property (decreased by 50%) but with superior corrosion resistance (improved by 68%), compared to the single MgCa alloy matrix. This bioactive HA/TCP scaffold was also proved to have slower degradation rate due to the presence of ceramic scaffolds, which is an important part to be considered in bone tissue engineering and cartilage tissue engineering.

Apart from metal, various inorganic and organic materials are also applied to the design of composite scaffolds. For example, chitosan is greatly used in the combination of polymers to fabricate the ideal composite scaffold, because of its important role in peripheral nerve regeneration [63]. Razavi et al. [64] fabricated the tissue-engineered scaffold with chitosan nanopowders and PLGA, and found that the high chitosan content and aligned-orientation of nanofibers in biocomposite scaffold can promote differentiation and myelinogenic capacity of Schwann-like cells induced from human adipose-derived stem cells (ADSCs), which is a critical issue in nerve regeneration medicine. Collagen, as a natural polymer found in ECM, is resorbable with high swelling ability, low antigenicity and cytocompatibility [65]. Due to its poor mechanical strength, pure collagen can't be directly used as a bone substitute material. Hence the composite scaffold of collagen and bioactive ceramics has drawn great attention [65]. Kuttappan et al. [43] summarized the current state of the field by outlining composite scaffolds made of gelatin/collagen in combination with bioactive ceramics, and Table 2.5 below shows commercially available collagen/gelatin based composites.

Table 2.5 Collagen/gelatin based composites available in the market [43]

Product	Company	Collagen/ gelatin	Additive	Application
Collagraft®	Zimmer	Bovine Collagen Type I (>95%), Type II (<5%)	HAP −65% β-TCP −35%	Acute long bone fractures and traumatic osseous defects, bone void filler
GingivAid®	Maxigen Biotech Inc.	Type I collagen	HAP/β-TCP	Dental implant surgeries, sinus lift and alveolar ridge augmentation
Formagraft®	Maxigen Biotech Inc.	Type I collagen	HAP/β-TCP	Bone graft substitute
CopiOs Sponge and Paste	Zimmer	Type I Bovine Collagen	Calcium phosphate	Bone void filler
Ossigen®	Exactech	Collagen	Bone mineral	Bone void filler
OsseoFit®	DSM Biomedical	Type I bovine collagen	PLLA, β-TCP	Bone voids or bone defects in the pelvis and extremities
CONDUCT® Matrix	GLOBUS Medical	Type I collagen	Carbonate apatite mineral	Bone void filler
Puros® DBM Putty	Zimmer	Demineralized bone matrix		Bone void filler
CONFORM® Demineralized Cancellous Bone	Depuy Synthes	Demineralized cancellousbone		Bone void filler, use within posterolateral gutters of the spine
DBX®	Depuy Synthes	Gelatin-Sodium hyaluronate	Demineralized bone matrix	Bone void filler, treatment of oral/ maxillofacial and dental intraosseous defects
Integra Mozaik™	Integra Life Sciences	Type I collagen	TCP	Bone void filler
INFUSE® Bone Graft	Medtronic	Type 1 bovine collagen	rhBMP-2	Bone void filler
Vitoss®	Stryker	Collagen	β-TCP, Bioactive glass	Bone void filler
OP-1 Implant	Stryker	Type I bovine collagen	BMP-7	Bone void filler
MASTERGRAFT® Putty	Medtronic	Type I bovine collagen	β-TCP	Bone void filler
RegenOss®	JRI Orthopaedics	Type Icollagen fibers	Magnesium-enriched HAP	Long bone fractures, Spinal fusion

(continued)

Table 2.5 (continued)

Product	Company	Collagen/ gelatin	Additive	Application
Orthoss® Collagen	Geistlich Surgery	Porcine Collagen	Bovine HAP	Bone void filler, reconstruction in orthop-aedic and in spinal surgery, volume extender for composite bone grafting
Bio-Oss Collagen®	Geistlich Biomaterials	Porcine Collagen	Geistlich Bio-Oss® particles	Sinus Floor Elevation, Peri-implantitis, Periodon-tal Regeneration, Ridge Augmentation

Adapted with permission from Ref. [43]. Copyright 2014 Elsevier Ltd

2.3 Mechanism of Reinforcements

Regarding the scaffold, it is generally agreed that a highly porous microstructure with interconnected pores and a large surface area is conducive to growth of tissues. In this way, a satisfactory scaffold should possess not only appropriate porosity but also ideal mechanical properties [66]. The scaffold can only retain its shape and other related characteristics after a cell-seeding procedure and after following being embedded in the body. A number of solutions have been applied to reach this biomedical goal, while researchers found that the use of reinforced scaffolds, containing substrate materials as a basic platform and reinforcement materials, such as particles, fibers and nanotubes, can achieve the object referred above. Thus, in the composite material, the reinforcement part is essentially dedicated to increasing the mechanical properties of the entire material system. All of the different reinforcements in the composites have different properties and thus affect the properties of the composite in different ways.

The reinforced action requires that the generated structure should have a higher stiffness and strength than the matrix and be well bonded to the matrix polymer. As a consequence, the stress can be transferred from the "weak" matrix to the "strong" reinforcement structure, according to the "working principle" of all composites. The reinforcing phase in a composite can exist in a number of different forms. To date, several reinforcements, including particles, fibers and tubes, have been investigated and used for biocomposites. Particle-reinforced composites were once used as the cheapest and most widely used composite materials for industrial applications. However, in recent years, it has been seen that significant advances in the field of scaffold reinforcements for tissue repair strategies are the use of fiber/tube reinforcements in cell-seeded gel carrier systems [67, 68]. This section will provide a brief summary of the fiber-, tube- and self-reinforced biocomposite scaffolds.

2.3.1 Fiber

Particulate systems have dimensions that are approximately equal in all directions, so they remain predominantly isotropic in their physical behavior. Technically, the most important composites are those in which the dispersed phase is in the form of fibers. Fiber-based structures represent a wide range of morphological and geometric possibilities that can be customized for each specific tissue engineering application [69, 70]. Materials with fiber reinforcemens have long been used in the repair or replacement of tissues, ranging from their use to more complicated applications such as vascular grafts, hernia replacement tissue and musculoskeletal tissues such as bone or cartilage [9, 71].

Fibers are a continuous material structure with an extremely high aspect ratio that can be supplied directly from nature or produced from synthetic polymers [72, 73]. The fibers are fairly adequate materials as reinforcements, which can be utilized as controllable elements to make the mechanical and physical properties of biomaterials suitable for a particular application. Natural polymers as fiber reinforcements are fibrous proteins, including collagen, such as gelatin, elastin, silk fibroin, and fibrinogen, and carbohydratebased polymers, such as chitin, chitosan, and hyaluronic acid [74–76], and most of them have been successfully fabricated and evaluated for various tissue applications. As for synthetic fiber reinforcement, the polymers include but not limited to poly(α-hydroxy esters) [74], such as PLA, PGA, PCL, poly(hydroxyalkenoates) (PHAs), poly(propylene fumarate) (PPF) and poly(3-hydroxybutyrate-*co*-3 -hydroxyvalerate) (PHBV) [77].

In addition to conventional textile techniques, such as knitting, weaving, or braiding, various methods and strategies, such as electrospinning, wetspinning, meltspinning, interfacial complexation, direct writing and 3D printing, have been implemented to fabricate fiber reinforced materials with varying geometries, mechanical properties, porosity, anisotropy. The four main factors that govern the contribution of fibers are the basic mechanical properties, the surface interaction, the amount and the orientation of those fibers in the composite. The resulting properties of fiber-reinforced composites depend on the properties of the fiber and matrix as well as the fiber-matrix interactions [1]. Technically, the fiber reinforced systems typically have a preferred orientation to provide optimum mechanical properties [78]. As shown in Fig. 2.4, CaP biocomposites can be enhanced with different fiber dispositions, and the result biocomposites present different mechanical properties. For different types of matrix materials, mechanical properties are considered to be different.

This option for the reinforcement of the scaffold in tissue engineering should be determined primarily by the type of host tissue as well as its mechanical properties and anatomical structure. Suitable reinforcing materials generally have harder and stronger mechanical properties than those of the substrate, and their desired surface properties, porosity, morphology and biocompatibility are also important factors for their successful applications [1]. In fact, the mechanical properties of the reinforced composites can be affected by a large number of material parameters, as shown in

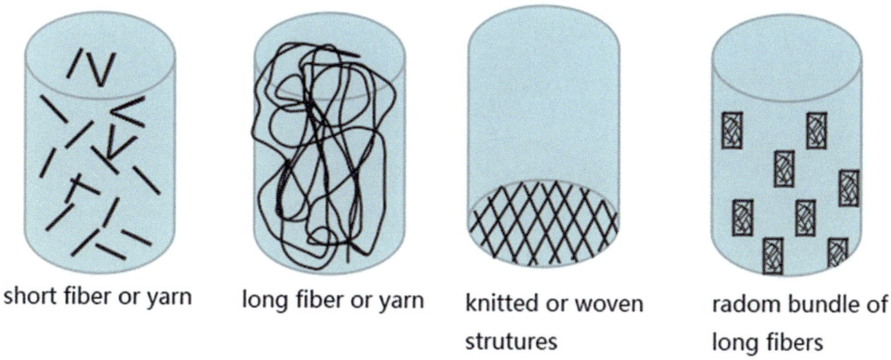

Fig. 2.4 Fibre-reinforced CPCs: different fiber dispositions

Fig. 2.5 Material parameters which influence the time dependent mechanical behavior of the FRCPC composite. [31] (Adapted with permission from Ref. [31]. Copyright 2012 Elsevier Ltd)

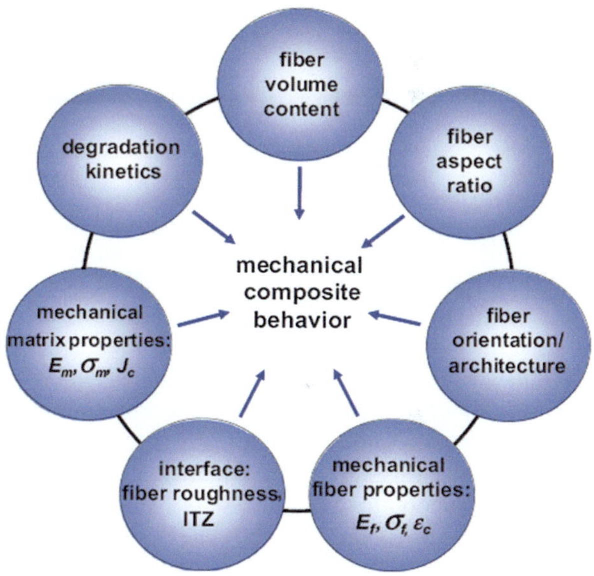

Fig. 2.5. In addition to the complex interaction between the above factors, scaffolds in clinical applications require biodegradation over time, making the scaffold behavior time-dependent [31].

Fiber-based materials or structures, having been widely used as reinforcing additives in tissue engineering, are be designed to have high porosity, good mechanical properties based on the arrangement of the fibers and suitable conformability for various tissues and organs [1, 75]. There are a number of studies conducting on the mechanical properties of the polymeric matrix polymers reinforced by fibers. Caves and cooperators [79] combined EMC protein analogs to produce multilamellar, elastin-like protein sheets reinforced with collagen microfibers. The manufacturing process of an angle-ply composite was suitable for varying fiber orientation and

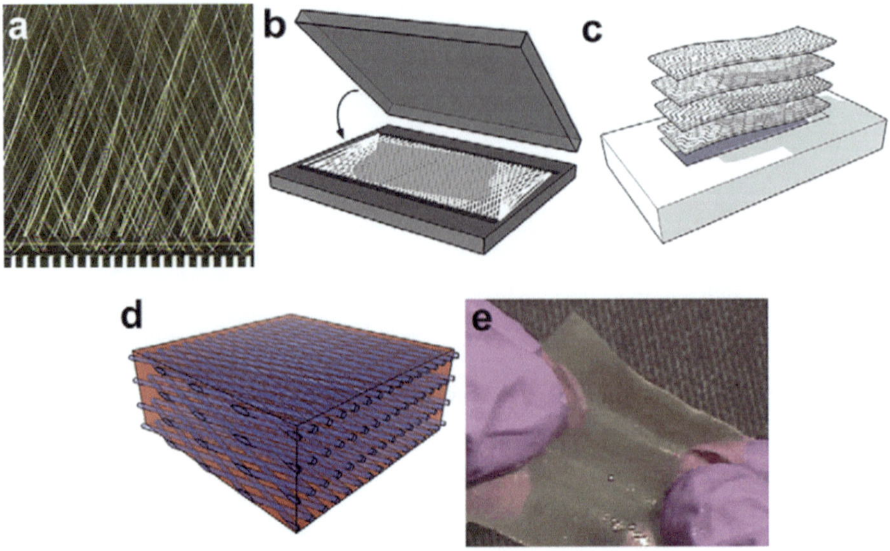

Fig. 2.6 Fabrication of a collagen microfiber reinforced elastin-like protein sheet. (**a**) Collagen microfiber is wound about rectangular frames to obtain the desired orientation and average spacing. (**b**) A cooled protein polymer solution is distributed over the microfiber layout and molded into a thin membrane through a temperature-driven solegel process. (**c**) Stacked membranes are laminated by a temperature transition to yield multilamellar, angle-ply composites (**d**, **e**). [79] (Adapted with permission from Ref. [79]. Copyright 2011 Elsevier Ltd)

volume fraction within and between individual sheets of a planar sheet, leading to an adjustable biomechanical property similar to that of the natural tissue (Fig. 2.6). The addition of reinforced collagen fibers extends the range of mechanical properties of elastomeric protein polymer composites to carry a wide range of suitability for the load bearing soft tissue. Significantly, these composites provided excellent mechanical support for a full thickness abdominal wall defect (Fig. 2.7). In the other study reported by Slivka et al. [80], the PGA fiber-reinforced PLGA scaffold had an increasing compressive modulus from 10 to 50 MPa with the amount PGA fibers from 0 to 20 wt.%. The resulted strength was approximately five fold higher than the cartilage, which might be beneficial for tissues with high loads. Guarino et al. [30] prepared the PLA fiber-reinforced PCL scaffolds and revealed that the composite scaffolds exhibited a higher activity of the human osteoblasts and human marrow stromal cells. The synergic addition of PLA fiber and TCP particle reinforcements in PCL scaffolds resulted in a high porosity (80%) structure, and the elastic modulus increased up to a maximum value of 2.21 ± 0.24 MPa [81]. Cont et al. [82] designed a novel composite scaffolds, consisting of long continuous bidirectional fibers embedded in an electrospinning matrix. The long bidirectional continuous PDO fibers were embedded in a PLA matrix to induce anisotropic behavior and increase the mechanical strength of the scaffolds along the direction(s) of the PDO fibers. The highest ultimate tensile load applied to PLA fibers (17.9 N) was almost one

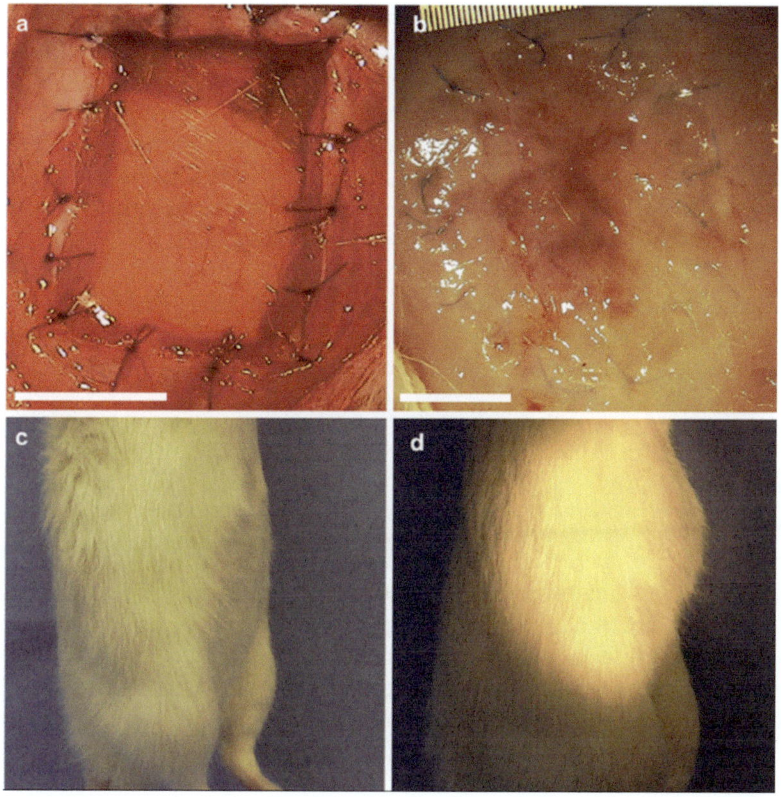

Fig. 2.7 Abdominal defect repair. Appearance of the multilamellar elastin-composite immediately following implant (**a**) and at 8 weeks (**b**). None of the repaired defects demonstrated hernia formation for the duration of the study (**c**), as compared to unrepaired defects (**d**). Scale 10 mm. [79] (Adapted with permission from Ref. [79]. Copyright 2011 Elsevier Ltd)

quarter of the lowest ultimate load applied on the PLA-PDO composite (70.6 N) along the fibers.

Many research studies have used strong fibers to enhance hydrogels, which are highly desirable but difficult for medical applications. This reinforced composite will incorporate sufficiently robust properties to provide the strength and transport of the liquid phase inside the hydrogel. Illeperuma et al. [83] successfully fabricated a composite using an alginate-polyacrylamide hydrogel reinforced with a random network of stainless steel fibers. The tests showed that both stiffness and strength could be significantly increased by adding the fibers to the hydrogel. Yodmuang et al. [84] investigated a silk-based hydrogel composite made from silk microfibers and silk hydrogels, having the potential to be used as a support material for engineered cartilage. Calvert et al. [85] described a hydrogel scaffold reinforced with elastic fibers to achieve high strength and high fracture strain. Agrawal et al. [86] demonstrated a new class of hydrogel composites reinforced with elastic fibers had a cartilage-like structure by using a 3D rapid prototyping technique. In addition, Visser et al. [87] also reinforced soft hydrogels with highly organized, high-porosity microfiber networks to fabricate a high-porosity scaffold by combining the 3D-printing and melt-electrospinning writing technology.

The fiber reinforcement with its unique characteristics and functions has played a critical role in improving biomechanics, biocompatibility, bioactivity, integration, and degradation of artificial composite scaffolds in tissue repair and regeneration. The properties of fiber-reinforced systems are influenced by the fiber's organization, volume fraction, mechanical properties, structure etc. Despite the current studies with significant achievements, there is still a need to develop advanced theories and satisfactory processing techniques of fiber-reinforced scaffolds to get the homogeneous structure, properties and composition similar to native ECM.

2.3.2 Nanotubes

As one of the most interesting studies in nanomaterials, nanotube materials have a structure similar to small straws with large inner volumes, which makes it possible to fill with a variety of chemicals and biomolecules, ranging from small molecules to proteins [88, 89]. Nanotubes as a young class of material are usually composed of carbon nanotubes (CNTs) and non-carbonic nanotubes. Similar to nanofibers, nanotubes are also a type of one-dimensional macromolecules with a larger surface area, and normally exhibit similar strengthening mechanism as nano-fiber reinforcements.

Since the CNTs was first reported by Iijima [90] and Bacon [91] in 1991, they have become a biomaterial with great application prospects, especially for prostheses positioned in contact with bone, like joint replacement prostheses and plates or screws for fracture fixation, as well as drug delivery systems [92]. Because of outstanding electrical, mechanical, and surface properties of CNTs, a great deal of research has been initiated in this field [93–95]. After the CNTs had been found, tungsten disulfide nanotubes (WSNTs), one of the noncarbonic nanotubes, were discovered by Rothschild et al. in 2000[10d]. Up to now, there are various noncarbonic nanotubes, such as ZnO nanotube [96], ZnS nanotube [97], NbS_2 nanotube [98], boron nitride nanotubes (BNNTs) [99], $NbSe_2$ nanotube [100], Bi_2S_3 nanotube [101], GaN nanotube [102], InP nanotube [103], SiO_2 nanotube [104], AIN nanotube [105], and MS_2 nanotube[11d]. As an important member of quasi-one-dimensional family, noncarbonic nanotubes have a high volume percentage of surface area, which shows a high chemical activity and unique physical properties. In recent years, nanotubes have been synthesized from graphite and boron nitride compounds with layered structure by using various methods, such as arc discharge, chemical vapor deposition, laser ablation, carbothermal reduction, pyrolysis method, ball milling method, and hydrothermal synthesis method [92]. A new mechanism for the preparation of nanotubes from the non-layer material has developed. For example, Rothschild et al. prepared nano-diameter WS_2 nanotubes by vaporsolid growth method [106]. There are two main types of nanotubes existing: the singlewalled nanotubes (SWNTs) consisting of coiled monolayer graphene sheets and the multi-walled nanotubes (MWNTs) with several graphite concentric layers. In the high-temperature methods, multi-walled carbon nanotubes (MWCNTs) can be produced by pure carbon evaporation, but the synthesis of single-walled carbon nanotubes (SWCNTs) requires the presence of metallic catalysts.

CNTs [107] has their unique structural (cylindrical nanostructure) and physical properties, such as high aspect ratio [108], strength (Young's modulus ~1 TPa and tensile strength ~150 GPa) [109, 110], high conductivity (metallic nanotubes can carry an electric current density of 4 × 109 A/cm^2) [111], and chemical stability. In recent years, some reports indicate that functionalized CNTs can improve cell compatibility of the matrix material, promote tissue regeneration and inhibit the formation of glial scar and fibrous tissue [112]. For instance, Shi et al. studied three different highly porous materials of scaffolds: poly (propylene fumarate (PPF) polymer, an ultrashort single-walled CNT (US tube) nanocomposite and a dodecylated US-tube (F-US tube) nanocomposite. The functionalized ultra-short SWCNTs nanocomposite with tunable porosity and mechanical properties possessed good performance among the composite materials, which could be a promising candidate as the ideal scaffolds materials for the bone tissue engineering applications. CNTs constitute a new type of materials and have been recognized as the most commonly used tubes reinforced materials, like fibers, in tissue engineering [113]. Because of its extraordinary properties, CNTs have been introduced into many host materials (polymer, ceramics, metal) to improve the overall properties of the reinforced composite system [114]. Remarkably, the alignment of CNTs can provide a crucial advantage in terms of directional conductivity, which is often required for the damage tissues or organs with the transmission of electrical signals [115].

Researchers have study the properties of CNT reinforced scaffold in the field of tissue repair and regeneration. Zawadzak et al. [116] deposited CNTs on the surfaces of PU foams using electrophoretic deposition (EPD) to create a new family of functional bone scaffolds. The CNTs coating was thought to promote the scaffold's osteoconductivity and mineralization and provided a nanostructured surface and a conductivity function. Mikael et al. [117] found that the compressive strength and modulus of CNT reinforced PLGA scaffold were significantly increased (35 MPa, 510.99 MPa) compared to pure PLGA scaffolds (19 MPa and 166.38 MPa) by adding only 3% MWNTs. In addition to the random dispersion of the CNT, high electrical conductivity and mechanical strength can be achieved at low CNTs concentrations by aligning CNTs to form a connective pathway and obtain an effective charge transfer. This scaffold can be beneficial for the regeneration of nerve tissue to promote neurite outgrowth, orientation and branching [118]. Shin et al. [115] prepared the reinforced CNT-gelatin methacrylate (GelMA) hybrid, and a dose-dependent increase in mechanical properties was observed with the increase of the amount of CNTs. The tensile modulus increased about three times and the compressive modulus increased about two times by the adding 0.5 mg/ml CNTs in GelMA hybrids. Zhang et al. [119] found that the mechanical and biological properties of MWNT-reinforced PLGA improved significantly, making them potentially for tissue engineering applications, particularly for bone tissue regeneration. Compared with pure PLGA scaffolds, the tensile stress of the PLGA scaffold with 0.25% MWNTs was increased by 54% (from 5.88 to 9.08 MPa), the Young's modulus was increased by 8% (from 163.53 to 176.83 MPa), and the elongation at break was increased by 49% (from 27.14% to 40.39%). Vozzi et al. [120] microfabricated 3D PLLA/MWCNTs nanocomposite scaffolds for bone tissue engineering. The results suggested that the presence of CNTs increased the elastic modulus, and the cells on the PLLA/MWCNTs scaffold were more viable. Chen et al. [121] synthesized chitosan-multiwalled CNTs/HA nanocomposites (CHIMWCNTs/HA) for bone tissue engineering. These elastic modulus and compressive strength of the composites were increased sharply from 509.9 to 1089.1 MPa and from 33.2 to 105.5 MPa respectively, when multiwalled carbon/chitosan weight ratios were increased from 0 to 5%. And the cells on the surface of the composites with different weight ratios were spread out and had some filamentous pseudopods. Abarrategiet et al. [122] also proved that MWCNTs/CHI scaffolds with a well-defined micro channel porous structure had biocompatible and biodegradable supports for culture growth. Yildirim et al. [123] studied the preparation, characterization and biocompatibility of SWCNT-reinforced alginate composite scaffolds by free formation technique. In addition, Hu et al. [124] used CNTs as a substrate for culturing the hippocampal neuron growth. Mattson et al. [125] succeeded to cultivate embryonic rat-brain neurons on nanotubes, and the neurons survived and grew for 8 days in culture. Shin et al. [115] used CNTs as reinforcing agents to produce a porous hydrogel with adjustable mechanical properties.

As for noncarbonic nanotubes, their manufacturing method and properties have attracted extensive research and discussion, while their applications as reinforced materials in tissue engineering have few reports. Only a few of materials has been investigated as reinforcements in composites. Boron nitride nanotubes (BNNTs) are synthesized from boron nitrid with a stable hexagonal structure analogous to that of graphite [126]. BNNTs as an important material of noncarbonic nanotubes have excellent mechanical properties and thermal conductivity similar to CNTs, which makes it a potential candidate as reinforcements. Moreover, BNNTs have its own characteristics. BNNTs have chemical inertness, stable structure and high temperature oxidation resistance [127], so there is little effect on the ductility of the scaffolds when they reinforce polymer composites. As a kind of wide band gap semiconductors, the electrical property of BNNTs is independent from geometry. BNNTs are noncytotoxic to osteoblasts, macrophages, human embryonic kidney cells and human neuroblastoma cell line.

As a relatively new type of reinforcements, few researchers have studied the performance of BNNTs in the composite for tissue engineering. But BNNTs reinforced composites have been proved to have good thermal, mechanical, and optical properties, which makes them a promising candidate material for obtaining ideal composite material for tissue engineering scaffolds [128–130]. Lahiri et al. [131] reported that BNNTs reinforced PLC copolymer composite increased the tensile strength from 2.67 MPa in PLC to 5.59 MPa. BNNTs exhibited better expression levels of the Runx2 gene, which was the main regulator of osteoblast differentiation, and the composite had favorable biocompatibility to osteoblast and macrophages in the cytotoxicity assay. Meanwhile, Lahiri et al. [132] also proposed BNNTs as reinforcing agents to enhance HA for the stronger orthopedic implant applications. The results showed that the HA-4wt% BNNTs composite possessed excellent mechanical properties compared with HA. And the elastic modulus, hardness, wear resistance and fracture toughness were improved by 120%, 129%, 75% and 86%, respectively. BNNTs reinforcements also showed no adverse effect on osteoblast proliferation and cell viability. Zhi et al. [133] used BNNTs as reinforced agents to prepare poly(methyl methacrylate) (PMMA), polystyrene (PS), poly(vinyl butyral) (PVB), and poly(ethylene vinyl alcohol) (PEVA) composite materials to evaluate the thermal, electrical, and mechanical properties of reinforced composites. The BNNTs reinforced polymers maintained good electrical insulation and achieved higher thermal conductivity, an increase of more than 20 times. The thermal expansion coefficient of the BNNTs-loaded polymers was significantly reduced, and all the polymers remained insulated and possessed a high breakdown electric field in any case. The Vickers micro hardness tests results indicated that there was no obvious significant effect on the mechanical properties of the composites.

The inorganic WSNT is a closed polyhedron and a cylindrical crystal of a tungsten disulphide semiconductor compound having a high mechanical properties (Young's modulus ≈150 GPa, bending modulus ≈217 GPa) [134, 135] and functional groups, which make it possible to disperse in organic solvents, polymers, epoxy, and resins. Like BNNTs, there is little research on the mechanical properties of WSNTs used as reinforcing materials for scaffolds. Elad et al. [136] evaluated the effect of embedding inorganic nanotubes of tungsten disulfide in an epoxy matrix, and the fracture toughness, shear strength, and peel strength of the resulting nanocomposites were increased by about 49%, 39% and 85%. Reddy et al. [137] found that the mechanical properties of polymeric nanocomposites could be significantly improved at very low load concentrations of WSNTs. The elastic modulus of electrospun PMMA fiber composites at 2 wt.% loading of WSNTs improved 22-fold, and their tensile strength increased 30–35% compared with pristine PMMA fibers. Lalwani et al. [138] reported that PPF nanocomposites reinforced by WSNTs at low loading concentrations (0.01–0.2 wt.%) possessed a significant enhancement (up to 28–190%) in the mechanical properties, including compressive modulus, compressive yield strength, flexural modulus, and flexural yield strength. And the efficacy of WSNTs as reinforcing agents was better than (up to 127%) or equivalent to that of carbon nanotubes (SWCNTs and MWCNTs). In general, WSNTs have the potential to be a new reinforcing material for the manufacture of scaffolds in tissue engineering, and there is still great space for further research in the future.

Titania (TiO_2) is known as a class of biomaterials with excellent biocompatibility for orthopedic applications, and has been widely used as a coating for bone implants. TiO_2 nanotube is highlighted as an implantation material due to its unique properties such as high specific surface area and the ability to exhibit positive cellular response. TiO_2 nanotube might also be a good candidate as nanotube reinforcements in scaffolds for tissue engineering. Recently, TiO_2 nanotube surfaces have been proposed as alternative architectures to enhance the interaction between the implant and tissues for orthopedic implant considerations, as well as the drug delivery systems. To date, a great number of studies have demonstrated the TiO_2 nanotube surfaces exhibit a positive effect on cell behavior such as accelerated apatite formation, enhanced osteoblast adhesion, proliferation and differentiation [139]. The diameter of TiO_2 nanotube shows significant influence on cell response. For instance, a recent study by Minagar et al. [140] indicated that TiO_2-ZrO_2-$ZrTiO_4$ nanotubes with inner diameter (Di) ~40 nm exhibited the highest level of SaOS2 cell adhesion (41.0%), compared to 25.9%, 33.1% and 33.5% cell adhesion for nanotubes with Di 59 nm, 64 nm and 82 nm. The studies *in vivo* demonstrated that the TiO_2 nanotube surfaces significantly improved the bone bonding strength, about nine fold compared with TiO_2 gritblasted surfaces [141]. However, there is little literature on the use of TiO_2 nanotubes to enhance tissue engineering scaffolds.

This nanotube structure can provide pographical cues and signals to guide the cell functions of different cell lines. It is therefore foreseen that nanotubes, both CNTs and noncarbonic nanotubes, are potential candidates as reinforcements for future biomedical applications. The future growth of interest in the research of reinforcing materials should be expected in the field of nanotubes. More future work should focus on making directional nanotubes by advanced manufacturing techniques to further improve the performance of composite materials.

2.3.3 *Self-Reinforcement*

There is a growing interest for the development of self-reinforced materials, which are a substitute for traditional fiber-reinforced and nanotube-reinforced composites. This research topic has been focused because both the reinforcing and continuous phases involve the composites with the same chemical composition, which are easier to convert and facilitate mass production. Self-reinforced composite (SRC) is an interesting concept that a matrix is reinforced with oriented fibers and tapes, or particles, where both the reinforcing and matrix phases are given by the same suitable material but have different mechanical properties. Moreover, in the published literature, the matrix and reinforcing materials of the composites, which are different material but belong to the same family, are also referred to as SRCs [142].

SRCs are particularly important in biomaterials applications, since any additives composed of different chemicals can affect biocompatibility and biodegradability. As a new development field, the fundamental research and processing technologies are still far from mature, but the recent research results have demonstrated the great potential of SRCs as biomaterials in tissue engineering. More and more R&D works have been devoted to the field biomaterials [143, 144].

As a common phenomenon in nature, some natural materials are SRCs. For example, wood is formed by cellulose-based SRC, and muscle is composed of self-reinforced protein. In the recent, the development of SRCs has emphasized on natural materials. The SRCs of cellulose have also been widely used to reinforce other polymers of the same family. Li et al. [145] used regenerated silk fibroin SRC fabrics to enhance the interfacial adhesion between the silk fibers and the fibroin matrix, which significantly improved the overall mechanical properties and thermal stability of silk fiber/fibroin composites. Lu et al. [146] reported the development of the self-reinforced composite of sisal prepared by melt and hot pressing, which had inherent interfacial compatibility and full biodegradability. Similarly, Nishino et al. [40] reported that the whole cellulosic composite exhibited excellent mechanical, thermal and biodegradable properties in application.

Reinforcing a synthetic matrix by embedding the same fibers also provides a possible means of meeting the high strength and stiffness requirements for many tissue applications. Niiranen et al. [58] studied the properties and degradation behavior of self-reinforced bioabsorbable polymer/bioactive glass composites, which consisted of the same poly($_{\text{L/DL}}$)lactide (PLA) 70:30 and bioactive glass (BaG) 13–93 by conventional melt-extrusion [147]. The bioactive glass fillers increased the hydrophilic nature of the composite material at the interface, and they also played a role in providing the macropores at the die stretching. The structure and quantity of glass fillers on a macro-scale determined the hydrolysis degradation of the composite, which was more significant *in vivo* than *in vitro*. Thus, the self-reinforced composites with a porous structure can provide an advantageous substrate for creating the interlocking mechanism stability and guiding cell adhesion, proliferation and deposition to form bone tissue. In the other study, Majola et al. [148] indicated that the initial bending and shear strengths of PLLA SRCs at 250–271 MPa and 94–98 MPa, respectively, were higher than those of non-reinforced PLLA with the same molecular mass. Similarly, Wrightcharlesworth et al. [149] also reported that a oriented PLA fiber SRCs prepared by the hot press method had higher initial mechanical properties. As for the study of other artificial polymers, Deng et al. [147] indicated that the ultra-high-molecular-weight polyethylene (UHMWPE) matrix reinforced with the same fibers was a good potential candidate of SRCs, which could be used in load-bearing biomedical applications. Compared with pure UHMWPE, the tensile modulus, double notch impact strength and creep resistance of the SRCs were increased, although there was no obvious difference in the wear properties of the two materials. In addition, Wright et al. [150–154] evaluated mechanical properties of three different types of wove PMMA SRCs, and found that each one had its own fracture processes different from others, which were caused by the different orientation of fibers and the fracture toughness. The best combined PMMA SRC, absorbing the same amount of water as bone cement, was equal to or better than the bone cement in the tests, which made the PMMA SRC worthy of further consideration as a candidate biomaterial in tissue engineering. Amer et al. [155] investigated optimum processing conditions of high-density PE (HDPE) SRCs, including processing temperature, pressure and duration, and found that the composite modulus increased with the decrease of the composite cooling rate. And the authors held that PE SRCs was promising candidate materials for the preparation of a number of scaffolds with wear-resistance in tissue engineering.

Self-reinforced materials are one of the most promising reinforcing biomaterials in tissue engineering, because they possess special characteristics that are different from conventional polymers. Thus, it is expected that the self-reinforced material as reinforcements will provide an interesting method to increase the strength, degradability and thermal stability of the composites for the future development in various biomedical applications. The growing interest in the self-reinforced material will continue to rise.

2.4 Fabrication Techniques

Cells and tissues in the body are organized into physiological 3D architecture. Tissue engineering scaffolds are aimed to engineer these functional tissues and organs by the different methodologies to facilitate the cell distribution and guide their growth into suitable 3D space. As for an ideal reinforced scaffold, it is important not only to select a material, but also to develop a satisfactory processing technique so as to achieve uniform structure and composition throughout the reinforced scaffolds and to have good adhesive strength between the matrix and the fibers or tubes. The processing technique should also be able to orient the fibers or tubes in the main trajectories, leading to an expected final shaping of the reinforced scaffold.

The selection of suitable manufacturing techniques that are compatible with the biomaterial properties is as important as choosing the biomaterial for the development of tissue scaffolds. Over the past decade, the development of manufacturing technology for porous scaffolds has been an intensive area of research. Here, we discuss the main techniques of scaffold manufacturing and emphasize the application of these technologies in the field of the reinforced scaffold.

2.4.1 Textile Technique

Textile techniques have been developed to fabricate the complex tissue-like constructs with the structure and properties similar to native tissues. Conventional textile techniques, including weaving, knitting, and braiding (Fig. 2.8), have been used to making porous structures with good mechanical properties as tissue scaffolds or reinforcements [1, 73].

Weaving, as one of the widely used textile techniques, is totally based upon two distinct sets of warps or wefts, which interlace at right angles to form fabric with controlled strength, porosity, morphology, and geometry [67, 156]. The weaving structures are lightweight, strong, and flexible and have been used to prepare 3D scaffolds. However, the disadvantage of this technique is the poor resistance towards forces applied in the through-plane direction, because the structure is usually two dimensions. Recently, different layers have been connected by interlocking all the layers to overcome this problem and create a structure with a low shear resistance in the in-plane direction.

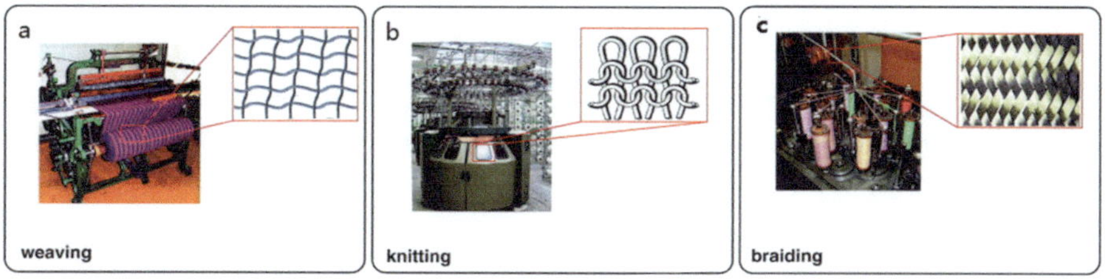

Fig. 2.8 Overview of the instruments and techniques used for fabricating fibrous structures and scaffolds and representative microstructures. (**a**) weaving of scaffolds by interlacing warp and weft fibers in perpendicular directions; (**b**) knitting of yarns in a series of connected loops; (**c**) braiding of fibers achieved by intertwining three or more fiber strands [73] (Adapted with permission from Ref. [73]. Copyright 2013 Elsevier Ltd)

Knitting is formed by intertwining yarns or threads in a series of symmetric loops as compared to weaving or other textile technique with straight threads. The knitted textile substrate includes fibers in the through-plane directions to offer higher through-plane mechanical properties [157, 158]. In addition, the knitted structure generally has a higher porosity but a lower thickness than that made by weaving. However, the disadvantage of the knitting structure is the weak mechanical properties in the in-plane direction. For their suitable mechanical properties and easy manufacture of 3D geometries, the knitted structure has been combined with other materials to provide sufficient and adjustable mechanical properties for various tissue engineering applications.

Braiding is good at forming complex structures or patterns by intertwining three or more fiber strands [159], which can make cylinder and rod structures suitable for engineering tissues with the similar shapes, like blood vessels etc. The multi-step braiding structure is the only textile construct that is resistant to bending, shear, in-plane and through-plane loads, and they can also withstand twisting and internal pressure and offer the highest axial strength [160]. But the porosity of this technology is lower than that of knitting. The braid has the potential for the preparation of scaffolds with high mechanical strength and has also been developed to combine with other manufacturing techniques.

Textile technique is a conditional tissue engineering fabrication system that has been employed to fabricate the matrix or reinforcements in composite scaffolds and plays an important role among the fabrication techniques [161]. There are many studies on these techniques used for preparing reinforced scaffolds. For example, McCullen et al. [1] stated the use of traditional textile techniques to prepare fiber-reinforced scaffolds for tissue engineering and regenerative medicine. Huang [162] et al. investigated the mechanical properties of composites reinforced with woven and braided fabrics. Xu et al. [163] reinforced small-diameter PU vascular grafts by weft-knitted tubular fabric to improve the mechanical properties of the grafts. Ananta et al. [164] fabricated a 3D hybrid scaffold for cartilage regeneration by incorporating a type I collagen within a PLGA knitted mesh to improve cell adhesion. In another study, the knitted structure was used to enhance PU blood vessel grafts. The results were shown that the strength and elasticity of the graft made from the composite material were higher than that of the implant made from pure PU. Over the next few years, this technology will continue to be one of the main processing methods used to prepare reinforced scaffolds, especially fiber-based composites, and attract extensive research and attention in the field of tissue engineering.

2.4.2 *Solution Casting*

Solution casting and particulate leaching are widely used in the manufacturing of porous polymer scaffolds [165]. The solution casting is based entirely on the evaporation of some solvent to form scaffolds through one of the two routes. With any large equipment, it is a very simple, easy and inexpensive way to prepare scaffolds. One method is to immerse the mold into polymeric solution with sufficient time to produce a polymer film. Another one is to add the polymeric solution into a mold with sufficient time until the solvent evaporates, which can also produce a layer of a polymeric membrane that is adhered to the mold [166]. Recent studies have shown that that solution casting is not only commonly used in the manufacture of 3D but also abled to form reinforced composite scaffolds. Fan et al. [167] synthesized graphene-reinforced chitosan composites as scaffold materials in tissue engineering by solution casting method. Sadat-Shojai et al. [168] designed nano-HA (HAp) reinforced PHB composites prepared by solution casting. Lin et al. [169] evaluated PCL reinforced type II collagen scaffolds as tissue-engineered trachea by solution casting and freeze drying the PCL in a mold container. Julkapli et al. [170] evaluated thermal properties of 4,4-oxydiphathalic anhydride chitosan reinforced chitosan bio-composites by solution casting method. In addition, Govindasamy et al. [171] produced chitosan-cellulose-halloysite membranes by using solvent casting method. Kong et al. [172] manufactured natural fiber-reinforced polymer composites by solvent casting method.

The main drawback of the technique is that the designed scaffold may retain some toxicity from the toxic solvent. The method of removing toxic solvent is completely dry by a vacuum process, but it will cost a lot of time. To overcome these problems, solvent casting usually combines with particulate leaching techniques to fabricate scaffolds, namely solution casting/particulate leaching technique [14].

2.4.3 *Particulate-Leaching Techniques*

Particulate leaching is one of the popular and easy techniques to fabricate scaffolds with the pore size (~500 μm), percentage porosity (around 94–95%) and desired crystallinity for tissue engineering applications [173, 174]. In short, the salt is first grounded into small particles, and those particles with a desired size are poured into a mold [166, 175]. A polymer solution is then cast into the salt-filled mold. After the evaporation of the solvent, the salt crystals are leached away using water to form the pores of the scaffold. Particulate leaching is a widely used and mature technique in the manufacturing of porous polymer scaffolds, and also has a good performance in the preparation of the reinforced scaffolds, as some research results shown. Guarino et al. [30] prepared the PLLA fibers embedded in a porous PCL matrix by co-use of particulate leaching technique and filament winding technology. Jiao et al. [52] fabricated PLLA–chitosan hybrid scaffolds by particulate-leaching and subsequently freeze-drying, which combined the individual advantages of synthetic and natural polymers. Mou et al. [176] prepared silk fibroin-collagen-HA composite scaffold by particulate leaching method.

The pore size can be controlled by the porogen/polymer ratio and the size of the porogen, such as salt, wax and sugar. The advantage of this method is the requirement of very less amount of polymer to fabricate the scaffold. However, certain critical variables such as pore shape and inter-pore openings are not controlled.

2.4.4 Phase-Separation

Phase separation is a homogeneous multicomponent system, just like a polymer-water emulsion, which becomes thermodynamically unstable under certain conditions and tends to separate into more than one phase for lowering the system free energy [177]. A polymer solution separates into two phases, one having low polymer concentration (polymer lean phase) and other having the high polymer concentration (polymer rich phase). The polymer is dissolved in phenol or naphthalene, and then the biologically active molecules are dispersed in these solutions. By lowering the temperature, the liquid-liquid phase is separated by lowering the temperature and quenching the mixture below the freezing point of the solvent to form a two-phase solid, followed by freeze-drying to produce porous scaffolds, in which the biologically active molecule is integrated into the structure [166, 178, 179]. The architecture of the scaffold could be controlled by various thermodynamic and kinetic parameters, such as the selection of solvent phase separation temperature. Phase-separation technique can be used to fabricate scaffolds from many types of polymers and polymeric composite materials [179]. Phase separation, as one of the common technologies for fabricating various composites, is also considered to be an effective means of developing a reinforcing material to provide adequate topographical architecture and mechanical properties of scaffolds for tissue regeneration. There are many studies on the applications of reinforced scaffolds prepared by phase separation. Bonnaud et al. [175] studied the effect of reinforcing glass fibers on morphology and properties of thermoplastic modified epoxy-aromatic diamine matrix prepared by the phase separation. Ma et al. [180] prepared the highly porous CNTs reinforced thermoplastic PLLA nanocomposite scaffold by a thermally induced phase separation method. Nakhoda et al. [181] designed a new class of PU biocomposites reinforced with green biocellulose nanofibers (BC), and the porosity

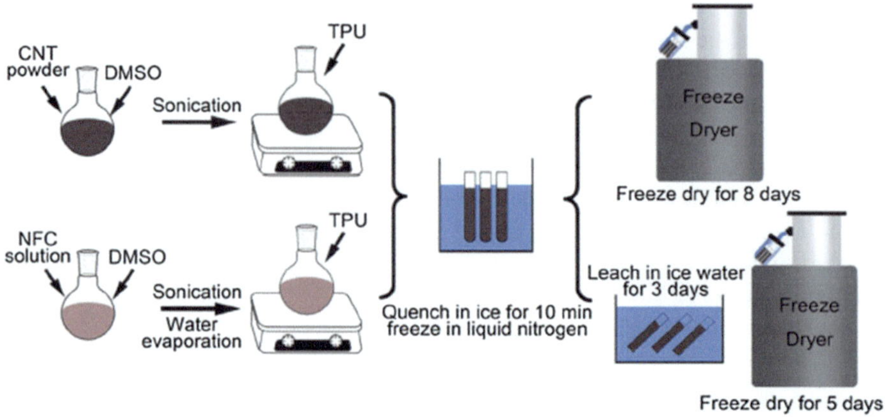

Fig. 2.9 Fabrication of TPU/CNTandTPU/NFC scaffolds using thermally induced phase separation via different solvent removal methods [182] (Adapted with permission from Ref. [182]. Copyright 2016 Elsevier Ltd)

was introduced in the PU matrix using a combination of salt leaching and thermally induced phase separation (TIPS). Mi et al. [182] prepared the TPU/CNT and TPU/NFC scaffold by using thermally induced phase separation based on different solvent removal methods (Fig. 2.9). Tian et al. [183] built TiO_2 nanoparticle reinforced PLLA/TiO_2 porous scaffolds by using thermally induced phase separation (TIPS) and further examined the influence of TiO_2 contents on the physicochemical properties of porous scaffolds.

The advantage of the phase separation technique is that it can easily combine with other fabrication technology, such as particulate leaching or rapid prototyping, to create 3D scaffolds for tissue engineering applications [184]. However, the disadvantage of this technique is that the pores have a diameter on the order of a few to tens of microns and are usually unevenly distributed, which limits its applications in tissue engineering.

2.4.5 Porogen Leaching

Porogen leaching is a widely used, facile, convenient and cost-effective technique for the preparation of a wide range of polymer scaffolds to introduce porosity porous in tissue engineering. The porogen leaching method involves the casting of a polymer/porogen solution in a mold, drying the mixture, followed by solvent evaporation and leaching the incorporated porogen with water to generate the pores [185, 186]. Various porogens, including sodium chloride (NaCl), paraffin spheres, sugar crystals, gelatin, and polymers [187–189], have been successfully used to fabricate porous structures in practical manufacturing techniques. Porogen leaching has been already used to fabricate reinforced scaffolds in tissue engineering, especially for

Fig. 2.10 Three-dimensional micro-CT reconstructions showing the porosity and architecture of 2 mm thick sections from the center of 10 mm diameter PEKK scaffolds with 40 vol.% HA whisker reinforcement and (**a**) 75, (**b**) 82.5 and (**c**) 90% porosity, molded at 350 °C [188] (Adapted with permission from Ref. [188]. Copyright 2010 Elsevier Ltd)

bone repair and regeneration [190, 191]. Rajkhowa et al. [192] used porogen leaching technology to reinforce silk scaffolds with silk particles prepared from aqueous and HFIP silk solutions through the addition of silk particles. Gil et al. [193] systematically studied porogen leached scaffolds prepared from silk proteins through the addition of silk particles in combination with silk solution concentration to improve the mechanical properties. Converse et al. [188] prepared and characterized HA whisker-reinforced polyetherketoneketone (PEKK) bone ingrowth scaffolds by porogen leaching (Fig. 2.10).

The main advantage of this technique is easy and versatile to manipulate the pore structure by controlling the property and fraction of porogen to fabricate the highly porous scaffold with porosity up to 93% and pore diameter up to 500 μm [194, 195]. One of the main disadvantages of this technique is the realization of larger pore size and the increase of accurate pore interconnectivity [196].

2.4.6 *Gas-Forming*

Gas foaming uses high pressure carbon dioxide gas for the fabrication of highly porous scaffolds. A polymer solution is saturated with carbon dioxide at a high pressure (800 psi) [47]. When the dissolved carbon dioxide becomes unstable under this condition, it will separates from the polymer, and its molecules become cluster to minimize the free energy to create pore nucleation (foaming) [197]. After completing the foaming process, a 3D porous polymeric structure of scaffold is formed with expanded polymeric volume and decreased polymer density [198]. In the recent, Gravel et al. [199] described the first attempt in using natural coralline biomaterials as reinforcing gas-forming agents for the development of 3D macroporous hybrid biomatrices of chitosan and natural coralline material. Yun et al. [200] used a gas forming method to obtain micro-sized pores in the electrospun fiber bundles for designing a dual-porosity structure of Montmorillonite reinforced PLLA nanocomposite scaffold. And the robust dual-porosity scaffold structure was used for a facile transport of metabolic nutrients and wastes through the nano-sized pores, as well as, for the cell implantation and blood vessel invasion through the micro-sized pores. Therefore, the gas foaming can be utilized as a step in the manufacture of highly porous reinforced scaffolds, in combination with other fabrication techniques.

The great advantage of gas foaming is that the technique doesn't require the use of organic solvents compared to many manufacturing techniques requiring high temperature and organic solvents, where the residue will damage cells and nearby tissues. Such techniques widely used in industry are rarely used in the manufacture of scaffolds. Because it is hard to control the size and interconnection of pores, and large closed pores can be created inside the polymer structure during the foaming process, which is unsuitable for cells [201]. Thus, the gas foaming is mainly used to form bone scaffolds [201–203]. Recently, many modifications of the gas foaming method have been explored, such as the use of polymer-salt mixtures [204], or ammonium bicarbonate salt instead of high pressure carbon dioxide with [205].

2.4.7 Melt Molding

A large number of techniques are developed for the manufacturing of scaffolds to control the pore interconnectivity and geometry, which are provided channels to the exchange of nutrient/wast used for cell survival and growth. Melt molding is commonly used for the preparation of porous ceramics and metals [206]. In addition, it is also used to fabricate polymeric materials [207]. Melt molding process is that the Teflon molds are filled with PLGA powder and specific diameter gelatin microspheres, and then heated at a polymer's glass transition temperature (Tg) under the elevated pressure [185]. This action causes the PLGA particle to be bonded together to form a mold-shaped scaffold. Once the mold is removed, gelatin microspheres are leached out by immersing the mixture into water, and the scaffold in the shape of the mold is then dried. The mold decides the shape of the scaffolds produced by this technique. The melt molding is the most convenient and economical method of making porous scaffolds, for it allows the rapid production of polymer scaffolds with various shapes and sizes and doesn't require the use of organic solvents to create scaffolds [186]. The shape and size of pores can be controlled by selecting porogen. As for the manufacture of scaffolds, Liao et al. [208] incorporated MWNTs of 0.1 and 0.3 wt.% and hydoxyapatite nanorods (nHAs) of 8–20 wt.% into polypropylene (PP) to form biocomposites using melt molding techniques. Although there are few reported about this application, this technique still has the potential to be developed as a new processing technique for forming the reinforced scaffolds with suitable structures and compositions. This method can also be combined with other methods, like the solvent casting technique, to provide a route for the preparation of the reinforced scaffolds.

However, there are some disadvantages of this method, such as the presence of nonporous layers on the surface, the complexity of the design and construction of the mold, the difficulty of leaching out the porogen compound, and the unfavorable high-temperature process [209, 210]. The melt molding technique is mostly used for the preparation of bone scaffolds as it allows the combination of ceramic HA and PU [211, 212].

2.4.8 Freeze Drying

Freeze drying has been extensively used to prepare porous structures with many different polymers, including silk proteins [213, 214], PGA, PLLA, PLGA and PLGA/PPF blends, for tissue engineering and biological applications. Freeze drying is a slow batch process based on the principle of sublimation to extract dry product from an aqueous solution. This technique first dissolves the polymer in a solvent to form a solution of desired concentration. Then the solution is usually frozen, and the solvent is removed by lyophilization under the high vacuum to prepare a the scaffold with high porosity and interconnectivity [215]. Freeze drying technique is also a common and effective method that is widely used in a suitable reinforced scaffold for successful transplantation therapy. Kane et al. [216] studied the freeze-dried HA reinforced collagen scaffolds to evaluate the effects of HA reinforcement on the architecture and mechanical properties of freeze-dried collagen scaffolds. Mobini et al. [217] fabricated 3D scaffolds by embedding of natural degummed silk fibers in a matrix of regenerated fibroin, followed by freeze-drying. Hiraoka et al. [218] obtained PGA fiber-reinforced collagen sponges by freeze-drying a collagen solution with PGA fibers and the dehydrothermal cross-linking technique.

When compared to the other methods [219], the main advantage of this technique is that it doesn't require a high temperature and is conducive to maintaining the activity of biomacromolecules and pharmaceuticals [220, 221]. The freeze-drying process doesn't bring any impurities into the samples, so that a separate leaching step is not required. The drawback of this technique is smaller pore size and long processing time [221]. Despite some of the shortcomings of this technique, it is often applied for the manufacture of tissue-engineered scaffolds, even in the present study.

2.4.9 Emulsion Freeze Drying

Emulsion freeze-drying is mainly used for the preparation of soft tissue scaffolds [222, 223]. The basic principle of emulsion freeze drying is to form a water-in-oil emulsion by homogenisation of two immiscible phases, in which the continuous phase contains the polymer-rich solvent and the dispersed phase is water [224]. Next, the homogeneous emulsion is quenched in liquid nitrogen to maintain a liquid structure and then freeze-dried to produce a porous structure with a pore size of 20–200 μm and a porosity of more than 90% [221, 225]. The porosity and pore size depend on the concentration and viscosity of the polymer solution and the amount of aqueous phase dispersed in the emulsion. The average pore size also decreases as the volume of the aqueous phase decreases. However, the dispersed aqueous phase is exposed to higher shear forces when the polymer concentration increases in the continuous phase, which reduces the size of the small water domains, resulting in materials with lower porosity and smaller pores. More uniform porous scaffolds, including the reinforcing scaffold, can be manufactured using a constant cooling rate freezing techniques. For example, Shemesh et al. [226] developed and studied the bioresorbable hybrid structures that combined a synthetic porous drug-loaded top layer with a spongy collagen sublayer made by emulsion freeze drying. As for the reinforced composite materials, this technology provides a promising way, though there are few research reports.

The advantage of this technique is that the use of toxic solvents is reduced and it doesn't eliminate the drying and leaching process of the porogen component. In order to maintain the stability of the emulsion, the method requires the addition of suitable surfactants, like PU etc.

2.4.10 Electrospinning

Among different fiber manufacturing techniques, electrospinning has been widely used in the manufacture of tissue scaffolds with an average fiber diameter at micrometer or nanometer scales due to its relatively simple and easy control of critical process parameters [227, 228]. In the past decade, different natural materials, like collagen and chitosan, and synthetic polymers, like PLA, polyglycolic acid, polycaprolactone, etc. can be electrospun into fibers with different diameters from nanoscale to microscale for the development of tissue scaffolds [229, 230]. As the most commonly used technique, it also plays an important role in the manufacture of scaffolds for tissue engineering, and its application examples are numerous.

Basically, in electrospinning, a high voltage is used to produce an electrically charged jet of polymer solution or melt that forms polymer fibers after drying or solidification [73, 230, 231]. This process is controlled by a high intensity electric field between two electrodes having an opposite polarity charge. One electrode is placed in the polymer solution, and the other is placed in a collector. Typically, the polymer solution is pumped as a result to form a drop of solution. Afterward, an electric field is generated and intended to produce a force caused by the droplet to overcome the surface tension of the solution. A jet of the polymer is ejected to produce the fibers while the solvent begins to evaporate due to the jet formation and continues after the nanofibers are deposited to the collector (Fig. 2.11). The electrospinning can produce fibers with diameters in the range of 100 nm to several micrometers, and the fibers have special orientation, high aspect ratio, high surface area and controllable pore geometry, which are favorable for better cell adhesion, cell expression, and transportation of oxygen and nutrients to the cells. The factors that influence the electrospinning process are the physical properties of polymer

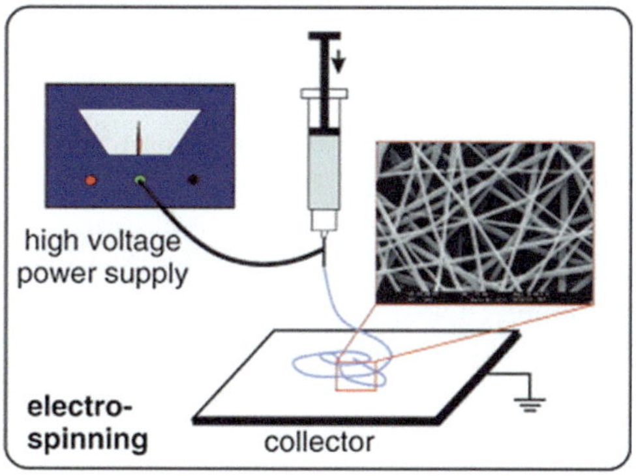

Fig. 2.11 Electrospinning of polymeric fibers that are drawn under an applied electric field and deposited on a surface to form a fibrous scaffold [73] (Adapted with permission from Ref. [73]. Copyright 2013 Elsevier Ltd)

(e.g., viscosity, electrical conductivity and surface tension), the applied electric field, the polymer flow rate and the distance between the jet and the collector.

This technique, as one of the main processing methods, has been applied to fabricate the reinforced composites for various scaffolds in tissue engineering. There are countless examples of this technique used for the reinforced scaffolds. Kang et al. [113] found that MWCNTs embedded in the silk fibroin nanofibers by electrospinning were well aligned along the nanofibers axis. Cont et al. [82] designed a novel composite scaffolds consisting of long continuous bidirectional fibers embedded in an electrospun matrix for the purpose of using them in soft tissue engineering applications. He et al. [232] prepared all-cellulose nanocomposite nanofibers reinforced with cellulose nanocrystals by electrospinning. Yang et al. [233] developed an electrospun-aligned nanoyarn-reinforced scaffold for tendon tissue engineering to improve mechanical strength and cell infiltration. Rui et al. [234] fabricated anisotropically aligned nanofibrous scaffolds reinforced with cellulose nanocrystals for tendon tissue engineering by electrospinning. Eslami et al. [235] integrated electrospun poly(glycerol sebacate) (PGS)- PCL microfiber scaffolds for heart valve engineering.

The advantages of electrospinning for scaffold fabrication are relative simple and have the possibility of enlarging and effectively controlling critical process parameters such as flow rate and voltage. However, the disadvantages of the technique are that it is difficult to create complex 3D scaffold shapes and internal pore networks, and the average fiber diameter is usually on the larger side of the ECM fibers and sometimes falls within micrometer range.

2.4.11 *Self-Assembly*

Self-assembly is another exciting area of research, especially in nanometer size and patterned biological structures. Self-assembly is now loosely defined as the autonomous organization of the molecule into an ordered structure without human intervention [236, 237]. The self-assembly of natural or synthetic molecules has been used to produce nanofiber materials with the potential as tissue engineering scaffolds [229].

A common method of self-assembly is the hydrophobic interaction of amphiphilic peptide sequences, because molecules are clustered together for the manufacture of 3D nanofiber structures in tissue engineering. The hydrophobic and hydrophilic domains within these peptides in aqueous solution interact together with the help of weak non covalent bonds, such as hydrogen bond, Van der Waals interactions, ionic bond and hydrophobic interaction, resulting in different rapid recovery of hydrogen [238]. Peptide self-assembling has been used to produce 3D cell culture systems and feature fibers diameters on the order of 10 nm with pore sizes between 200 and 500 nm [236]. The modification of the peptide ampholyte structure involves various self-assemblies including layered and lamellar structures, and its reversibility properties provide flexibility for the system [239]. Diblock polymer (AxBy) is another self-assembly method for the preparation of synthetic polymer nanofibers, which is not belong to the peptide. when the two blocks separate from each other in large quantities due to their incompatibility, the volume formation of A and B can be controlled to obtain B cylindrical domain with nanoscale diameter embedded into matrix A [165]. Hosseinkhani et al. [240] designed the hybrid scaffold consisting of two biomaterials, that is a hydrogel formed through self-assembly of peptide-amphiphile and a collagen sponge reinforced with PGA fiber. Zhou et al. [241] developed a versatile approach using layer-by-layer self-assembly to incorporate cell adhesion and spatial representation of neurotrophic factors into complex nanofibrous scaffolds. Therefore, self-assembly will be a good candidate for the formation of reinforced scaffolds both in the matrix and reinforcements.

Self-assembly has several advantages over the electrospinning. The fabricated nanofiber is much thinner [242] and have amino acid residues that can be chemically modified by adding bioactive moieties. The technique is carried out in aqueous salt solution or physiological media without the use of organic solvent, which reduces cytotoxicity [243]. Its main disadvantage is a complicated and elaborated process. In general, the self-assembly techniques show the potential to be used in the design of novel scaffolds for tissue engineering applications.

2.4.12 Rapid Prototyping

As a more advanced technique for scaffold manufacturing, rapid prototyping (RP), also known as solid free-form technique (SFF), is computer-controlled fabrication technique that can rapidly produce 3D architectures by using a layer-by-layer additive method [244, 245]. Since biomedical devices were developed by using a custom-built 3D printing (3DP) machine in the early 1990s, numerous SFF technologies, such as fused deposition modeling [246], stereolithography [247], and selective laser sintering [248] have been utilized to fabricate scaffolds. By the year 2000, computer-aided tissue engineering (CATE) has emerged to produce

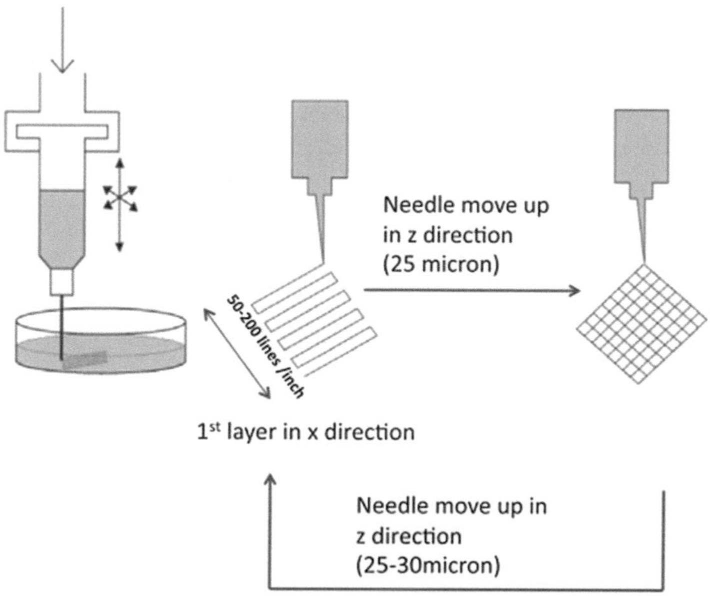

Fig. 2.12 Schematic of the experimental setup [86] (Adapted with permission from Ref. [86]. Copyright 2012 Elsevier Ltd)

multi-functional scaffolds. PR technology uses the computer added design (CAD) software to express the designed scaffold as a series of cross-sections [249], then lays down a layer of material from the bottom and moves up a layer at a time to build complex 3D scaffolds that correspond to each cross section. According to the state of materials before forming, the RP technology can be divided into the liquid-based, solid sheet and discrete particle technologies.

RP is an efficient to produce the scaffolds with desired properties. At present, metal, polymer and ceramic can be used as materials to create 3D models by this technology. The advantage of RP is that not only the matrix architecture of scaffolds, such as size, shape, inter connectivity, branching, geometry and orientation, but also the mechanical properties, biological effects and degradation kinetics of scaffolds can be designed similar to the original structure with different material composition [250, 251]. In addition, it is also used for the preparation of reinforced scaffolds. For example, Agrawal et al. [86] used the 3D RP technology to form crossed "log-piles" of elastic fibers that were then impregnated with an epoxy-based hydrogel in order to form the fiber-reinforced gel, as shown in Fig. 2.12. Lian et al. [252] developed a method based on RP technology to produce chitosan fiber CPC composites (CF/CPC) as scaffold materials for bone tissue engineering applications. Alge et al. [253] also used RP technology to prepare poly(propylene fumarate) reinforced dicalcium phosphate dihydrate cement composites for bone tissue engineering. Yong et al. [254] utilized RP technology to form collagen scaffolds reinforced with PCL/beta-TCP nanofibres to obtain the high efficiency of cell seeding and enhanced mechanical properties for bone tissue regeneration.

RP technique can be easily combined with imaging techniques to produce the scaffolds. The main inherent shortcoming of this technique is the limited material selection and inadequate resolution. Most of RP systems are too expensive, which is an important factor confining the use of RP technique. In addition, educing product costs and manufacturing time is also the long-term aim of the project.

2.4.13 *Three-Dimensional Printing*

Solid freeform fabrication (SFF) technology creates a 3D structure in laminated fashion by numerical models [255]. 3D printing also uses a layer-by-layer deposition technique. The molten polymers or ceramics are extruded through a small jet head, which is combined with the material on the previous layer. In fact, 3DP employs ink jet printing technique to eject a printing liquid binder from a jet head, which moved to the polymer powder bed of the porogen and polymer particles based on the CAD cross-sectional data [256, 257], as shown in Fig. 2.13 [258]. The mechanical manipulation of the x-y position of the jet head controls the pattern for each layer according to the computer parameters, which can be different or arbitrary for each deposited layer [259]. The solvent will dissolve the polymer and evaporate, and the polymer will re-precipitate to form a solid structure. The final porosity is achieved after particulate leaching and solvent removal.

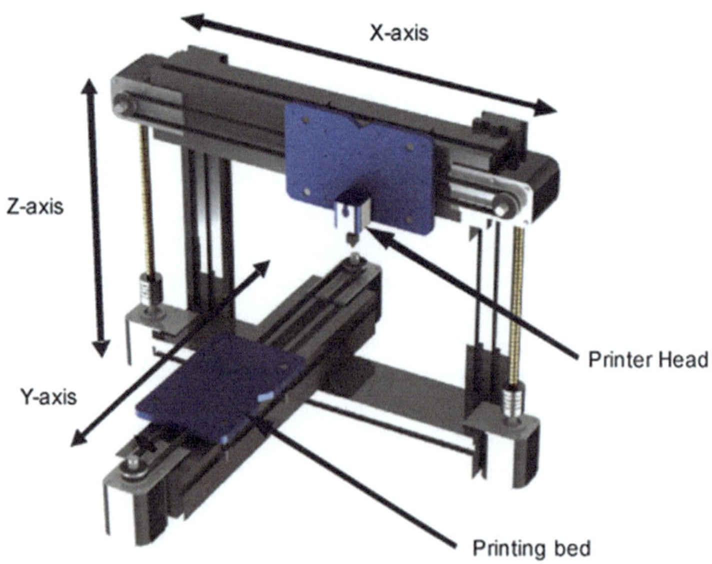

Fig. 2.13 Schematic of a 3D positioning system incorporating a print head and printing bed system. [258] (This article is available under the terms of the Creative Commons Attribution License)

As one of the advanced technologies in tissue engineering, 3DP technology has been successfully applied for the fabrication of complex reinforced 3D scaffolds by using both direct and indirect techniques. Bakarich et al. [260] combined digital modeling and 3D printing to prepare fiber reinforced hydrogels in a single-step process. Suwanprateeb et al. [261] designed a double infiltration technique combined with 3DP technology to increase the mechanical properties of natural polymers. Christ et al. [257] also introduced the process of 3DP in detail for preparing fiber reinforcements.

The main advantage of this technique is to fabricate 3D complex scaffolds with large pore sizes and fine features that can better simulate the native tissue structure as a template for cell growth and ECM formation. 3DP also has the potential to fabricate zero-order release formulations with complex geometries. However, the drawback of the 3DP is that it is difficult to remove the organic solvent and the support powder from complex architectural features deep within the scaffold. The solvent-based process of 3DP technology results in a significant increase in manufacturing time and poor mechanical properties. Synthetic thermoplastic polymers are the only type of materials that can be processed by this method, excluding many natural biomaterials or thermosetting synthetic polymers.

2.4.14 Microsyringe Deposition

Microsyringe deposition is also similar to a layer-by-layer deposition technique in practice, that a polymer is patterned onto a precision-controlled surface using a continuous stream of material via layer-by-layer deposition to fabricate 3D structures with various shapes for tissue engineering [262]. The deposition based on pressure-assisted microsyringe takes use of the compressed air to eject a solution of a polymer in volatile solvent through a narrow capillary needle with ranging a diameter from 10 to 20 μm [47, 263]. Microsyringe deposition takes advantage of the micropositioning system with the lateral precisions of 0.1 μm to manipulate the position of the solution in the *x*-*y* plane. The resulting 3D structure has a physical dimensions between 5 and 600 μm, depending upon various processing parameters [120]. The 3D PLGA scaffolds have been fabricated by this technique, so that any synthetic or natural polymer, that is soluble in a volatile solvent, can be used as a material for this method in theory [264].

Microsyringe deposition technique as a young advanced technique offers a convenient method to fabricate multiple 3D scaffolds with micron-scale precision. Low temperature is also conducive to maintaining protein and other biomolecule activity, which is required by creating a favorable microenvironment for tissue repair. Therefore, this technique will be a potential method for creating the reinforced scaffolds in tissue engineering.

2.5 Conclusion and Future Trends

Biomaterial scaffolds play a crucial role as a temporary support for cell life activities and directly determine the success of tissue transplantation in tissue engineering. By using various preparation techniques to remold and regenerate different tissues and organs in the body, numerous scaffolds are produced from a variety of biocompatible and bioresorbable materials. According to the characteristics of the repair organization, there are several key considerations for the development of scaffolds in tissue engineering, such as biocompatibility, biodegradability, mechanical properties, structure and manufacturing techniques. Choosing the right biomaterial is the basis for the scaffold design, which is the first step in the process. As described in the previous section, there are four typical individual groups of biomaterials, metals, ceramics, polymers and composites, used in the manufacture of tissue engineering scaffolds. Although these single-phase homogeneous materials have been widely accepted for the development of tissue scaffolds, each of these materials has particular advantages, not to mention shortcomings. Thus, there are still significant challenges in the development of scaffolds with desired pore structure, mechanical and bioactive properties, degradation and tissue in-growth rates and no inflammatory response in the host tissue. Recent studies have shown that the composite scaffolds containing heterogeneous or multifaceted materials are an effective and viable concept to improve the biomechanical and biological properties of scaffolds and promote the regeneration of functional engineered tissues.

This chapter describes the basic information of the research status that numerous potential matrix and reinforcement materials have been developed for the preparation of the scaffolds reinforced by fibers or tubes in tissue engineering. This research of the reinforced scaffolds has opened a new step for the development of bioengineered composites. As we have described in this chapter, the addition of reinforcements can optimize the chemical composition and distribution of the composites and improve their morphology, degradation and mechanical properties. Thus, additive agents in tissue engineering with a broad application prospect are an effective tool for engineering a variety of composite scaffolds with specific mechanical properties. Although these reinforced materials can better mimic the natural ECM of damage tissues or organs, their applications also cause a series of new problems that need further study when used in clinical applications.

From a material science point of view, the current challenge in tissue engineering is to design and manufacture reproducible bioactive and bioabsorbable scaffolds with 3D structures, which can maintain their structural and performance integrity at the expected time even under load-bearing conditions. Recently, the most interesting development of tissue engineers seems to develop new types of advanced composite biomaterial scaffolds and evaluate their important properties. Therefore, the integration of composite technology to fabricate suitable synthetic materials still remains great interest to all researchers and engineers. In future research on biological scaffold materials, the focus should be on the use of materials that may activate the body's own repair mechanism in clinical applications. Thus, further research should concentrate on the systematic study of molecular biology levels of cell-material interactions and the design of a new generation of bioactive materials to activate cell proliferation and differentiation for enhancing the rapid formation of ECM and tissue growth in situ. In the design of these scaffolds, bioactive molecules, such as growth factors, enzymes, ECM protein and DNA, are another important consideration.

The research so far has proven the potential of various materials as the matrix and reinforcements, however, the techniques used to produce these materials are far from mature. The manufacturing techniques are directly related to the properties and proposed functions of the tissue engineered scaffolds, as well as the success of tissue repair and reconstruction. Due to the diversity of materials, each technique has its own relative merits and different applications for manufacturing scaffolds, and thus the appropriate selection of the methodologies must satisfy the micro-environmental requirements of the repaired or engineered tissues. There is a need to develop new technologies and manufacturing systems for different biomaterials and further study the advantages and disadvantages of their use. In addtion, the existing techniques must be combined with other methods for implementing in a wider range of applications. The frequently used conventional techniques should be developed in depth to overcome their limitations for fabricating the next generation of tissue engineered scaffolds. In general, utilizing developed methods and strategies to fabricate scaffolds, that can better mimic the natural properties of native tissue, continues to attract wide attention and in-depth research.

Interdisciplinary research in the development of tissue engineering will continue to be a major concern for the researchers in this field. For a successful application of the scaffold in tissue repair or regeneration, the proper material with the suitable porous size, degradation rate and surface morphology can provide a good beginning, while choosing an adequate processing route is also contributed a lot. Furthermore, there is a need for closer ties between materials science, life sciences, and engineering researchers to develop the reinforced materials for specific applications in tissue engineering, which is critical to achieving success in the coming years.

References

1. McCullen SD, Haslauer CM, Loboa EG. Fiber-reinforced scaffolds for tissue engineering and regenerative medicine: use of traditional textile substrates to nanofibrous arrays. J Mater Chem. 2010;20:8776–88.
2. Wang H, Chang X, Qiu G, Cui F, Weng X, Zhang B, et al. Axial vascularization of nano-HA/collagen/PLA composites by arteriovenous bundle. J Nanomater. 2013;2013:3061–9.
3. Mooney DJ, Mikos AG. Growing new organs. Sci Am. 1999;280:60–5.
4. Vimal SK, Ahamad N, Katti DS. A simple method for fabrication of electrospun fibers with controlled degree of alignment having potential for nerve regeneration applications. Mat Sci Eng C-Mater. 2016;63:616–27.

5. Habib FN, Nikzad M, Masood SH, Saifullah ABM. Design and development of scaffolds for tissue engineering using three-dimensional printing for bio-based applications. 3d Print Addit Manuf. 2016;3:119–27.
6. Ma SH, Yu B, Pei XW, Zhou F. Structural hydrogels. Polymer. 2016;98:516–35.
7. Janowska-Wieczorek A, Marquez-Curtis LA, Shirvaikar N, Ratajczak MZ. The role of complement in the trafficking of hematopoietic stem/progenitor cells. Transfusion. 2012;52:2706–16.
8. Bezlaj. Še o rodbinskem imenu Tavčar. J Biomed Mater Res A. 2014;102:1580–94.
9. Li X, Yang Y, Fan Y, Feng Q, Cui FZ, Watari F. Biocomposites reinforced by fibers or tubes as scaffolds for tissue engineering or regenerative medicine. J Biomed Mater Res A. 2014;102:1580–94.
10. Murugan R, Ramakrishna S. Design strategies of tissue engineering scaffolds with controlled fiber orientation. Tissue Eng. 2007;13:1845–66.
11. Levine B. A new era in porous metals: applications in orthopaedics. Adv Eng Mater. 2008;10:788–92.
12. Nover AB, Lee SL, Georgescu MS, Howard DR, Saunders RA, Yu WT, et al. Porous titanium bases for osteochondral tissue engineering. Acta Biomater. 2015;27:286–93.
13. Rummo H, Veinthal R, Aruniit A. Pultruding of metal powder filled glass fiber reinforced polymer composites. Key Eng Mater. 2016;674:48–53.
14. Suh SW, Shin JY, Kim JH, Kim JG, Beak CH, Kim DI, et al. Effect of different particles on cell proliferation in polymer scaffolds using a solvent-casting and particulate leaching technique. ASAIO J. 2002;48:460–4.
15. Gu XN, Zheng YF, Zhong SP, Xi TF, Wang JQ, Wang WH. Corrosion of, and cellular responses to Mg-Zn-Ca bulk metallic glasses. Biomaterials. 2010;31:1093–103.
16. Li ZJ, Gu XN, Lou SQ, Zheng YF. The development of binary Mg-Ca alloys for use as biodegradable materials within bone. Biomaterials. 2008;29:1329–44.
17. Jiang GF, Li QY, Wang CL, Dong J, He G. Characterization and investigation of the deformation behavior of porous magnesium scaffolds with entangled architectured pore channels. J Mech Behav Biomed Mater. 2016;64:139–50.
18. Witte F, Ulrich H, Palm C, Willbold E. Biodegradable magnesium scaffolds: part II: peri-implant bone remodeling. J Biomed Mater Res A. 2007a;81A:757–65.
19. Witte F, Ulrich H, Rudert M, Willbold E. Biodegradable magnesium scaffolds: Part I: Appropriate inflammatory response. J Biomed Mater Res A. 2007;81:748–56.
20. Zhang X, Li XW, Li JG, Sun XD. Preparation and mechanical property of a novel 3D porous magnesium scaffold for bone tissue engineering. Mat Sci Eng C-Mater. 2014;42:362–7.
21. Cheng MQ, Wahafu T, Jiang GF, Liu W, Qiao YQ, Peng XC, et al. A novel open-porous magnesium scaffold with controllable microstructures and properties for bone regeneration. Sci Rep. 2016;6:24134.
22. Teixeira S, Fernandes H, Leusink A, Blitterswijk CV, Ferraz MP, Monteiro FJ, et al. In vivo evaluation of highly macroporous ceramic scaffolds for bone tissue engineering. J Biomed Mater Res A. 2010;93:567–75.
23. Woodard JR, Hilldore AJ, Lan SK, Park CJ, Morgan AW, Eurell JAC, et al. The mechanical properties and osteoconductivity of hydroxyapatite bone scaffolds with multi-scale porosity. Biomaterials. 2007;28:45–54.
24. Oliveira JM, Rodrigues MT, Silva SS, Malafaya PB, Gomes ME, Viegas CA, et al. Novel hydroxyapatite/chitosan bilayered scaffold for osteochondral tissue-engineering applications: Scaffold design and its performance when seeded with goat bone marrow stromal cells. Biomaterials. 2006;27:6123–37.
25. Tarafder S, Balla VK, Davies NM, Bandyopadhyay A, Bose S. Microwave-sintered 3D printed tricalcium phosphate scaffolds for bone tissue engineering. J Tissue Eng Regen M. 2013;7:631–41.
26. Cox SC, Thornby JA, Gibbons GJ, Williams MA, Mallick KK. 3D printing of porous hydroxyapatite scaffolds intended for use in bone tissue engineering applications. Mat Sci Eng C-Mater. 2015;47:237–47.

27. Descamps M, Duhoo T, Monchau F, Lu J, Hardouin P, Hornez JC, et al. Manufacture of macroporous beta-tricalcium phosphate bioceramics. J Eur Ceram Soc. 2008;28:149–57.
28. Farzadi A, Waran V, Solati-Hashjin M, Rahman ZAA, Asadi M, Abu Osman NA. Effect of layer printing delay on mechanical properties and dimensional accuracy of 3D printed porous prototypes in bone tissue engineering. Ceram Int. 2015;41:8320–30.
29. Dorozhkin SV. Calcium orthophosphates as bioceramics: state of the art. J Func Biomat. 2010;1:22–7.
30. Guarino V, Causa F, Taddei P, di Foggia M, Ciapetti G, Martini D, et al. Polylactic acid fibre-reinforced polycaprolactone scaffolds for bone tissue engineering. Biomaterials. 2008;29:3662–70.
31. Kruger R, Groll J. Fiber reinforced calcium phosphate cements – on the way to degradable load bearing bone substitutes? Biomaterials. 2012;33:5887–900.
32. Xu HH, Eichmiller FC, Giuseppetti AA. Reinforcement of a self-setting calcium phosphate cement with different fibers. J Biomed Mater Res B Appl Biomater. 2000;52:107–14.
33. Canal C, Ginebra MP. Fibre-reinforced calcium phosphate cements: a review. J Mech Behav Biomed Mater. 2011;4:1658–71.
34. Santos LAD, Oliveira LCD, Rigo ECDS, Carrodéguas RG, Boschi AO, Arruda ACFD. Fiber reinforced calcium phosphate cement. Artif Organs. 2000;24:212–6.
35. Gunatillake PA, Adhikari R. Biodegradable synthetic polymers for tissue engineering. Eur Cell Mater. 2003;5:405–6.
36. Janas VF, Tenhuisen KS. Viscous suspension spinning process for producing resorbable ceramic fibers and scaffolds. US:US6451059. 2002.
37. Smoak M, Hogan K, Kriegh L, Chen C, Terrell LKB. Modulation of mesenchymal stem cell behavior by nano and micro-sized beta-tricalcium phosphate particles in suspension and composite structures. J Nanopart Res. 2015;17:1–14.
38. Wang Z, Li N, Li R, Li Y, Ruan L. Biodegradable intestinal stents: a review. Prog Nat Sci Mat Int. 2014;24:423–32.
39. Howard D, Buttery LD, Shakesheff KM, Roberts SJ. Tissue engineering: strategies, stem cells and scaffolds. J Anat. 2008;213:66–72.
40. Nishino T, Peijs T. All-cellulose composites. Macromolecules. 2014;37:7683–7.
41. Nidhin M, Vedhanayagam M, Sangeetha S, Kiran MS, Nazeer SS, Jayasree RS, et al. Fluorescent nanonetworks: a novel bioalley for collagen scaffolds and tissue engineering. Sci Rep. 2014;4:5968.
42. Karam JP, Muscari C, Montero-Menei CN. Combining adult stem cells and polymeric devices for tissue engineering in infarcted myocardium. Biomaterials. 2012;33:5683–95.
43. Kuttappan S, Mathew D, Nair MB. Biomimetic composite scaffolds containing bioceramics and collagen/gelatin for bone tissue engineering – a mini review. Int J Biol Macromol. 2016;93:1390–401.
44. Malafaya PB, Santos TC, van Griensven M, Reis RL. Morphology, mechanical characterization and in vivo neo-vascularization of chitosan particle aggregated scaffolds architectures. Biomaterials. 2008;29:3914–26.
45. Chen JP, Chang GY, Chen JK. Electrospun collagen/chitosan nanofibrous membrane as wound dressing. Colloid Surface A. 2008;313:183–8.
46. Cheung HY, Lau KT, Lu TP, Hui D. A critical review on polymer-based bio-engineered materials for scaffold development. Compos Part B-Eng. 2007;38:291–300.
47. Rana D, Arulkumar S, Vishwakarma A, Ramalingam M. Chapter 10–Considerations on designing scaffold for tissue engineering. Stem Cell Biol Tiss Eng Dental Sci. 2015;133–48.
48. Xiong Y, Zeng YS, Zeng CG, Du BL, He LM, Quan DP, et al. Synaptic transmission of neural stem cells seeded in 3-dimensional PLGA scaffolds. Biomaterials. 2009;30:3711–22.
49. Chen J, Wang C, Lü S, Wu J, Guo X, Duan C, et al. In vivo chondrogenesis of adult bone-marrow-derived autologous mesenchymal stem cells. Cell Tissue Res. 2005;319:429–38.
50. Kim GH. Electrospun PCL nanofibers with anisotropic mechanical properties as a biomedical scaffold. Biomed Mater 2008;3:025010.

51. Okamoto M, John B. Synthetic biopolymer nanocomposites for tissue engineering scaffolds. Prog Polym Sci. 2013;38:1487–503.
52. Jiao YP, Liu ZH, Zhou CR. Fabrication and characterization of PLLA-chitosan hybrid scaffolds with improved cell compatibility. J Biomed Mater Res A. 2007;80:820–5.
53. Collins MN, Birkinshaw C. Hyaluronic acid based scaffolds for tissue engineering – a review. Carbohydr Polym. 2013;92:1262–79.
54. Ben-Nissan B, Pezzotti G. Bioceramics: processing routes and mechanical evaluation. J Ceram Soc Jpn. 2002;110:601–8.
55. Fresno F, Portela R, Suarez S, Coronado J. Photocatalytic materials: recent achievements and near future trends. J Mater Chem. 2014;2:2863–84.
56. John MJ, Thomas S. Biofibres and biocomposites. Carbohydr Polym. 2008;71:343–64.
57. Boccaccini AR, Verrier S. Bioactive composite materials for bone tissue engineering scaffolds. Expert Rev Med Device. 2005;2:303–17.
58. Niiranen H, Pyhältö T, Rokkanen P, Kellomäki M, Törmälä P. In vitro and in vivo behavior of self-reinforced bioabsorbable polymer and self-reinforced bioabsorbable polymer/bioactive glass composites. J Biomed Mater Res A. 2004;69:699–708.
59. Yunos DM, Bretcanu O, Boccaccini AR. Polymer-bioceramic composites for tissue engineering scaffolds. J Mater Sci. 2008;43:4433–42.
60. Miao X, Tan LP, Tan LS, et al. Porous calcium phosphate ceramics modified with PLGA–bioactive glass. Mater Sci Eng C. 2007;27:274–9.
61. Marsich E, Bellomo F, Turco G, Travan A, Donati I, Paoletti S. Nano-composite scaffolds for bone tissue engineering containing silver nanoparticles: preparation, characterization and biological properties. J Mater Sci Mater Med. 2013;24:1799–807.
62. Gu XN, Wang X, Li N, Li L, Zheng YF, Miao X. Microstructure and characteristics of the metal–ceramic composite (MgCa-HA/TCP) fabricated by liquid metal infiltration †. J Biomed Mat Res Part B App Biom. 2011;99:127–34.
63. Hall SM. The Schwann cell: a reappraisal of its role in the peripheral nervous system. Neuropathol Appl Neurobiol. 2008;4:165–76.
64. Razavi S, Zarkesh-Esfahani H, Morshed M, Vaezifar S, Karbasi S, Golozar MA. Nanobiocomposite of poly(lactide- co -glycolide)/chitosan electrospun scaffold can promote proliferation and transdifferentiation of Schwann-like cells from human adipose-derived stem cells. J Biomed Mater Res A. 2015;103:2628–34.
65. Jia L, Duan Z, Fan D, Mi Y, Hui J, Chang L. Human-like collagen/nano-hydroxyapatite scaffolds for the culture of chondrocytes. Mater Sci Eng C. 2013;33:727–34.
66. Li X, Feng Q, Wang W, Cui F. Chemical characteristics and cytocompatibility of collagen-based scaffold reinforced by chitin fibers for bone tissue engineering. J Biomed Mat Res B Appl Biomater. 2006;77:219–26.
67. Moutos FT, Freed LE, Guilak F. A biomimetic three-dimensional woven composite scaffold for functional tissue engineering of cartilage. Nat Mater. 2007;6:162–7.
68. Moutos FT, Guilak F. Composite scaffolds for cartilage tissue engineering. Biorheology. 2008;45:501–12.
69. Tuzlakoglu K, Reis RL. Biodegradable polymeric fiber structures in tissue engineering. Tissue Eng Part B Rev. 2009;15:17–27.
70. Grinstaff MW. Dendritic macromers for hydrogel formation: tailored materials for ophthalmic, orthopedic, and biotech applications. J Polym Sci Part A Polym Chem. 2008;46:383–400.
71. Mauck RL, Baker BM, Nerurkar NL, Burdick JA, Li WJ, Tuan RS, et al. Engineering on the straight and narrow: the mechanics of nanofibrous assemblies for fiber-reinforced tissue regeneration. Tissue Eng Part B Rev. 2009;15:171–93.
72. Byers BA, Baksh D. Translation of new tissue engineering materials to clinical application. In: Biomaterials for Tissue Engineering Applications. 2011. p. 541–61.
73. Tamayol A, Akbari M, Annabi N, Paul A, Khademhosseini A, Juncker D. Fiber-based tissue engineering: progress, challenges, and opportunities. Biotechnol Adv. 2013;31:669–87.

74. Noh HK, Lee SW, Kim JM, Oh JE, Kim KH, Chung CP, et al. Electrospinning of chitin nanofibers: degradation behavior and cellular response to normal human keratinocytes and fibroblasts. Biomaterials. 2006;27:3934–44.
75. Michael CM, Boland ED, Simpson DG, Barnes CP, Bowlin GL. Electrospun fibrinogen: feasibility as a tissue engineering scaffold in a rat cell culture model. J Biomed Mater Res A. 2007;81:299–309.
76. Wan-Ju Lia, JA Cooper Jr, RLM, Rocky ST. Fabrication and characterization of six electrospun poly(alpha-hydroxy ester)-based fibrous scaffolds for tissue engineering applications. Acta Biomater. 2006;2:377–385.
77. Choi JS, Lee SW, Jeong L, Bae SH, Min BC, Ji HY, et al. Effect of organosoluble salts on the nanofibrous structure of electrospun poly(3-hydroxybutyrate- co -3-hydroxyvalerate). Int J Biol Macromol. 2004;34:249–56.
78. Palmer S, Hall W. Objective surface evaluation of fiber reinforced polymer composites. Appl Compos Mater. 2013;20(4):627–37.
79. Caves JM, Cui WX, Wen J, Kumar VA, Haller CA, Chaikof EL. Elastin-like protein matrix reinforced with collagen microfibers for soft tissue repair. Biomaterials. 2011;32:5371–9.
80. Slivka MA, Leatherbury NC, Kieswetter K, Niederauer GG. Porous, resorbable, fiber-reinforced scaffolds tailored for articular cartilage repair. Tissue Eng. 2001;7:767–80.
81. Guarino V, Ambrosio L. The synergic effect of polylactide fiber and calcium phosphate particle reinforcement in poly ε-caprolactone-based composite scaffolds. Acta Biomater. 2008;4:1778–87.
82. Cont L, Grant D, Scotchford C, Todea M, Popa C. Composite PLA scaffolds reinforced with PDO fibers for tissue engineering. J Biomater Appl. 2013;27:707–16.
83. Illeperuma WRK, Sun JY, Suo Z, Vlassak JJ. Fiber-reinforced tough hydrogels. Extreme Mech Letter. 2014;1:90–6.
84. Yodmuang S, McNamara SL, Nover AB, Mandal BB, Agarwal M, Kelly TAN, et al. Silk microfiber-reinforced silk hydrogel composites for functional cartilage tissue repair. Acta Biomater. 2015;11:27–36.
85. Calvert P, Agrawal A, Chalivendra NRV. Tough, strong hydrogels with elastomeric fiber reinforcement. Mrs Online Proc Lib. 2012;1420. DOI:10.1557/opl.2012.686.
86. Agrawal A, Rahbar N, Calvert PD. Strong fiber-reinforced hydrogel. Acta Biomater. 2013;9:5313–8.
87. Visser J, Melchels FP, Jeon JE, van Bussel EM, Kimpton LS, Byrne HM, et al. Reinforcement of hydrogels using three-dimensionally printed microfibres. Nat Commun. 2015;6:6933.
88. Mitchell DT, Lee SB, Trofin L, Li N, Nevanen TK, Söderlund H, et al. Smart nanotubes for bioseparations and biocatalysis. J Am Chem Soc. 2002;124:11864–5.
89. Sang BL, Mitchell DT, Trofin L, Nevanen TK, SÖderlund H, Martin CR. Antibody-based bio-nanotube membranes for enantiomeric drug separations. Science. 2002;296:2198–200.
90. Iijima S. Helical microtubules of graphitic carbon. Nature. 1991;354:56–8.
91. Bacon R. Growth, structure, and properties of graphite whiskers. J Appl Phys. 1960;31:283–90.
92. Wang W, Liao S, Liu M, Zhao Q, Zhu Y. Polymer composites reinforced by nanotubes as scaffolds for tissue engineering. Int J Polym Sci. 2014;2014:1–14.
93. Li X, Liu H, Niu X, Yu B, Fan Y, Feng Q, et al. The use of carbon nanotubes to induce osteogenic differentiation of human adipose-derived MSCs invitro and ectopic bone formation invivo. Biomaterials. 2012;33:4818–27.
94. Kim JS, Song KS, Yu IJ. Evaluation of in vitro and in vivo genotoxicity of single-walled carbon nanotubes. Toxicol Ind Health. 2013;31:747–57.
95. Barrientos-Durán A, Carpenter EM, Zur Nieden NI, Malinin TI, Rodríguez-Manzaneque JC, Zanello LP. Carboxyl-modified single-wall carbon nanotubes improve bone tissue formation in vitro and repair in an in vivo rat model. Int J Nanomedicine. 2014;9:4277–91.
96. Xing YJ, Xi ZH, Xue ZQ, Zhang XD, Song JH, Wang RM, et al. Optical properties of the ZnO nanotubes synthesized via vapor phase growth. Appl Phys Lett. 2003;83:1689–91.

97. Dloczik L, Engelhardt R, Ernst K, Fiechter S, Sieber I, Konenkamp R. Hexagonal nanotubes of ZnS by chemical conversion of monocrystalline ZnO columns. Appl Phys Lett. 2001;78:3687–9.
98. Nath M, Rao CNR. ChemInform abstract: new metal disulfide nanotubes. ChemInform. 2010;32:4841–2.
99. Lourie OR, Jones CR, Bartlett BM, Gibbons PC, Ruoff RS, Buhro WE. ChemInform abstract: CVD growth of boron nitride nanotubes. ChemInform. 2000;31:1808–10.
100. Galvan DH, Jun-Ho K, Maple MB, Avalos-Borja M, Adem E. Formation of NbSe2 nanotubes by electron irradiation. Fuller Sci Technol. 2000;8:143–51.
101. Kim H, Jin C, Park S, Wan IL, Chin IJ, Lee C. Structure and optical properties of Bi_2S_3 and Bi_2O_3 nanostructures synthesized via thermal evaporation and thermal oxidation routes. Chem Eng J. 2013;s215–216:151–6.
102. Wang Z, Wang S, Li J, Gao F, Weber WJ. Structure and electronic properties of saturated and unsaturated gallium nitride nanotubes. J Phys Chem C. 2009;113:19281–5.
103. Chandiramouli R. Band structure and transport studies on InP nanotube – a first-principles investigation. Superlattice Microst. 2015;83:193–209.
104. Fan R, Wu Y, Li D, Yue M, Arun Majumdar A, Yang P. Fabrication of silica nanotube arrays from vertical silicon nanowire templates. J Am Chem Soc. 2003;125:5254–5.
105. Wu Q, Hu Z, Wang X, Chen X, Xu H, et al. Synthesis and characterization of faceted hexagonal aluminum nitride nanotubes. J Am Chem Soc. 2003;125:10176–7.
106. Rothschild A, Sloan J, Tenne R. Growth of WS2 nanotubes phases. J Am Chem Soc. 2000;122:5169–79.
107. Gupta P, Sharan S, Roy P, Lahiri D. Aligned carbon nanotube reinforced polymeric scaffolds with electrical cues for neural tissue regeneration. Carbon. 2015;95:715–24.
108. Wang X, Li Q, Xie J, Zhong J, Wang J, Li Y, et al. Fabrication of ultralong and electrically uniform single-walled carbon nanotubes on clean substrates. Nano Lett. 2009;9:3137–41.
109. Saito R, Dresselhaus G, Dresselhaus MS. Physical properties of carbon nanotubes. Int Winterschool Electron. 1998;442:467–80.
110. Demczyk BG, Wang YM, Cumings J, Hetman M, Han W, Zettl A, et al. Direct mechanical measurement of the tensile strength and elastic modulus of multiwalled carbon nanotubes. Microsc Microanal. 2002;334:173–8.
111. Hong S, Myung S. Nanotube electronics: a flexible approach to mobility. Nat Nanotechnol. 2007;2:207–18.
112. Li X, Gao H, Uo M, Sato Y, Akasaka T, Abe S, et al. Maturation of osteoblast-like SaoS2 induced by carbon nanotubes. Biomed Mater. 2009;4:015005.
113. Kang M, Chen P, Jin HJ. Preparation of multiwalled carbon nanotubes incorporated silk fibroin nanofibers by electrospinning. Curr Appl Phys. 2009;9:S95–7.
114. Thostenson ET, Ren Z, Chou TW. Advances in the science and technology of carbon nanotubes and their composites: a review. Compos Sci Technol. 2001;61:1899–912.
115. Shin SR. Carbon nanotube reinforced hybrid microgels as scaffold materials for cell encapsulation. ACS Nano. 2012;6:362–72.
116. Zawadzak E, Bil M, Ryszkowska J, Nazhat SN, Cho J, Bretcanu O, et al. Polyurethane foams electrophoretically coated with carbon nanotubes for tissue engineering scaffolds. Biomed Mater. 2008;4:71–9.
117. Mikael PE, Amini AR, Basu J, Arellano-Jimenez MJ, Laurencin CT, Sanders MM, et al. Functionalized carbon nanotube reinforced scaffolds for bone regenerative engineering: fabrication, in vitro and in vivo evaluation. Biomed Mater. 2014;9:035001.
118. Edwards SL, Werkmeister JA, Ramshaw JAM. Carbon nanotubes in scaffolds for tissue engineering. Expert Rev Med Devices. 2009;6:499–505.
119. Zhang HL. Electrospun poly (lactic-co-glycolic acid)/multiwalled carbon nanotubes composite scaffolds for guided bone tissue regeneration. J Bioact Compat Polym. 2011;26:347–62.
120. Vozzi G, Corallo C, Daraio C. Pressure-activated microsyringe composite scaffold of poly(L-lactic acid) and carbon nanotubes for bone tissue engineering. J Appl Polym Sci. 2013;129:528–36.

121. Chen L, Hu JX, Shen XY, Tong H. Synthesis and characterization of chitosan-multiwalled carbon nanotubes/hydroxyapatite nanocomposites for bone tissue engineering. J Mater Sci-Mater M. 2013;24:1843–51.
122. Abarrategi A, Gutierrez MC, Moreno-Vicente C, Hortiguela MJ, Ramos V, Lopez-Lacomba JL, et al. Multiwall carbon nanotube scaffolds for tissue engineering purposes. Biomaterials. 2008;29:94–102.
123. Yildirim ED, Yin X, Nair K, et al. Fabrication, characterization, and biocompatibility of single-walled carbon nanotube-reinforced alginate composite scaffolds manufactured using freeform fabrication technique. J Biomed Mater Res B Appl Biomater. 2008;87:406–14.
124. Hu H, Ni YC, Mandal SK, Montana V, Zhao N, Haddon RC, et al. Polyethyleneimine functionalized single-walled carbon nanotubes as a substrate for neuronal growth. J Phys Chem B. 2005;109:4285–9.
125. Mattson MP, Haddon RC, Rao AM. Molecular functionalization of carbon nanotubes and use as substrates for neuronal growth. J Mol Neurosci. 2000;14:175–82.
126. Ma R, Bando Y, Sato T. Controlled synthesis of BN nanotubes, nanobamboos, and nanocables. Adv Mater. 2002;14:366–8.
127. Chen X, Wu P, Rousseas M, Okawa D, Gartner Z, Zettl A, et al. Boron nitride nanotubes are noncytotoxic and can be functionalized for interaction with proteins and cells. J Am Chem Soc. 2009;131:890–1.
128. Liu HK, Li B, Zhang CR, Wang SQ, Liu K. Microstructure and mechanical properties of BN nanotubes reinforced Si3N4 porous composites. Mater Sci Forum. 2015;816:64–70.
129. Arani AG, Shajari AR, Amir S, et al. Electro-thermo-mechanical nonlinear nonlocal vibration and instability of embedded micro-tube reinforced by BNNT, conveying fluid. Physica E. 2012;45:109–21.
130. Arani AG, Shajari AR, Atabakhshian V, Amir S, Loghman A. Nonlinear dynamical response of embedded fluid-conveyed micro-tube reinforced by BNNTs. Compos Part B Eng. 2013;44:424–32.
131. Lahiri D, Rouzaud F, Richard T, Keshri AK, Bakshi SR, Kos L, et al. Boron nitride nanotube reinforced polylactide–polycaprolactone copolymer composite: mechanical properties and cytocompatibility with osteoblasts and macrophages in vitro. Acta Biomater. 2010;6:3524–33.
132. Lahiri D, Singh V, Benaduce AP, Seal S, Kos L, Agarwal A. Boron nitride nanotube reinforced hydroxyapatite composite: mechanical and tribological performance and in-vitro biocompatibility to osteoblasts. J Mech Behav Biomed Mater. 2011;4:44–56.
133. Zhi C, Bando Y, Tang C, Golberg D. Immobilization of proteins on boron nitride nanotubes. J Am Chem Soc. 2005;127:17144–5.
134. Kaplan-Ashiri I, Cohen SR, Gartsman K, Ivanovskaya V, Heine T, Seifert G, et al. On the mechanical behavior of WS2 nanotubes under axial tension and compression. Proc Natl Acad Sci U S A. 2006;103:523–8.
135. Wang MS, Kaplan-Ashiri I, Wei XL, Rosentsveig R, Wagner HD, Tenne R, et al. In situ TEM measurements of the mechanical properties and behavior of WS2 nanotubes. Nano Res. 2008;1:22–31.
136. Elad Z, Sharon B, Mark S, Hanna D, Samuel K, Reshef T, et al. The effect of WS2 nanotubes on the properties of epoxy-based nanocomposites. J Adhes Sci Technol. 2011;25:1603–17.
137. Reddy CS, Zak A, Zussman E. WS2 nanotubes embedded in PMMA nanofibers as energy absorptive material. J Mater Chem. 2011;21:16086–93.
138. Lalwani G, Henslee AM, Farshid B, Parmar P, Lin L, Qin YX, et al. Tungsten disulfide nanotubes reinforced biodegradable polymers for bone tissue engineering. Acta Biomater. 2013;9:8365–73.
139. Tan AW, Pingguan-Murphy B, Ahmad R, Akbar SA. Review of titania nanotubes: fabrication and cellular response. Ceram Int. 2012;38:4421–35.
140. Minagar S, Li Y, Berndt CC, Wen C. The influence of titania–zirconia–zirconium titanate nanotube characteristics on osteoblast cell adhesion. Acta Biomater. 2015;12:281–9.
141. Bjursten LM, Rasmusson L, Oh S, Smith GC, Brammer KS, Jin S. Titanium dioxide nanotubes enhance bone bonding in vivo. J Biomed Mater Res A. 2009;92:1218–24.

142. Gao C, Yu L, Liu H, Chen L. Development of self-reinforced polymer composites. Prog Polym Sci. 2012;37:767–80.
143. Alcock B, Peijs T. Technology and development of self-reinforced polymer composites. Adv Polym Sci. 2013;251:H1775–84.
144. Alcock B, Peijs T. Technology and development of self-reinforced polymer composites. Adv Polym Sci. 2013;251:1–76.
145. Li MZ, Lu SZ. Material for repairing defective tissue of body and its preparing process. CN:1179757. 2004.
146. Lu X, Zhang MQ, Rong MZ, Shi G, Yang GC. Self-reinforced melt processable composites of sisal. Compos Sci Technol. 2003;63:177–86.
147. Deng M, Shalaby SW. Properties of self-reinforced ultra-high-molecular-weight polyethylene composites. Biomaterials. 1997;18:645–55.
148. Majola A, Vainionpää S, Rokkanen P, Mikkola HM, Törmälä P. Absorbable self-reinforced polylactide (SR-PLA) composite rods for fracture fixation: strength and strength retention in the bone and subcutaneous tissue of rabbits. J Mater Sci Mater Med. 1992;3:43–7.
149. Wrightcharlesworth DD, Miller DM, Miskioglu I, King JA. Nanoindentation of injection molded PLA and self-reinforced composite PLA after in vitro conditioning for three months. J Biomed Mater Res A. 2005;74:388–96.
150. Wright DD, Lautenschlager EP, Gilbert JL. Bending and fracture toughness of woven self-reinforced composite poly(methyl methacrylate). J Biomed Mater Res. 1997;36:441–53.
151. Wright DD, Lautenschlager EP, Gilbert JL. Interfacial properties of self-reinforced composite poly(methyl methacrylate). J Biomed Mater Res. 1998;43:153–61.
152. Wright DD, Gilbert JL, Lautenschlager EP. The effect of processing temperature and time on the structure and fracture characteristics of self-reinforced composite poly(methyl methacrylate). J Mater Sci Mater Med. 1999;10:503–12.
153. Wright DD. Optimizaton and thermomechanical characterization of highly oriented PMMA fibers and self-reinforced composite PMMA. 1999;60:6227–96.
154. Wright-Charlesworth DD, Peers WJ, Miskioglu I, Loo LL. Nanomechanical properties of self-reinforced composite poly(methyl methacrylate) as a function of processing temperature. J Biomed Mater Res A. 2005;74A:306–14.
155. Amer MS, Ganapathiraju S. Effects of processing parameters on axial stiffness of self-reinforced polyethylene composites. J Appl Polym Sci. 2001;81:1136–41.
156. Abrahamsson CK, Yang F, Park H, Brunger JM, Valonen PK, Langer R, et al. Chondrogenesis and mineralization during in vitro culture of human mesenchymal stem cells on three-dimensional woven scaffolds. Tissue Eng A. 2010;16:3709–18.
157. Barber JG, Handorf AM, Allee TJ, Li WJ. Braided nanofibrous scaffold for tendon and ligament tissue engineering. Tissue Eng A. 2013;19:1265–74.
158. Ellis-Behnke RG, Liang YX, You SW, Tay DK, Zhang S, So KF, et al. Nano neuro knitting: peptide nanofiber scaffold for brain repair and axon regeneration with functional return of vision. Proc Natl Acad Sci. 2006;103:5054–9.
159. Laurent CP, Durville D, Mainard D, Ganghoffer JF, Rahouadj R. A multilayer braided scaffold for anterior cruciate ligament: mechanical modeling at the fiber scale. J Mech Behav Biomed Mater. 2012;12:184–96.
160. Freeman JW, Woods MD, Laurencin CT. Tissue engineering of the anterior cruciate ligament using a braid–twist scaffold design. J Biomech. 2006;40:2029–36.
161. Aytemiz D, Sakiyama W, Yu S, et al. Small-diameter silk vascular grafts (3 mm diameter) with a double-raschel knitted silk tube coated with silk fibroin sponge. Adv Healthc Mater. 2013;2(2):361–8.
162. Huang ZM. The mechanical properties of composites reinforced with woven and braided fabrics. Compos Sci Technol. 2000;60:479–98.
163. Xu W, Zhou F, Ouyang C, Ye W, Yao M, Xu B. Mechanical properties of small-diameter polyurethane vascular grafts reinforced by weft-knitted tubular fabric. J Biomed Mater Res A. 2010;92A:1–8.

164. Ananta M, Aulin CE, Hilborn J, Aibibu D, Houis S, Brown RA, et al. A poly(lactic acid-co-caprolactone)-collagen hybrid for tissue engineering applications. Tissue Eng A. 2009;15:1667–75.
165. Subia B, Kundu J, Kundu SC. Biomaterial scaffold fabrication techniques for potential tissue engineering applications. Tissue Engineering. InTech, 2010.
166. Meitl MA, Zhou Y, Gaur A, Jeon S, Usrey ML, Strano MS, et al. Solution casting and transfer printing single-walled carbon nanotube films. Nano Lett. 2004;4:1643–7.
167. Fan HL, Wang LL, Zhao KK, Li N, Shi ZJ, Ge ZG, et al. Fabrication, mechanical properties, and biocompatibility of graphene-reinforced chitosan composites. Biomacromolecules. 2010;11:2345–51.
168. Sadat-Shojai M, Khorasani MT, Jamshidi A, Irani S. Nano-hydroxyapatite reinforced polyhydroxybutyrate composites: a comprehensive study on the structural and in vitro biological properties. Mat Sci Eng C-Mater. 2013;33:2776–87.
169. Lin CH, Su JM, Hsu SH. Evaluation of type II collagen scaffolds reinforced by poly(epsilon-caprolactone) as tissue-engineered trachea. Tissue Eng Pt C-Meth. 2008;14:69–77.
170. Julkapli NM, Akil HM, Ahmad Z. Thermal properties of 4,4-oxydiphathalic anhydride chitosan filled chitosan bio-composites. J Therm Anal Calorim. 2012;107:365–76.
171. Govindasamy K, Muhammad Hafiz R, Pooria P, Vengidesh P, Babak S. Chitosan/cellulose/halloysite membranes produced using solvent casting method. Polym Polym Compos. 2015;23:325–32.
172. Kong I, Tshai KY, Hoque ME. Manufacturing of natural fibre-reinforced polymer composites by solvent casting method. Springer International Publishing; 2015. p. 331–49.
173. Ma PX, Langer R. Fabrication of biodegradable polymer foams for cell transplantation and tissue engineering. Methods Mol Med. 1999;18:47–56.
174. Lu L, Peter SJ, Lyman MD, Lai HL, Leite SM, Tamada JA, et al. In vitro and in vivo degradation of porous poly(dl -lactic- co -glycolic acid) foams. Biomaterials. 2000;21:1837–45.
175. Bonnaud L, Pascault JP, Sautereau H, Zhao JQ, Jia DM. Effect of reinforcing glass fibers on morphology and properties of thermoplastic modified epoxy-aromatic diamine matrix. Polym Compos. 2004;25:368–74.
176. Mou ZL, Duan LM, Qi XN, Zhang ZQ. Preparation of silk fibroin/collagen/hydroxyapatite composite scaffold by particulate leaching method. Mater Lett. 2013;105:189–91.
177. Whang K, Thomas CH, Healy KE, Nuber G. A novel method to fabricate bioabsorbable scaffolds. Polymer. 1995;36:837–42.
178. Sachlos E, Czernuszka JT. Making tissue engineering scaffolds work. Review: the application of solid freeform fabrication technology to the production of tissue engineering scaffolds. Eur Cell Mater. 2003;5:29–39.
179. Nam YS, Park TG. Porous biodegradable polymeric scaffolds prepared by thermally induced phase separation. J Biomed Mater Res. 1999;47:8–17.
180. Ma H, Xue L. Carbon nanotubes reinforced poly(L-lactide) scaffolds fabricated by thermally induced phase separation. Nanotechnology. 2015;26:025701.
181. Nakhoda HM, Dahman Y. Novel biodegradable polyurethanes reinforced with green nanofibers for applications in tissue engineering. Synthesis and characterization. Can J Chem Eng. 2014;92:1895–902.
182. Mi HY, Jing X, Salick MR, Cordie TM, Turng LS. Carbon nanotube (CNT) and nanofibrillated, cellulose (NFC) reinforcement effect on thermoplastic polyurethane (TPU) scaffolds fabricated via phase separation using dimethyl sulfoxide (MOO) as solvent. J Mech Behav Biomed Mater. 2016;62:417–27.
183. Tian F, Niu XF, Li XM, Feng QL, Fan YB. Porous poly(L-lactic acid) scaffold reinforced by titanium dioxide nanoparticles. J Biomater Tiss Eng. 2016;6:478–83.
184. Cai Q, Yang J, Bei J, Wang S. A novel porous cells scaffold made of polylactide-dextran blend by combining phase-separation and particle-leaching techniques. Biomaterials. 2003;23:4483–92.

185. Thomson RC, Wake MC, Yaszemski MJ, Mikos AG. Biodegradable polymer scaffolds to regenerate organs. Berlin/Heidelberg: Springer; 1995.
186. Hou Q, Grijpma DW, Feijen J. Porous polymeric structures for tissue engineering prepared by a coagulation, compression moulding and salt leaching technique. Biomaterials. 2003;24:1937–47.
187. Capes JS, Ando HY, Cameron RE. Fabrication of polymeric scaffolds with a controlled distribution of pores. J Mater Sci Mater Med. 2005;16:1069–75.
188. Converse GL, Conrad TL, Merrill CH, Roeder RK. Hydroxyapatite whisker-reinforced polyetherketoneketone bone ingrowth scaffolds. Acta Biomater. 2010;6:856–63.
189. Johnson T, Bahrampourian R, Patel A, Mequanint K. Fabrication of highly porous tissue-engineering scaffolds using selective spherical porogens. Biomed Mater Eng. 2010;20:107–18.
190. Yang Y, Jin Z, Zhao Y, Liang W, Yuan X, Fan Y. Formation of porous PLGA scaffolds by a combining method of thermally induced phase separation and porogen leaching. J Appl Polym Sci. 2008;109:1232–41.
191. Salerno A, Fernández-Gutiérrez M, Barrio JSRD, Domingo C. Bio-safe fabrication of PLA scaffolds for bone tissue engineering by combining phase separation, porogen leaching and scCO 2 drying. J Supercrit Fluid. 2015;97:238–46.
192. Rajkhowa R, Gil ES, Kluge J, Numata K, Wang L, Wang X, et al. Reinforcing silk scaffolds with silk particles. Macromol Biosci. 2010;10:599–611.
193. Gil ES, Kluge JA, Rockwood DN, Rajkhowa R, Wang L, Wang X, et al. Mechanical improvements to reinforced porous silk scaffolds †. J Biomed Mater Res A. 2011;99:16–28.
194. Mikos AG, Sarakinos G, Leite SM, Vacant JP, Langer R. Laminated three-dimensional biodegradable foams for use in tissue engineering. Biomaterials. 1993;14:323–30.
195. Mather ML, Crowe JA, Morgan SP, White LJ, Kalashnikov AN, Ivchenko VG, et al. Ultrasonic monitoring of foamed polymeric tissue scaffold fabrication. J Mater Sci Mater Med. 2008;19:3071–80.
196. Moore MJ, Jabbari E, Ritman EL, Lu L, Currier BL, Windebank AJ, et al. Quantitative analysis of interconnectivity of porous biodegradable scaffolds with micro-computed tomography. J Biomed Mater Res A. 2004;71:258–67.
197. Xin HZ, Lai YL, Jie SHJ, Tong YW, Wang CH. Characterization of porous poly(D,L-lactic-co-glycolic acid) sponges fabricated by supercritical CO2 gas-foaming method as a scaffold for three-dimensional growth of Hep3B cells. Biotechnol Bioeng. 2008;100:998–1009.
198. Nam YS, Yoon JJ, Park TG. A novel fabrication method of macroporous biodegradable polymer scaffolds using gas foaming salt as a porogen additive. J Biomed Mater Res. 2000;53:1–7.
199. Gravel M, Vago R, Tabrizian M. Use of natural coralline biomaterials as reinforcing and gas-forming agent for developing novel hybrid biomatrices: microarchitectural and mechanical studies. Tissue Eng. 2006;12:589–600.
200. Yun HL, Lee JH, An IG, Chan K, Lee DS, Lee YK, et al. Electrospun dual-porosity structure and biodegradation morphology of Montmorillonite reinforced PLLA nanocomposite scaffolds. Biomaterials. 2005;26:3165–72.
201. Kim J, Hollinger JO. Recombinant human bone morphogenetic protein-2 released from polyurethane-based scaffolds promotes early osteogenic differentiation of human mesenchymal stem cells. Biomed Mater. 2012;7:45008–16.
202. Gorna K, Gogolewski S. Preparation, degradation, and calcification of biodegradable polyurethane foams for bone graft substitutes. J Biomed Mater Res A. 2003;67A:813–27.
203. Puglia D, Ceccolini R, Fortunati E, Armentano I, Morena F, Martino S, et al. Effect of processing techniques on the 3D microstructure of poly (L-lactic acid) scaffolds reinforced with wool keratin from different sources. J Appl Polym Sci. 2015;132:42890.
204. Armani DK, Liu C. Microfabrication technology for polycaprolactone, a biodegradable polymer. J Micromec Microeng. 2000;10:80–4.
205. Ji C, Annabi N, Khademhosseini A, Dehghani F. Fabrication of porous chitosan scaffolds for soft tissue engineering using dense gas CO2. Acta Biomater. 2011;7:1653–64.

206. Oh SH, Park SC, Kim HK, Koh YJ, Lee JH. Degradation Behavior of 3D Porous Polydioxanone-b-Polycaprolactone Scaffolds Fabricated Using the Melt-Molding Particulate-Leaching Method. J Biomater Sci Polym Ed. 2010;22:225–37.
207. Botchwey EA, Pollack SR, Levine EM, Laurencin CT. Bone tissue engineering in a rotating bioreactor using a microcarrier matrix system. J Biomed Mater Res. 2001;55:242–53.
208. Liao CZ, Li K, Wong HM, Tong WY, Yeung KWK, Tjong SC. Novel polypropylene biocomposites reinforced with carbon nanotubes and hydroxyapatite nanorods for bone replacements. Mat Sci Eng C Mat Biol Appl. 2013;33:1380–8.
209. Hutmacher DW. Scaffolds in tissue engineering bone and cartilage. Biomaterials. 2000;21:2529–43.
210. Leong KF, Cheah CM, Chua CK. Solid freeform fabrication of three-dimensional scaffolds for engineering replacement tissues and organs. Biomaterials. 2003;24:2363–78.
211. Bretcanu O, Samaille C, Boccaccini AR. Simple methods to fabricate Bioglass ® -derived glass–ceramic scaffolds exhibiting porosity gradient. J Mater Sci. 2008;43:4127–34.
212. Shokrolahi F, Mirzadeh H, Yeganeh H. Fabrication of poly(urethane urea)-based scaffolds for bone tissue engineering by a combined strategy of using compression moulding and particulate leaching methods. Iran Polym J. 2011;20:645–58.
213. Vepari C, Kaplan DL. Silk as a biomaterial. Prog Polym Sci. 2007;32:991–1007.
214. Altman GH, Diaz F, Jakuba C, Calabro T, Horan RL, Chen J, et al. Silk-based biomaterials. Biomaterials. 2003;24:401–16.
215. Mandal BB, Kundu SC. Osteogenic and adipogenic differentiation of rat bone marrow cells on non-mulberry and mulberry silk gland fibroin 3D scaffolds. Biomaterials. 2009;30:5019–30.
216. Kane RJ, Roeder RK. Effects of hydroxyapatite reinforcement on the architecture and mechanical properties of freeze-dried collagen scaffolds. J Mech Behav Biomed Mater. 2012;7:41–9.
217. Mobini S, Hoyer B, Solati-Hashjin M, Lode A, Nosoudi N, Samadikuchaksaraei A, et al. Fabrication and characterization of regenerated silk scaffolds reinforced with natural silk fibers for bone tissue engineering. J Biomed Mater Res A. 2013;101A:2392–404.
218. Hiraoka Y, Hiroki U, Kimura Y, Tabata Y. Fabrication and characterization of mechanically reinforced collagen sponge. Key Eng Mater. 2005;288–289:240–7.
219. Mandal BB, Kundu SC. Cell proliferation and migration in silk fibroin 3D scaffolds. Biomaterials. 2009;30:2956–65.
220. Zamanian A, Ghorbani F, Nojehdehian H. Morphological comparison of PLGA/gelatin scaffolds produced by freeze casting and freeze drying methods. Appl Mech Mater. 2013;467:108–11.
221. Clyne AM. Thermal processing of tissue engineering scaffolds. J Heat Transf. 2011;133:174–94.
222. Spaans CJ, Groot JHD, Belgraver VW, Pennings AJ. A new biomedical polyurethane with a high modulus based on 1,4-butanediisocyanate and epsilon-caprolactone. J Mater Sci Mater Med. 1998;9:675–8.
223. Finoli A, Ostrowski N, Schmelzer E, Nettleship I, Gerlach J. Multiscale porous ceramic scaffolds for culturing of primary human cells. Adv Appl Ceram. 2013;111:262–8.
224. Niu X, Li X, Liu H, Zhou G, Feng Q, Cui F, et al. Homogeneous chitosan/poly(L-Lactide) composite scaffolds prepared by emulsion freeze-drying. J Biomater Sci Polym Ed. 2012;23:391–404.
225. Jung SH, Jang JW, Kim SH, Hong HH, Oh AY, Rhee JM, et al. Articular cartilage regeneration using hyaluronic acid loaded PLGA scaffold by emulsion freeze-drying method. Tissue Eng Regen Med. 2008;5:643–9.
226. Shemesh M, Zilberman M. Structure–property effects of novel bioresorbable hybrid structures with controlled release of analgesic drugs for wound healing applications. Acta Biomater. 2013;10:1380–91.
227. Li W-J, Laurencin CT, Caterson EJ, Rocky ST, Ko FK. Electrospun nanofibrous structure: a novel scaffold for tissue engineering. J Biomed Mater Res. 2002;60:613–21.

228. Ramakrishna S, Fujihara K, Teo WE, Yong T, Ma Z. Electrospun nanofibers: solving global issues. Mater Today. 2006;9:40–50.
229. Smith LA, Ma PX. Nano-fibrous scaffolds for tissue engineering. Colloids Surf B Biointerfaces. 2004;39:125–31.
230. Doshi J, Reneker DH. Electrospinning process and applications of electrospun fibers. J Electrost. 1995;35:151–60.
231. Doshi J, Reneker DH. Electrospinning process and applications of electrospun fibers. Ind Appl Soc Meet. 1993;3:1698–703.
232. He X, Xiao Q, Lu C, Wang Y, Zhang X, Zhao J, et al. Uniaxially aligned electrospun all-cellulose nanocomposite nanofibers reinforced with cellulose nanocrystals: scaffold for tissue engineering. Biomacromolecules. 2014;15:618–27.
233. Yang CW, Deng GY, Chen WM, Ye XJ, Mo XM. A novel electrospun-aligned nanoyarn-reinforced nanofibrous scaffold for tendon tissue engineering. Colloid Surface B. 2014;122:270–6.
234. Rui MAD, Chiera S, Gershovich P, et al. Enhancing the biomechanical performance of anisotropic nanofibrous scaffolds in tendon tissue engineering: reinforcement with cellulose nanocrystals. Adv Healthc Mater. 2016;5:1364–75.
235. Eslami M, Vrana NE, Zorlutuna P, Sant S, Jung S, Masoumi N, et al. Fiber-reinforced hydrogel scaffolds for heart valve tissue engineering. J Biomater Appl. 2014;29:399–410.
236. Zhang S. Fabrication of novel biomaterials through molecular self-assembly. Nat Biotechnol. 2003;21:1171–8.
237. Whitesides GM, Grzybowski B. Self-assembly at all scales. Science. 2002;295:2418–21.
238. Joshi KB, Singh P, Verma S. Fabrication of platinum nanopillars on peptide-based soft structures using a focused ion beam. Biofabrication. 2009;1:1831–7.
239. Hartgerink JD, Beniash E, Stupp SI. Peptide-amphiphile nanofibers: a versatile scaffold for the preparation of self-assembling materials. Proc Natl Acad Sci. 2002;99:5133–8.
240. Hosseinkhani H, Hosseinkhani M, Tian F, Kobayashi H, Tabata Y. Ectopic bone formation in collagen sponge self-assembled peptide–amphiphile nanofibers hybrid scaffold in a perfusion culture bioreactor. Biomaterials. 2006;27:5089–98.
241. Zhou K, Thouas G, Bernard C, Forsythe JS. 3D presentation of a neurotrophic factor for the regulation of neural progenitor cells. Nanomedicine. 2014;9:1239–51.
242. Ma PX. Biomimetic materials for tissue engineering. Adv Drug Deliv Rev. 2008;60:184–98.
243. Ma Z, Kotaki M, Inai R, Ramakrishna S. Potential of nanofiber matrix as tissue-engineering scaffolds. Tissue Eng. 2005;11:101–9.
244. Li XY. Multi-object optimal design of rapid prototyping based on uniform experiment. Tsinghua Sci Technol. 2009;14:206–11.
245. Kim MS, Hansgen AR, Wink O, Quaife RA, Carroll JD. Rapid Prototyping. Circulation. 2008;117:2388–94.
246. Zein I, Hutmacher DW, Tan KC, Teoh SH. Fused deposition modeling of novel scaffold architectures for tissue engineering applications. Biomaterials. 2002;23:1169–85.
247. Melchels FPW, Feijen J, Grijpma DW. A review on stereolithography and its applications in biomedical engineering. Biomaterials. 2010;31:6121–30.
248. Williams JM, Adewunmi A, Schek RM, Flanagan CL, Krebsbach PH, Feinberg SE, et al. Bone tissue engineering using polycaprolactone scaffolds fabricated via selective laser sintering. Biomaterials. 2005;26:4817–27.
249. Naing MW, Chua CK, Leong KF. Computer aided tissue engineering scaffold fabrication. In: Virtual prototyping bio manufacturing in medical applications. Boston: Springer; 2008. p. 67–85.
250. Kai H, Wang X, Madhukar KS, Qin L, Yan Y, Zhang R, et al. Fabrication of a two-level tumor bone repair biomaterial based on a rapid prototyping technique. Biofabrication. 2009;1:025003.
251. Hutmacher DW. Scaffold design and fabrication technologies for engineering tissues–state of the art and future perspectives. J Biomater Sci Polym Ed. 2001;12:107–24.

252. Lian Q, Li D, Jin Z, Wang J, Li A, Wang Z. Fabrication and in vitro evaluation of calcium phosphate combined with chitosan fibers for scaffold structures. J Bioact Compat Polym. 2009;24:113–24.
253. Alge DL, Bennett J, Treasure T, Voytik-Harbin S, Goebel WS, Chu TMG. Poly(propylene fumarate) reinforced dicalcium phosphate dihydrate cement composites for bone tissue engineering. J Biomed Mater Res A. 2012;100:1792–802.
254. Yong BK, Kim GH. Rapid-prototyped collagen scaffolds reinforced with PCL/β-TCP nanofibres to obtain high cell seeding efficiency and enhanced mechanical properties for bone tissue regeneration. J Mater Chem. 2012;22:16880–9.
255. Min L, Dunn JCY, Wu BM. Scaffold fabrication by indirect three-dimensional printing. Biomaterials. 2005;26:4281–9.
256. Min L, Wu BM. Recent advances in 3D printing of tissue engineering scaffolds. Methods Mol Biol. 2012;868:257–67.
257. Christ S, Schnabel M, Vorndran E, Groll J, Gbureck U. Fiber reinforcement during 3D printing. Mater Lett. 2015;139:165–8.
258. Munaz A, Vadivelu RK, St. John J, Barton M, Kamble H, Nguyen N-T. Three-dimensional printing of biological matters. J Sci Adv Mat Devices. 2016;1:1–17.
259. Jung JW, Lee H, Hong JM, Park JH, Shim JH, Choi TH, et al. A new method of fabricating a blend scaffold using an indirect three-dimensional printing technique. Biofabrication. 2015;7:045003.
260. Bakarich SE, Gorkin R 3rd, in het Panhuis M, Spinks GM. Three-dimensional printing fiber reinforced hydrogel composites. Acs Applied Materials & Interfaces 2014;6:15998–16006.
261. Suwanprateeb J. Improvement in mechanical properties of three-dimensional printing parts made from natural polymers reinforced by acrylate resin for biomedical applications: a double infiltration approach. Polym Int. 2006;55:57–62.
262. Pegg DE. Principles of tissue engineering. Academic Press. 2013:378-379.
263. Vozzi G, Previti A, De RD, et al. Microsyringe-based deposition of two-dimensional and three-dimensional polymer scaffolds with a well-defined geometry for application to tissue engineering. Tissue Eng. 2002;8:1089–98.
264. Vozzi G, Flaim C, Ahluwalia A, Bhatia S. Fabrication of PLGA scaffolds using soft lithography and microsyringe deposition. Biomaterials. 2003;24:2533–40.

Chapter 3
The Mechanical Properties of the Scaffolds Reinforced by Fibres or Tubes for Tissue Repair

Sreekanth Pentlavalli, Helen O. McCarthy, and Nicholas J. Dunne

Abstract Tissue engineering is a broad technology for tissue repair of diseased or injured tissues and organs by using composite scaffolds. The scaffold material provides stability and creates natural environment to promote cell-seeded population to develop fully functional new tissue or organ by restoring original tissue qualities. Generally, these scaffolds are biocompatible, three dimensional and biodegradable frameworks that act as template for *ex-vivo* cell expansion. Fibres or tubes reinforced scaffolds are gaining importance due to high mechanical stability and eco-economic reasons. This tailored made technology is a combination of both natural and synthetic material which helps to design the scaffold based on needs. To achieve this, fibres or tubes have been used to provide suitable mechanical properties during entire duration of tissue repair. Ideally, fibres reinforced scaffolds must meet following criteria: (i) 3D network with interconnected pore network for cell attachment and growth; (ii) biocompatible and controlled degradation and resorption rate to match tissue growth; (iii) appropriate mechanical properties to match the native tissue at the site of implantation. In this article, we mainly discuss the impact of mechanical properties on various scaffolds such as hydrogel based, ceramic based composites with both natural and synthetic fibres. Also discussed are the various useful techniques used to measure mechanical properties of reinforced materials.

Keywords Mechanical properties • Tissue engineering • Scaffolds • Fibres

S. Pentlavalli • H.O. McCarthy
School of Pharmacy, Queen's University of Belfast, 97 Lisburn Road, Belfast BT9 7BL, UK

N.J. Dunne (✉)
School of Pharmacy, Queen's University of Belfast, 97 Lisburn Road, Belfast BT9 7BL, UK

Centre for Medical Engineering Research, School of Mechanical and Manufacturing Engineering, Dublin City University, 9 Stokes Building, Collins Avenue, Dublin, Ireland

Trinity Centre for Bioengineering, Trinity Biomedical Sciences Institute, Trinity College Dublin, Dublin 2, Ireland
e-mail: nicholas.dunne@dcu.ie

X. Li (ed.), *Tissue Repair*, DOI 10.1007/978-981-10-3554-8_3

3.1 Introduction

Tissue engineering (TE) uses scaffolds to provide temporal support for cells to regenerate new extracellular matrix (ECM) that has been destroyed by disease, injury, or congenital defects [1]. Natural ECM separates different tissues, forms a supportive meshwork around cells and provides anchorage to the cells. It is comprises of proteins like collagen and carbohydrate biopolymers such as glycosaminoglycans (GAGs). The purpose of a scaffold is to support the tissue and promote cell-seeded population for mature native tissue development. In general, scaffolds are three-dimensional (3D) that resemble the ECM, biocompatible and biodegradable. It should work like a template for *ex vivo* cell expansion and restoration of original tissue qualities with respect to the biochemical constituents, morphology, form and physiological function of the respective tissue. Scaffolds should be designed to degrade slowly in the body as the cells begin to regenerate. The rate of degradation must match the rate of tissue regeneration. To be considered biocompatible, the scaffold should be non-toxic to the cells or tissue in either the original and degraded form. The scaffold should also facilitate cell attachment, permit the diffusion of nutrients from the blood, and initially mimic the mechanical properties of the native tissue. Novel scaffold materials have been developed, which are not only 3D, but also to provide mechanical stability and replicate the functional properties of the tissue. For simple tissues, the scaffold materials are homogeneous in chemical composition and isotropic in mechanical properties. The need to utilise alternative strategies to fabricate scaffolds that better mimic the natural properties of native tissue with respect to its nanoscale morphology, gross anatomic form, localised mechanics, and biocompatibility continues to attract new methodologies for material fabrication.

Hydrogel materials [2–5] and fibrous materials/textile substrates [6–8] are two major categories of materials used as scaffolds for tissue engineering. Hydrogels have been used to create homogeneous 3D structure in the tissue or grafted to adjacent tissue for *in vivo* growth [5, 9]. These materials can create highly defined microenvironment based on their composition and the chemical moieties present and can be synthesised using both natural and synthetic polymers/materials [4]. Collagen Type-I, hyaluronic acid, agarose, keratin, alginates, fibrin, and gels derived from tissues or isolated cells can be classified as natural hydrogel materials. The main advantages of natural hydrogel materials are uniform cell distribution and high cellular viability. Most commonly used synthetic hydrogels are fabricated using poly(ethylene glycol) (PEG) and can be designed according to degradation profile (i.e. hydrolysis vs. enzymatic), bioactivity modification through the tethering of growth factors, chemical composition, cross-linking density for cell attachment and production of ECM [2]. However, the limiting factor for these materials is heterogeneity based on batch to batch processing, substantial contraction due to cellular remodelling and poor mechanical properties [2]. The increase in aggregate modulus typically decreases with permeability, limiting diffusion of nutrients to the entrapped cells. Various methods have been adopted to fabricate scaffold materials with

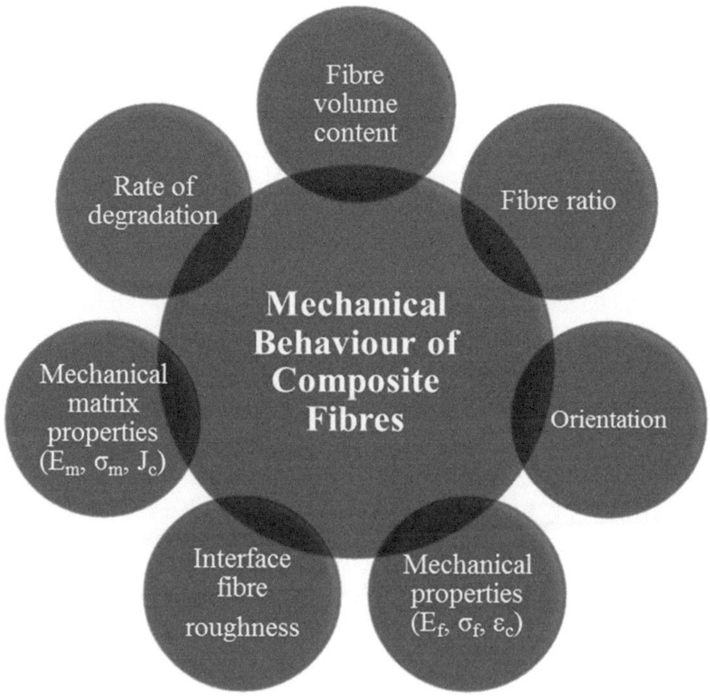

Fig. 3.1 Factors influencing the mechanical properties of fibre reinforced composites

varying geometries, mechanical properties, porosity and anisotropy. Interaction of cells with specific ECM is critical to maintain cell phenotype, cell-mediated ECM deposition and resultant properties of the cell-seeded constructs.

A number of material properties can influence the mechanical properties of fibres (Fig. 3.1). Introduction of fibres within the matrix is essential to improve the mechanical properties of composites as the fibres can take external load and increase the stability of the scaffold.

Fibre reinforcement within cell-seeded gel carrier systems has seen significant advancement in the field of scaffold fabrication for tissue engineering [10, 11]. Fibre based materials have long been used in tissue engineering for the repair or replacement of tissues, ranging from their use as sutures to more complicated applications such as vascular grafts, hernia tissue replacement and bone, cartilage tissue engineering [8]. Fibre reinforcement is a key variable to manipulate the mechanical and physical properties of biomaterials for targeted tissue engineering. These fibres can be organised as monofilaments or multifilament yarns [8, 12]. The mechanical properties of fibre reinforcement systems are based on various factors such as: fibre organisation (e.g. uniaxially aligned or randomly aligned), fibre volume, fibre-fibre and fibre-matrix interactions. These systems can have high porosity, which helps to contain gels and cells offering good mechanical properties and ease of handling for varying tissue and organ shapes.

Use of nanofibre-based composites is growing due to their high surface area, superior mechanical properties derived from highly aligned molecular chains minimising pre-mature crack formation and increasing biocompatibility. Many composites are being developed for use in biomaterials for drug delivery, wound healing and tissue engineering applications. Also, similarity between nanofibre mesh and natural collagen fibres exist in many tissues [13]. For tissue engineering applications – the most common fabrication method is coaxial electrospinning [14]. Various material properties affect the mechanical properties of nanofibres/mesh such as: fibre diameter, mesh porosity, fibres alignment and interconnection between fibres. Many researchers have attempted to correlate mesh stiffness with fibre diameter and mesh morphology. McManus et al., showed that the tensile strength of mesh is not only dependent on fibre diameter but also the pore size, interaction between fibres and conditions used to form mesh also play important roles [15].

3.2 Mechanical Characterisation

Fibre based composites demonstrate improved mechanical properties by sharing strain between the matrix and the fibre reinforcement [16]. The reinforcement can be calculated using the Hashin-Shtrikman model [17].

$$E_{C,U} = V_r E_r + (1 - V_r) E_g \tag{3.1}$$

$$E_{C,L} = (V_r / E_r + (1 - V_r) / E_g)^{-1} \tag{3.2}$$

V_r is volume fraction of reinforcement component; E_r is elastic moduli and E_g is gel matrix. $E_{C,U}$ and $E_{C,L}$ corresponds to the fibres aligned and perpendicular to the direction of loading. Randomly aligned fibre reinforced composites fall in the region between the two bounds. Fibres aligned in the direction of loading exhibit higher stiffness even at relatively low volume fractions. To quantify the effect of reinforcement an amplification factor, A, is introduced to facilitate comparison between studies, E_c is simple bounds on the modulus.

$$A = E_c / E_g \tag{3.3}$$

By combining Eqs. 3.1 and 3.3, the volume fraction V_r is influenced by A and there is modulus mismatch between the reinforcement and gel matrix, E_r/E_g.

It has been reported that fibre reinforcement polyvinyl alcohol (PVA) hydrogels demonstrate an increased tensile modulus from 0.23 MPa without any reinforcement to 260 MPa when reinforced with 29 wt.% ultra high molecular weight polyethylene (UHMWPE) fibres [18], whereas PVA covalently grafted onto the UHMWPE demonstrated increased interfacial shear strength from 11 kPa to over 220 kPa [19].

3.3 Tensile Testing

A well-known method for determining the mechanical properties of fibre reinforced composites is tensile testing; however, there is no standard protocol like in traditional fields such as metallurgy. Based on the composition and nature of the material – specific test conditions need to be applied such as wet, dry or fully swollen samples, geometry and range of applied strain (Fig. 3.2) [18, 20–25]. For example the elastic modulus of unreinforced alginate hydrogels has been shown to be 77.88 ± 18.67 kPa and the inclusion of gelatin electrospun fibres increased the tensile modulus to 0.50 ± 0.11 MPa. The failure strength increased by a factor of 17.6 times on incorporation of the gelatin electospun fibres [24]. In another study, the tensile modulus of a PCL-alginate composite increased from 30–200 kPa to 180–400 kPa depending on the level of fibre loading [25]. The effect of the fibres was high when 1 wt.% alginate gel was used, where a modulus mismatch between the gel and fibres was greatest. To evaluate the effectiveness of fibre reinforcement, tensile testing should be performed at a relatively low strain rate, which is considered as the gold standard for making comparisons between studies.

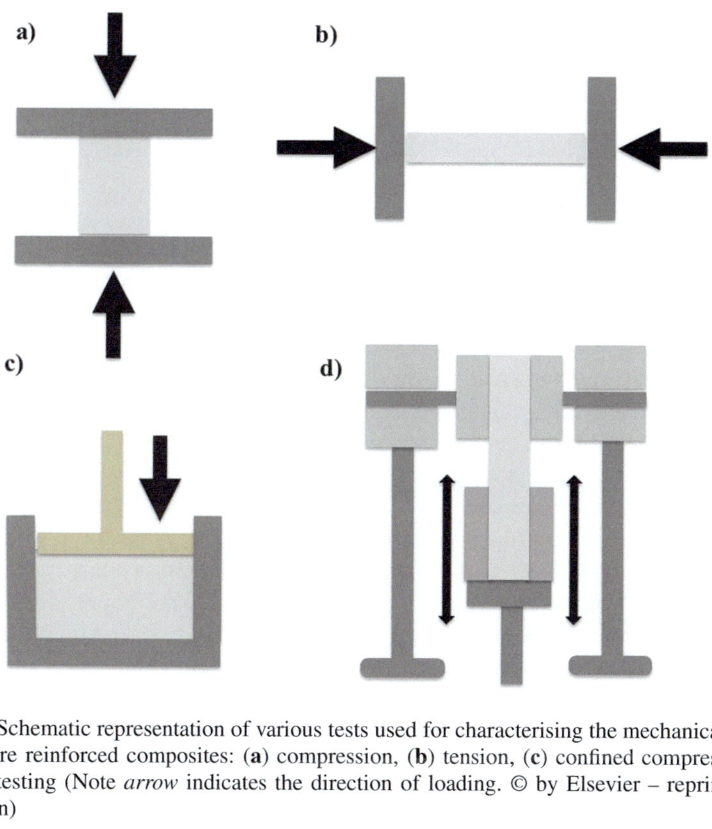

Fig. 3.2 Schematic representation of various tests used for characterising the mechanical properties of fibre reinforced composites: (**a**) compression, (**b**) tension, (**c**) confined compression, (**d**) dynamic testing (Note *arrow* indicates the direction of loading. © by Elsevier – reprinted with permission)

3.4 Uniaxial Tensile Testing

During uniaxial tensile testing, the load is applied along a single specified direction of the bulk scaffold (comprised of individual fibres with distinct mechanical properties), while the specimen is free to contract laterally. For rectangular or dumbbell-shaped samples, the load and displacement are measured. The stress vs. strain behaviour is calculated by normalising the force to the original cross sectional area (i.e. stress) and displacement to original gauge length (i.e. strain).

3.5 Biaxial Tensile Testing

For biological samples, adopting the uniaxial tensile test has limited value due to absence of freely contracting boundaries experienced by many tissues *in situ*. Many fibre reinforced soft tissues are subject to tensile loads along multiple directions. For this reason, biaxial tensile testing is more useful for measuring the mechanical properties of tissue-engineered constructs. Biaxial tensile testing allows for the measurement of material behaviour simultaneously along two orthogonal axes. This method is more accurate to characterise the mechanical function of scaffolds and tissues, which provides an approximate idea of *in vivo* loading [26, 27].

3.6 Compression Testing

The parameters for compression tests include confine or unconfined compression and creep vs. failure tests, but once again there is no standard protocol [28–31]. The effect of fibre inclusion has been shown to increase in stiffness and the ability to bear high stresss. Inclusion of electrospun silk fibrin fibres in chitosan hydrogel using various test methods resulted in different compressive stiffness values being determined. The modulus calculated from the linear portion (i.e. elastic region) of a stress vs. strain profiles under static compressive loading for a homogeneously dispersed chopped silk fibrous within a chitosan hydrogel and a layer by layer assembled hydrogel-fibre composite were approximately 2 and 3 times stiffer when compared to the stiffness for pure chitosan hydrogel [28]. A biocomposite consisting of polyacrylamide gel and short chitosan nanofibres sustained seven times higher stress when compared to pure polyacrylamide at 95% compressive strain and recovered to its original height on load removal [32]. These observed improvements in mechanical properties were due to the fibres preventing crack growth within the microstructure and the smooth transfer of stress between the hydrogel matrix to the fibres.

3.7 Dynamic Mechanical Analysis (DMA)

This technique is useful for evaluation of the mechanical properties under different modes of loading, e.g. under shear loading the complex shear modulus can be attained; and under tensile loading conditions the storage and loss modulus can be determined. Electrospun poly(l-lactide) (PLA) fibres with a poly(lactide-co-ethylene oxide fumarate) hydrogel was fabricated using multilayer laminates, which was tested using DMA at 37 °C. The compressive modulus of the composite hydrogel was determined under wet and dry conditions and compared with the pure equivalent. The composite hydrogel demonstrated a compressive modulus of 575 ± 14 MPa when wet, which was significantly higher than the pure hydrogel (A = 4.1) and slightly greater than the dry composite [33]. Various other methods have also been used, such as shear friction test [30], notched tension test [30], suture retention tests [22, 23], dielectric property tests [19], spherical indentation [28, 34], fibre pull-out tests [19] and monotonic and cyclic strain tests [35, 36]. Irrespective of test method, the amplification factor (A) allows for a relatively straightforward metric demonstrating the extent to which the inclusion of a reinforcement material has on the mechanical properties of a composite.

3.8 Fibre Reinforced Hydrogels

Hydrogels are primarily composed of synthetic or natural hydrophilic polymers. The 3D structure of hydrogels resembles that of natural ECM, absorbing large amounts of water or biological fluids, demonstrating a highly porous network that allows cell and nutrients migration and cell proliferation [5, 37–39]. Hydrogel composition must meet certain parameters in order to be successful as a tissue engineered scaffold with respect to its mechanical (e.g. Young's modulus), physical (e.g. mass transport), and biological (cell/matrix interaction) properties [40, 41]. The mechanical properties of the hydrogel largely depends on the cross-link density [42] and the nature of network (e.g. chemical or physical). Physical networks are reversible and can be synthesised using polymers such as collagen [43], chitosan [44] and aminated hyaluronic acid [45]. Whereas, chemical networks are achieved by covalent bonding and are non-reversible, e.g. chitosan/hyaluronic acid [46], PEG/fibrinogen [47], and alginate [48] based hydrogels. However, such hydrogels are weak isotropic materials and fail to provide any load-bearing capabilities *in vivo*. They also demonstrate low stability and survival following engraftment [49, 50].

To improve the mechanical integrity of hydrogels, different methods have been developed to allow fibre reinforcement. The most commonly used methods for fibre fabrication are electrospinning and electrospraying [51]. Researchers have developed highly injectable composite based hydrogels using small fibres (1 μm–1 mm) [20, 21, 28, 29, 32, 52, 53]. Fibre reinforced hydrogels can be tailored to demonstrate anisotropic mechanical properties [30, 33, 54, 55]. Moreover, fibre reinforced hydrogel systems containing electrospun fibres have been shown to exhibit good mechanical properties, high levels of cell growth and differentiation [53, 56–58].

3.9 Fibre Reinforced Ceramic Composites

In recent times, calcium phosphate (CaP) based materials have been applied for the augmentation of bone defects. They are available as allogenic, sintered or cementitious materials. Unfortunately these materials have very poor tensile and shear properties [59]. Calcium phosphate cements (CPCs) strength is less than that of bone, teeth, or sintered calcium orthophosphate bioceramics [60] due to their inherent brittleness, which restricts their use non-load bearing defects [61]. To overcome this problem, many researchers developed new CaP based materials to improve mechanical properties of composites using novel methods, including the addition of additives to the CaP composites either by using polymers that interpenetrate the porous matrix or via fibrous reinforcement. Interpenetration of polymer into porous matrix has been shown to improve the mechanical properties – however some limitations have been demonstrated, e.g. lack of biodegradability, rigidity, ductility [61, 64, 65]. The other major problem is the sterilisation requirement for biomaterials and medical devices, which may affect the polymer properties. Fibre reinforcement with ceramics overcomes these limitations [63]. The recent rise in the number of research publications clearly shows an increasing interest in fibre reinforcement of calcium phosphate cements (FRCPCs) as a viable alternative for bone tissue engineering applications. In this section, we will focus on mechanical properties of FRCPCs. They are classified into two types: (i) CPC fibres with non-resorbable fibres and (ii) CPC fibres with resorbable/biodegradable fibres (Fig. 3.3). The incorporation of the non-resorbable fibres provides mechanical stability to the cement. Whereas in the latter classification, the addition of the resorbable fibres provides two roles: (i) provision of mechanical reinforcement and (ii) the development of

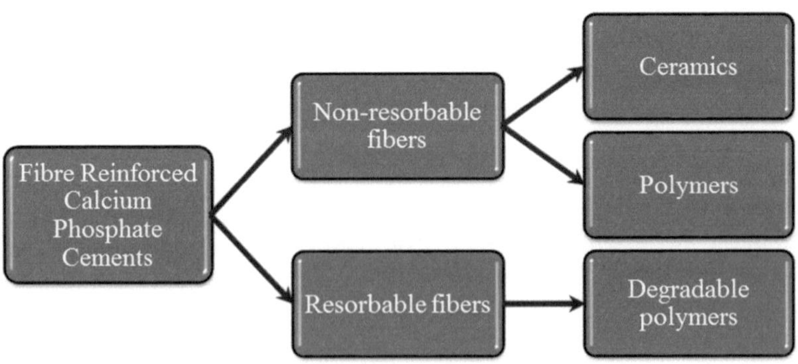

Fig. 3.3 Schematic diagram showing different classifications of fibre reinforced calcium phosphate cements

porosity within the cement microstructure following degradation of the fibres. Care should be taken when selecting the most appropriate type of resorbable fibres – as the rate of degradation should be closely match the rate of bone remodeling in order to maintain mechanical stability of the scaffold.

Fibres processed from polyamide (linear polymers), aramide (amide groups separated by phenylene groups) and glass fibres are typical non-resorbable fibres that have been used in CPC based systems. Most of polyesters such as polylactide (PLA), polyglycolide (PGA) and poly-ε-caprolactone (PCL) have been used as resorbable/biodegradable fibres. Polyester based polymers are largely insoluble in water but they can degrade via hydrolysis of ester bond under physiological conditions.

3.10 Mechanics of Fibre Reinforced Cement Systems

The mechanical behaviour of fibre reinforced cements is a result of the complex interaction between the all the constituents of the composite constitutes such as the macroscopic behaviour between the strength and stiffness of both fibre and cementitious matrix, matrix toughness, mechanical interaction between the fibres and matrix and supplementary effects of polymeric additives or aggregates [66, 68].

The stress vs. strain (σ vs. δ) curve generally defines the fibre bridging property across the matrix crack. That is the average tensile strength (σ) transmitted across the uniform crack opening (δ) in a uniaxial tensile specimen. The link between the composite (fibre, matrix and interface) and composite tensile ductility is gained from the σ vs. δ curve (Fig. 3.4).

Monolithic and fibre reinforcement σ vs. δ curves for different failure behaviour following loading under bending or tension conditions (Fig. 3.5a). The material strength or stress maximum before failure depends on two aspects: (i) strain to failure and (ii) the area below the curve or energy necessary to yield material failure. These properties will increase in the order of failure modes (i.e. brittle < tension softening

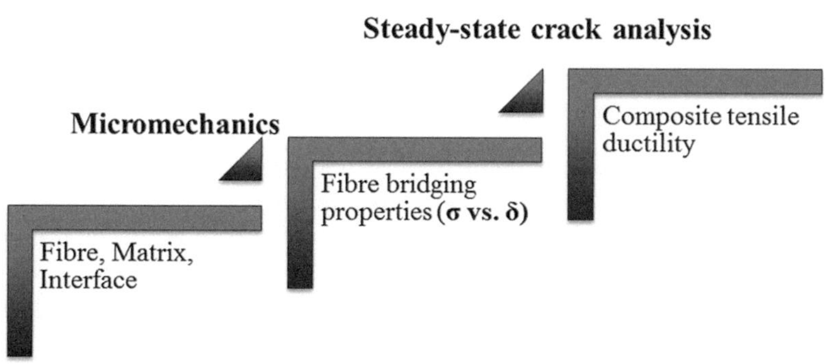

Fig. 3.4 Schematic representation of relation between composite material properties

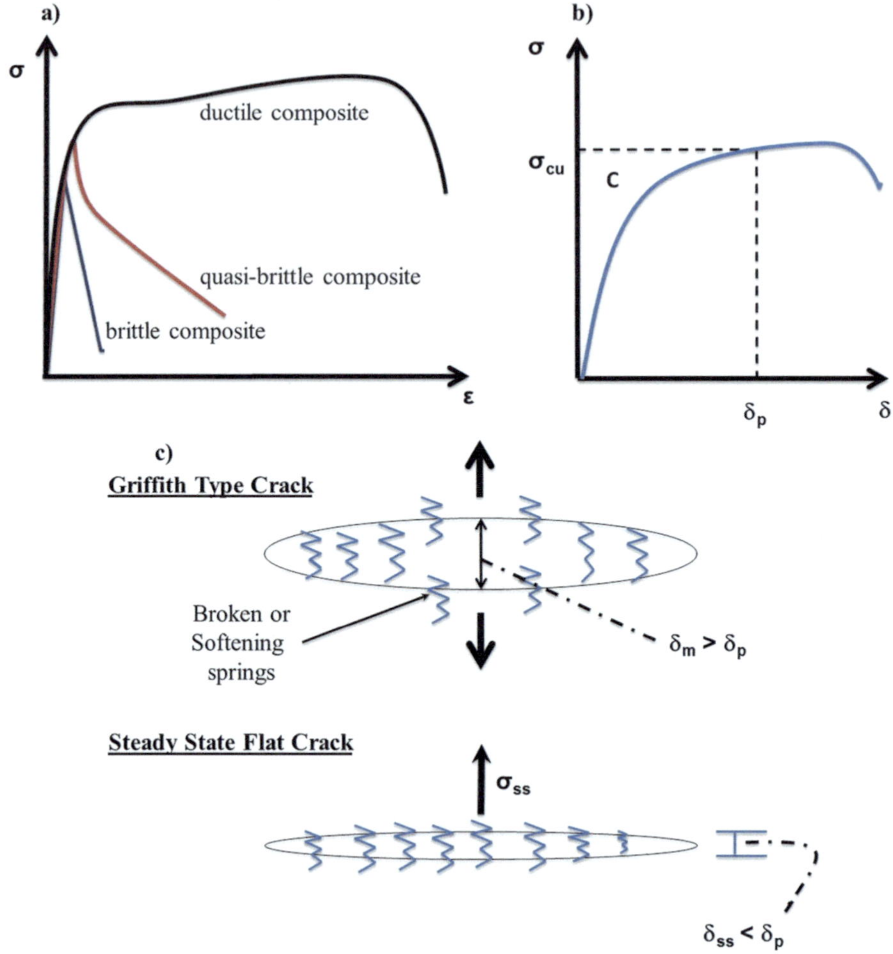

Fig. 3.5 Fracture of cement and composites. (**a**) Typical stress-strain curves of different composites. [68] © by RILEM reprint with permission from springer Science + Business Media B.V. (**b**) Complementary energy C originated from σ vs. δ curves (stress (σ) against crack opening displacement (δ)) [71]. (**c**) Schematic representation of cracks in brittle and quasi-brittle materials (Griffith crack, *top*) and strain hardening composites (steady state, *bottom*) [71] © by JCI – reprinted with permission

or quasi-brittle materials < strain hardening or strictly pseudo strain hardening; high ductile materials). High amounts of energy are absorbed during fracture of fibre reinforced cement based composites when compared with pure brittle failure, which leads to two or more fractures. Whereas for materials that undergo quasi-brittle composites, fracture is associated with a peak load and exhibits a tension softening behaviour, during this process the fibre bridging across the single crack opening delay the fracture via dissipation of the fracture energy – however the composite no longer demonstrates an ability to bear substantial load after the peak load reached.

Fibre reinforced composites with pseudo strain-hardening behaviour (elastic-plastic behaviour) demonstrates the highest ductility. For such composites even at peak load, the strain to failure can be several percent due to network of multiple cracks that develop and therefore, substantially more energy is dissipated during the fracture process [67, 68]. Energy dissipation happens at different sites and mechanisms during fracture. Some of the important sites are the frontal process zone (FPZ) where the fibres influence the fracture process before the crack tip, behind the crack and adjacent to the crack plane [62, 69, 70]. These behaviours of composite failure are based on different mechanisms. When tensile load is applied on a brittle solid, the stress intensity reaches a critical value where catastrophic failure occurs due to fast crack propagation from the initial crack (monolithic CPC) [80]. Whereas fibre reinforced composites that exhibit quasi-brittle behaviour, the linear elastic region ends with first crack of the composite and softening of the material is observed as crack propagates [67, 71]. The strength of the composite was determined by the behaviour of FPZ.

In pseudo strain-hardening composites, the composite can withstand substantial load after initial loss of stiffness in the elastic region. With increasing tensile loading multiple microcracks are formed, which could be visualised microscopically due to composite ductility [67]. The multiple cracking of fibre reinforced composites is based on strength of criterion and energy criterion. It is explained by the complementary energy on the σ vs. δ curve, which shows stress against crack opening displacement (Fig. 3.5b). Fibre pull-out occurs at a low peak strength when the interface is too weak, whereas fibre rupture takes place when the interface is strong with a small value of critical opening (δ_p). In both cases, the area between rising part of σ vs. δ curve and the σ–axis is very small (complementary energy). When complementary energy is large, the crack will remain flat as it propagates so that the steady state crack opening will be $< \delta_p$, and the composite maintain tensile load carrying capacity post-propagation (Fig. 3.5c). The shape of the σ vs. δ curve plays an important role in determining the uniaxial tensile load (strain-hardens or tension-softens) of the composite [71]. In composite based cements that undergo strain hardening, the main crack from initial notch is delayed and further loading can take place before real crack growth can begin [70]. High bending capacity of fibre-reinforced cements is a consequence of its high ductility. Also, ductility and fracture energy dissipation are higher in composites that undergo strain hardening when compared with composites that experience tension softening. Optimisations of these two variables are essential when attempting to improve the mechanical properties of fibre reinforced cements.

3.11 Fibre Length

The Young's modulus of the fibre is directly proportional to the external load, which can be calculated using following equation:

$$E_f = \sigma_f^{max} / \varepsilon_m \tag{3.4}$$

Where σ_f^{max} is the upper limit of tensile stress within the fibre, E_f is the fibre modulus and ε_m is the matrix strain [72]. In softer materials, the difference between the Young's moduli of fibre and matrix creates additional strain near their respective interface. Also, the interface area between fibre and matrix depends on the fibre diameter at a known fibre volume, which influences the homogeneity and processability of the composite. The mechanical stability of fibres within the matrix composite depends on fibre length and diameter. Substantial load must be transferred across the matrix to the fibre via interface to strengthen the composite. In general, the load transferred through shear stresses at the lateral surface of the fibre is high when compared with load transfer at the end face [63]. Fibre pull-out occurs before the fibre fracture if the fibre length is not sufficient or is small. To maximise the effect of fibres within composite material, the critical length (l_c) can be calculated using following equation:

$$l_c = (d_f \sigma_{f,B}) / 2\tau_i \tag{3.5}$$

Where d_f, $\sigma_{f,B}$ and τ_i represents fibre diameter, fracture strength and shear stress at the interface. To understand stress distribution with respect to fibre length along its axis a simple model has been proposed by many researchers. When the fibre length is greater than the l_c, then the shear stress in the fibre end is maximum and zero at the centre of the fibre [63, 72, 73].

3.12 Fibre Volume and Strength

Fibre volume plays an important role in composite strength and ductility. Critical fibre volume is necessary to induce pseudo strain hardening behaviour and multiple cracking. The following equation can be used to estimate the critical fibre volume in a composite containing parallel-aligned fibres:

$$V_f^{crit} = \sigma_{mu} / \sigma_{fu} + (1 - (E_f / E_m))\sigma_{mu} \tag{3.6}$$

Where σ_{fu} and σ_{mu} represent the ultimate strength of fibre and matrix. E_f and E_m refer to the Young's moduli of the fibre and matrix [67, 74]. The critical volume content of a composite that contains fibres organised in a random (isotropic) discontinuous manner can be expressed as follows:

$$V_f^{crit} = 12J_c / g\tau (L_f / d_f)\delta_o \tag{3.7}$$

Where J_c is the composite crack tip toughness, g is the snubbing factor, τ is interface frictional bond strength, L_f and d_f are the fibre length and diameter and δ_o is the crack opening when the fibre bridging stress reaches maximum. Compared with random oriented fibres, aligned fibres requires lower critical fibre content and low bonding strength requires with high fibre content at fibre-matrix interface [67, 71].

The interfacial strength (transfer of load between the fibre and matrix) can be determined by single fibre pull-out test or using a uniaxial tensile test where the tensile stress vs. the opening relationship is continuously monitored [72, 75–77]. Various methods have been developed to improve interfacial strength, e.g. coatings to the fibres and introducing functional groups on the fibre surface [78–80].

3.13 Mechanical Properties of Fibre Reinforced Calcium Phosphate Cements

Fibre reinforced calcium phosphate cements (FRCPCs) have been introduced in an effort to improve their tensile, flexural strength and ductility. Some of the important measuring techniques used to determine the mechanical properties of these reinforced cements have been: uniaxial compression, diametric compression, three point and four point bend and biaxial flexure tests. However, some of these techniques have limitations, e.g. toughness and flaw sensitivity cannot be characterised by compression due to stress concentration [81] and adoption of the diametric test configuration is not useful as this theory can only be applied to brittle materials. The most commonly used test method for determining the mechanical properties of fibre reinforced CPCs is either the three or four point bend test method. The mechanical properties of resorbable and non-resorbable fibre composite data are summarised in Table 3.1.

The mechanical properties of polyamide (linear polymers) fibres into CPCs demonstrated increased compressive strength (30%) [82]. When aramid fibres were added to CPCs, the flexural strength increased from 13 MPa (pristine cement) to 39 MPa and 65 MPa at 2 wt.% and 4 wt.% fibre loading [83, 84]. Inclusion of PLLA in CPC increased the work of fracture, however the flexural strength and Young's modulus decreased progressively (from 7 MPa to 2.6 MPa and from 49 GPa to 15 GPa). It was suggested that the reduction in flexural strength and Young's modulus was due to an increase in the pore volume created by inter-fibre porosity from the fibre bundle. Xu and Quin showed that adding 25 wt.% PLGA yarns created long cylindrical macropores within the cement and resulted in a three-fold increase in the flexural strength (25 MPa), the work of fracture increased by two-fold (up to 3.4 KJ/m^2) – however the Young's modulus did not change considerably (Fig. 3.6) [88]. Due to fibre degradation under physiological conditions, the mechanical properties of composites decreased over time. Calcium phosphate cements containing electrospun PCL fibres exhibited an increase in the work of fracture. However, the flexural strength and elastic modulus decreased on introduction of the fibre bundles, which were not well integrated within the cement microstructure due to the hydrophobic surface of the fibres [98].

Table 3.1 Overview of composition and mechanical properties of calcium phosphates cements containing resorbable and non-resorbable fibres

Fibres	Additives	Matrix	Fibre weight/ Fibre volume (wt.%/vol.%)	Fibre diameter (μm)	Fibre length (mm)	Strength (MPa)	Work to fracture (KJ/m^2)	Elastic modulus (GPa)	Refs.
FRCPC composites with non-resorbable fibres									
Polyamide	–	α-TCP	0.2–1.6	100	3	12.5–9.8	–	–	[82]
Aramide	Mannitol	DCPA: TTCP	6	15 ± 2	75	44–4	–	6.5 ± 1	[83]
Carbon	–	HA(DCPA:TTCP)	2–10	3–200					[84]
CNT	BSA	HA(β -TCP:DCPA)	0.2–0.5	20–30	30 nm	14	–	–	[85]
CNT	BSA	HA(β -TCP:DCPA)	0.25–1	30–50	30 nm	16.3	–	–	[86]
Bioglass	–	HA(TTCP:DCPD:Na_2HPO_4)	5–25	20	–	1.5–3.7	0.2–0.5	–	[84]
Bioglass	–	DCPA:TTCP	1.9–9.5	16 ± 2	3–200	20–35	2–4.5	4–5	[84]
FRCPC composites with resorbable/degradable fibres									
PLLA	–	α-TCP, $CaHPO_4$, PHA, $CaCO_3$	1–7	1–2	0.2	4–2.6	0.1–0.25	32–15	[87]
PLA/PGA	–	HA(DCPA: TTCP)	1.9–9.5	198 ± 64	3–75	12.5–28	2.5–5	2.2–4.5	[84]
PLA/PGA	–	HA(DCPA: TTCP)	25	322	8	14–25	2.5–5	5.5–2.2	[88]
PLA/PGA	Chitosan	HA(DCPA: TTCP)	45	322	8	41	11	3.2	[89]
PCL	–	HA(α-TCP, $CaHPO_4$, HA	1–7	1–2	0.2–3	3.5–7	0.1–0.5	15–47	[90]
PLA/PGA	Chitosan, alginate beads, hUCMSC	DCPA:TTCP	5–20	322	8	7–11	0.7–1.5	1–2	[91]
PLA/PGA	Chitosan, alginate hydrogel beads, hBMSC	DCPA:TTCP	20	322	8	11.7	1.7	2	[92]

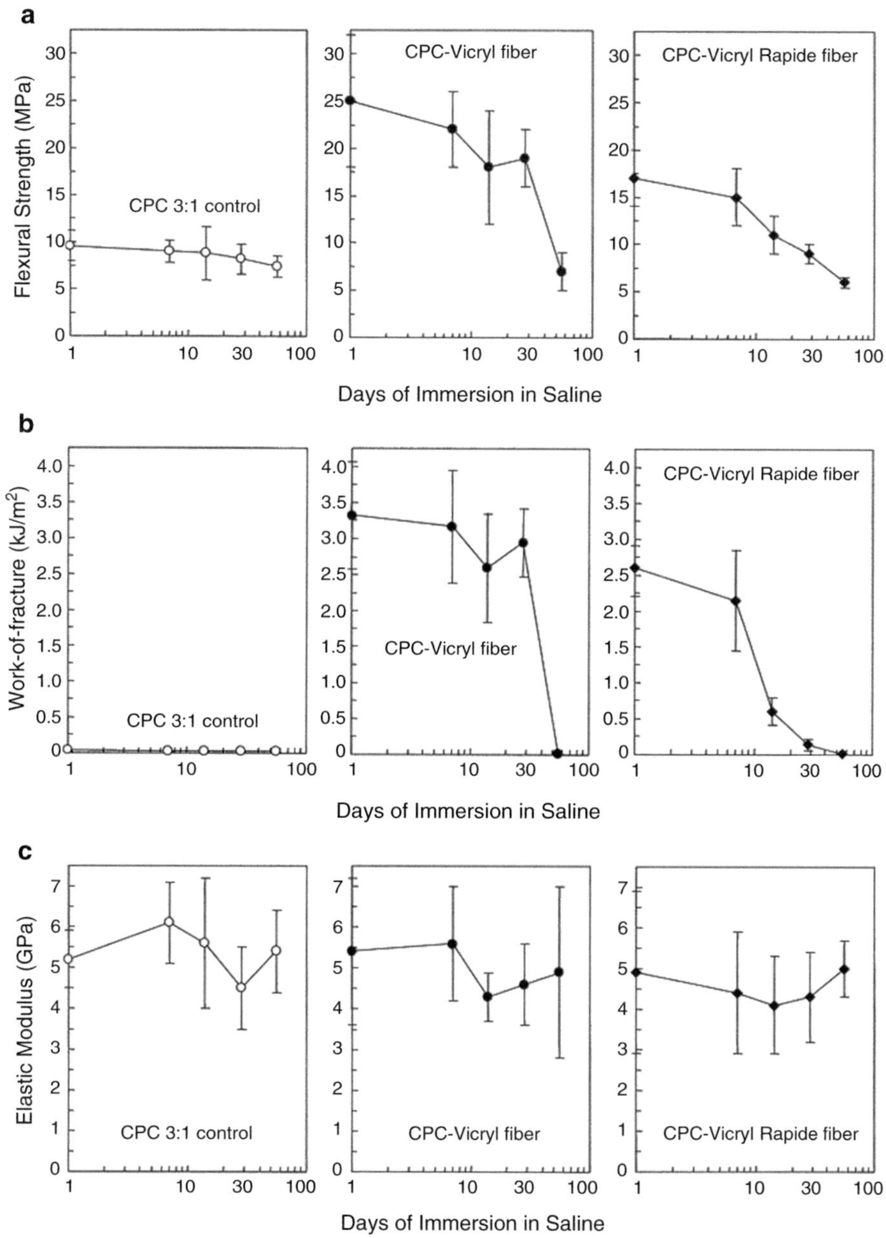

Fig. 3.6 (**a**) Flexural strength vs. days of immersion in saline for CPC control, CPC-Vicryl™ fiber composite, and CPC-Vicryl Rapide™ fiber composite, all at a CPC powder: water ratio of 2:1. Each datum is the mean value of six measurements ($n = 6$), with the error bar showing one standard deviation (SD). (**b**) Work-of-fracture for CPC control, CPC-Vicryl™ fiber composite, and CPC-Vicryl Rapide™ fiber composite, all at a CPC powder: water ratio of 2:1. (**c**). Elastic modulus for CPC control, CPC-Vicryl™ fiber composite, and CPC-Vicryl Rapide™ fiber composite, all at a CPC powder: water ratio of 2: 1 [88] © by Elsevier – reprinted with permission

3.14 Reinforcement Mechanisms of the Fibres or Tubes in the Scaffold

Engineering materials composed of two or more constitutes are called composites and mainly consist of a matrix and reinforcement [93]. Both these materials have different physical and chemical properties, but as a composite they are connected and distinct on a microscopic level [94, 95]. In polymer composite materials, the polymer acts as a matrix and the fibres acts as a reinforcement material. The properties of the scaffold depend on bond between matrix and fibres [96]. The main advantage of composite materials is the matrix holds/transfers the load along the reinforcement. In general, polymer materials are used as matrix and natural/synthetic fibres as the reinforcement. Polymer composites are classified into different types, fibre reinforced polymer composites where fibres are used as reinforcement either natural or synthetic and particle reinforced polymer composite, the reinforcement is converted into particle form of different dimensions [93, 97].

Based on the nature of polymer composites (renewable/non-renewable), they are categorised namely as renewable composites where both the matrix and reinforcement are based on 100% renewable materials. In partly renewable composites either the matrix or reinforcement material are obtained from renewable source. Both materials for non-renewable composites are obtained from non-renewable source [93].

Due to economic and environmental reasons, natural fibres are gaining much attention. Natural fibres have higher stiffness/strength compared with synthetic fibres. Although natural fibres show some disadvantages such as being hydrophilic in nature and less thermally stable [97]. However, to overcome these disadvantages, researchers have developed various techniques such as surface modification and pretreatment [98]. Many reviews have been published reporting detailed techniques and analysis for the chemical surface modifications of natural fibres [97, 99–102]. Natural fibres/polymer composites have been frequently used for many applications due to their mechanical properties [103, 104]. Some commercially important thermoset polymer composites are phenolic, epoxy, polyester, unsaturated polyester; vinyl ester composites, which are discussed herein.

In phenolic composites, the phenol-formaldehyde (PF) resin system is the most common. Using compression-moulding technique – natural fibre reinforced with PF resin system has been developed [105]. In this study, the authors analysed various mechanical properties such as tensile strength, flexural strength, and compressive strength and wear resistance with respect to fibre loading. The mechanical properties increased up to 30 wt.% of fibre loading and decreased beyond this loading. These composites demonstrated high mechanical strength. PF matrix grafted with grewia optiva fibres as the reinforcement showed better mechanical properties compared to the un-grafted composites. To improve the adhesion of the sisal fibres with the PF polymer matrix, the sisal fibres were modified via hydroxymethylation. These modified fibres showed highed impact strength and strong bond between the fibres and matrix [106]. PF polymer composites have been prepared using short sisal fibres using two different methods: (1) direct mixing and (2) polymerisation under com-

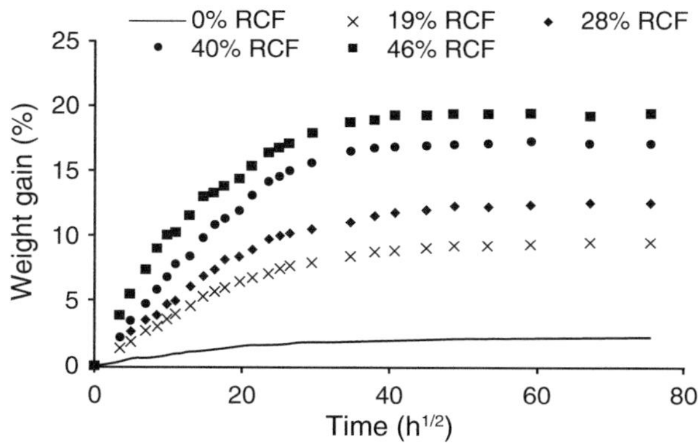

Fig. 3.7 Water absorption curves of RCF/epoxy composites [119] © by Elsevier – reprinted with permission

pression moulding conditions. It was observed that the polymer composites prepared via polymerisation showed better mechanical properties. Reinforcement of the phenol-furfural polymer matrix with sisal fibres - the polymer composite exhibited high mechanical properties, which is an alternative method to avoid the use of formaldehyde [107–109].

Different natural fibres like pine needles and banana fibres have been used as reinforcement in PF polymer composite with and without surface modification [110, 111]. Banana fibres were made into micro- and nano-sized fibres using a steam explosion process and reinforced with PF polymer matrix. Enhanced mechanical properties were observed for the nanofibre reinforced composite when compare with the microfibre reinforced system [112].

Different types of natural fibres based epoxy polymer composites have been developed [113–115]. The mechanical properties of composites prepared from palm tree fibre as a reinforcement and epoxy as a matrix have been studied using resin transfer molding (RTM). It was observed that chemical modification of fibres helps the bonding between fibre and matrix and increases the mechanical properties [116]. To improve interfacial adhesion between fibres and the matrix, short palm fibres were reinforced with polyester and epoxy matrices. Surface modification improved the bond between the fibre and matrix via esterification. It was concluded that epoxy based composites exhibited higher mechanical properties due to better interfacial adhesion when compared to unsaturated polyester matrix [117]. To understand the real potential of natural fibre-reinforced composites, [118] studied effect of surface modification on mechanical properties of composites. Natural fibres were aged in distilled water and alkali medium and used to reinforce epoxy and polyester matrices. Alkali treated fibres showed better adhesion and increased the mechanical properties with low water absorption [118]. The effect of fibre loading (19, 28, 40 and 46 wt.%) on mechanical properties such as flexural strength, flexural modulus, fracture toughness and impact strength was studied on recycled cellulose fibre (RCF) reinforced epoxy composites. The mechanical properties and water uptake were directly proportional to the fibre loading (Fig. 3.7) [119].

Polyester is another polymer matrix widely used for preparation of natural fibre reinforced polymer composites [93, 97]. To improve the fibre-matrix interface, the outer surface of Luffa fibres were completely removed and treated with 5% sodium hydroxide. Due to the high surface area and roughness of the fibres, high flexural mechanical properties were observed [120]. Bamboo derived fibres have been treated with various chemicals such as maleic anhydride, permanganate, benzoyl chloride, benzyl chloride. These fibres have been used as reinforcement of polyester based matrices and to improve the tensile, flexural, impact and water absorption properties. Irrespective of the type of chemical treatment, the mechanical properties improved significantly for the resultant composites [121]. To improve the adhesion between bacterial cellulose and unsaturated polyesters composites, surface functionalisation has been fibres used. The mechanical properties of silane functionalised fibre composites have been shown to exhibit higher fibre-resin adhesion strength when compares with untreated fibres [122]. Different types of fibres such as hardwood and softwood high yield pulp (HYP), Kraft pulp and whatman cellulose pulp have been used to reinforce hunsaturated polyester (UPE) and vinyl ester (VE) to evaluate interfacial bonding. HYP fibres were shown to be more compatible with UPE resin due to residual lignin content, which acted as a natural coupling agent and helped improve the fibre-matrix interfacial bond, whereas Kraft and whatman pulp were more compatible with the VE based resin [123].

3.14.1 Synthetic Fibres

The mechanical and physical properties of any biomaterial for tissue specific applications depends on the reinforcement material, i.e. fibres or tubes. The scaffold material must demonstrate high porosity with interconnected pores and a high surface area. However, a lack of good mechanical properties impedes their use for many load bearing tissue engineering applications. Natural origin scaffold materials (e.g. chitosan, collagen and hyaluronic acid), synthetic polymers such as poly(lactic acid) (PLA), poly(glycolic acid) (PGA), poly(lactic acid-co-glycolic acid) (PLGA) have been used for many tissue engineering applications. [124, 125] developed chitosan fibres reinforced with porous collagen in an effort to improve the mechanical properties of porous collagen scaffold. They studied the relationship between both volume and length of the fibre (which was greater than or equal to critical length) and the compressive properties of the scaffold. [125] proposed a modified empirical formula for (critical length) fibre reinforced porous materials by introducing a modifying factor (F_l), which represented the elasticity of the fibres as shown in the equation below:

$$(l_{cr} F_l) / d_r = \sigma_f / (2\tau_m) \tag{3.8}$$

Where l_{cr} is the critical length of the fibres, d_r the diameter of the fibres, σ_f the compressive strength of the fibres, τ_m the shear strength of the matrix material.

Also, for the relationship between volume content of the fibres and the compressive strength of the porous scaffold, they modified formula to calculate compressive strength of the scaffold as shown in the equation below:

$$\sigma_c = 2F_\sigma V_f \sqrt{[E_f E_m V_f O / 3(1 - V_f)]} + \sigma_m \tag{3.9}$$

Where σ_c, V_f, E_f, E_m, σ_m, F_σ, O represents the compressive strength, volume content of the fibres, compressive modulus of the fibres, compressive modulus of the matrix material, compressive strength of matrix material, modifying factor and porosity of the intensified material respectively [125]. Carbon microfibre and collagen nanofibre reinforcement in collagen Type I matrices showed that the interaction between collagen nanofibres with matrix was greater than that between carbon microfibres and matrix. This study indicated the importance of material choice and size of the fibre, which directly influenced the interaction between fibres and matrix and also the scaffold mechanical properties [126].

Incorporation of fibres not only improved the mechanical properties of the scaffold but also maintained the size and shape of the medical device or implant while maintaining the bioactivity. It is well known that collagen based gel contract while cellular remodelling and reduce in the dimensions. To overcome this problem, collagen I has been reinforced with poly(ester terephthalate) microfibres. Collagen Type I fibres were extruded into polymerising buffer and incorporated into Type I collagen gels as fibre bundles. The addition of fibres maintained the gel dimensions [127–129].

Pore size and surface area of composites play important roles with respect to fibre-reinforced scaffolds. The diameter of the fibres is inversely proportional to the surface area. Also, the fibre length and volume decreases the porosity of the scaffold due to intertwined and low solubility in pore forming reagents. Therefore, to reduce the porosity and pore size, the reinforcement material should be added in a limited way [124]. The mechanical integrity of fibres and matrix increases with chemical crosslinking when compared to conventional mechanical mixing. Chitin fibres have been integrated into a PLA matrix successfully when crosslinked and the crosslinking treatment did not invoke a cytotoxic effect or reduce the impact of the environment for efficient cell growth [130].

As discussed previously, the mechanical properties of reinforcing material depends on biocompatibility and biodegradability of the scaffold. The mechanical properties of the composite affect the adhesion of the cells. The softer the substrate then the better for bone marrow- and adipose tissue-derived mesenchymal stem cells attachment when compare to a stiffer substrate [131]. Biodegradable PLA decreases the pH rapidly over time due to degraded products, which affects the microenvironment. [132] reinforced PLA with chitin fibres (30 vol.%) and showed improved pH variation, which was maintained between 7.0 and 7.2 over 16 weeks *in vitro* by neutralising the acidic degradation by-products of PLA with chitin degradation by-products. Furthermore, the chitin reinforced PLA demonstrated better cell attachment and proliferation of human osteoblast (HOBt) like cells [132]. An alginate was reinforced with single-walled carbon nanotubes (SWCNT) via a free-form fabrication technique. The resultant composite exhibited improved biocompatibility and a higher strength (436 vs. 542 kPa) and secondary modulus (7.5 vs. 426 kPa) when compared with pure alginate [133].

3.15 Homogenisation of Natural Fibres in Production of Composite Reinforced Scaffolds

Natural fibres including flax and cellulose/biomass are attractive as possible alternatives to reinforcement of polymer matrix composites due to their high mechanical and biodegradable property. Also, these fibres are environmentally friendly, economical and biocompatible. However, major limitations include identification of natural raw materials and nanosizing. To overcome these problems, researchers have developed a number of techniques such as pre-treatment and surface modification of natural fibres. Furthermore, mechanical processes such as homogenisation have been employed to produce nano-sized fibres – however, it is a high-energy consumption process, which is uneconomical and therefore not commercially viable [134]. Alternatively, cryocrushing [135], enzyme pre-treatments [136] and chemical pre-treatments [137] are employed to reduce size of the fibres prior to homogenisation, which helps to reduce the energy consumption and makes the process more economical.

Cellulose is a natural biopolymer composed of β-D-glucopyranose units linked together by β-1-4-linkages and present in animals, plants and some bacteria [138]. Introduction of nano-sized cellulose fibres into other polymers helps to improve the mechanical properties of the reinforced scaffold. These nano-reinforced scaffolds showed low thermal expansion, high aspect ratio and good mechanical/optical properties. One of the widely used techniques to produce nano-sized cellulose fibre is high-pressure homogenisation (HPH). The process involves the pumping of cellulose slurry through a very small nozzle at high speed and the pressure into a vessel generates shear rate in the stream, which converts micro-sized cellulose fibres to nano-sized fibres. For example, [139] first produced nano-sized cellulose fibres from wood pulp using HPH [139]. They reported that the main drawback of HPH process was the clogging of the orifice due to is small size. To avoid this problem, the size of the fibres needs to be reduced prior to the HPH using one of the many mechanical pre-treatments available. Nanofibrillated cellulose from wheat straw and wood fibres has been pre-treated by milling and homogenised at 150 MPa [140]. Refining can also be used as a pre-treatment method, whereby the diluted fibre slurry is pumped between stator and rotor disks. The grooves of the rotor help to create sequential stress, which causes irreversible alterations and increases the bonding potentials [141].

To fibrillate nano-sized fibres from hard wood and soft wood, [142] used a refining and homogenisation technique at 50 MPa (Fig. 3.8) [142]. They reported that

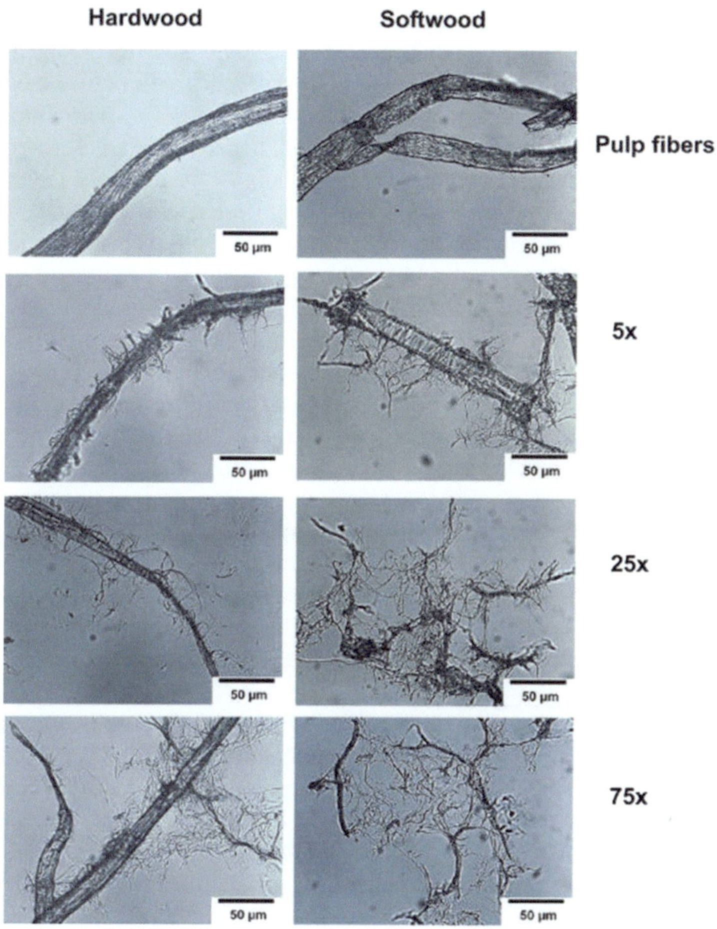

Fig. 3.8 Effect of mechanical refining on hard- and softwood cellulose fibres, before and after 5, 25, and 75 passes through the refiner. The presented images are regarded to be representative of the material after the applied number of processing steps [142] © by ACS – reprinted with permission

the diameter of nano-sized fibres was around 10–25 nm and increasing the number of cycles exhibited a decrease in the strain to failure and strength of softwood. Also they demonstrated that refining was more efficient and faster for softwood when compare to hard wood pulp. Due to internal and external fibrillation phenomenon of the cellulose, softwood broke after 25 passes, whereas even after 75 passes hardwood fibres remained intact.

Alternatively, microfluidisers are regularly used to produce nano-sized fibres. The principle behind a microfluidiser is like HPH. The fibres collide under high pressure to defibrillate the fibres using shear and impact forces against colliding streams and channel walls [143, 144]. Increasing the number of cycles (< 20) has been shown to cause agglomeration due to high surface area. Microfluidisers generate nano-sized fibres with a more homogeneous size distribution. Also, adoption of the microfluidising methodology does not significantly change the kappa number of nano-sized cellulose fibre compared with original cellulose. To reduce the cellulose to nano-size fibres, grinding is another potential method. The pulp slurry passes through static and rotating grinding stones, which helps to break cell wall structure using shear forces. Different pulp materials such as bleached eucalyptus pulp, bleached rice straw and bagasse pulps have been tested using this technology to reliably and efficiently produce nano-sized fibres [145, 146]. When the input energy increased from 5 to 30 W/kg, the fibres formed mainly in two forms: (1) highly kinked and untwisted fibrils and (2) entangled and twisted nanofibres. Extended fibrillation of untwisted nano-sized fibres could form nano-sized whiskers with high crystallinity. Also, refining was the main process for isolation of the nano-sized cellulose fibre, while HPH helped to form smaller and uniform size nanofibres. These processes do not affect the degree of polymerisation of the fibres, however the number of cycles demonstrated during HPH influenced the characterisation of nano-sized cellulose fibres.

Examples of other mechanical fibrillation methods include cryocrushing and high intensity ultrasonication (HIUS). Cryocrusing involves soaking cellulosic fibres in water and then immersing the fibres in liquid nitrogen – followed by crushing using a mortar and pestle. The cell walls of the frozen fibres rupture due to high impact force thus forming nano-sized fibres. For HIUS, the hydrodynamic force of ultrasound isolates the cellulose fibrils.

The aforementioned mechanical fibrillation methods need high consumption of energy in order to operate; therefore such techniques are not economical for producing nano-sized fibres of good yield and appropriate length. Consequently either pre-treatment and/or combination of multiple methods have been to nano-sized natural nanofibres. For example, refining and microfluidisation results in uniform nano-sized cellulose fibres with high fibrillation when compared to individual methods [147], combination of HIUS and HPH produced uniform nano-sized fibres [148]. Overall, homogenisation produces nano-sized cellulose fibres demonstrating a high surface area. However, grinding and microfluidiser showed nano-sized fibres exhibiting good mechanical, optical and physical properties with low energy consumption.

3.16 Methods to Analyse Adhesive Properties of Fibre/Matrix in Composite Reinforced Scaffold

The fibre reinforced composites consist of two phases: matrix phase and fibre phase. The adhesive strength of the composite mainly depends on characteristics of these materials. Fibres are divided into three groups based on their diameter and character, that is namely (i) whiskers, (ii) fibres and (iii) wires. Whiskers are very thin single crystals with large length-to-diameter ratio, and considered as the strongest materials due to their high strength and flaw free nature. However, whiskers are not widely used as a reinforcement medium due to high cost and practically not possible to include into a matrix. Materials like silicon nitride, graphite, silicon carbide and aluminium oxide can be included in the whisker group. Generally fibres are polycrystalline or amorphous materials with small diameter fibrous material. Polymer aramids, carbon, glass are classed within the fibre group. Finally, wires have relatively large diameter, typical materials include steel, molybdenum and tungsten and are considered most useful in the automotive industry.

The matrix phase mainly binds the fibres together and distributes the external applied stress to the fibres. A small amount of an applied load is sustained by the matrix phase. Moreover, the matrix phase should be ductile and protect individual fibres from surface damage due to mechanical abrasion or chemical reactions with the environment to avoid surface flaws capable of forming cracks, which may lead to failure at low tensile stress levels. Most importantly, the matrix phase acts as a barrier for crack propagation. It is essential that the adhesive bonding forces between fibre and matrix are high to minimise fibre pull-out. Also, adequate bonding helps to maximise the stress transmittance from weak matrix to the strong fibres. So choice of fibre-matrix combination dictates the bond strength of composite.

Fibre-matrix adhesion is crucial for competitive mechanical properties of natural fibre reinforced composites. The most commonly used test method to determine the fibre strength and adhesion is the single fibre fragmentation test (SFFT), which yields fibre break number as a function of the applied strain for a fibre embedded in matrix [149]. The adhesion between the reinforcing fibres and the matrix in composite materials determines the reinforcement efficiency as the stress transfers between the fibres and matrix. The mechanical adhesion of reinforcing fibres characterised by the interfacial shear strength (ISS) using various test methods such as the fibre pull-out, micro-debond test, fibre fragmentation test methods. The shear strength of fibre/polymer interface was evaluated using the SFFT method. The ISS, τ, is estimated as:

$$\tau = \{\sigma\} d / 2 l_c \tag{3.10}$$

Where d is the fibre diameter, l_c is the critical length related to the average fibre length at saturation of the fragmentation process, $\{l\}$, as:

$$l_c = (4/3)\{l\} \tag{3.11}$$

and $\{\sigma\}$ is the average fibre strength at critical length [150].

Some polymers like phenol-formaldehyde resins form chemical bond with natural fibres and modification of the fibre or matrix is needed to improve adhesion for other polymer matrices. In the case of flax fibres, several papers have been published in terms of the adhesion achieved [151–154].

Moreover, the mechanical properties of the composite depends on fibres orientation within the matrix and also the concentration and distribution. Basically two types of fibre orientations (i.e. parallel alignment and random alignment) influence

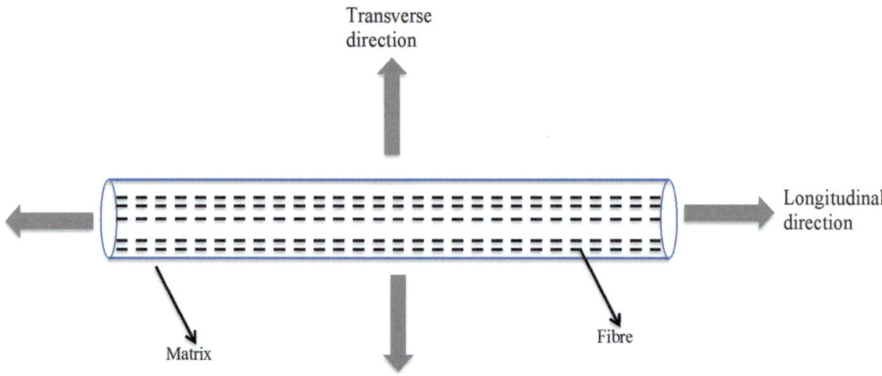

Fig. 3.9 Schematic representation of continuous fibre composite with different types of loads

the mechanical properties of the composite. Normally continuous fibres are aligned whereas discontinuous fibres may be aligned or randomly oriented.

For continuously aligned fibre composites, the mechanical properties depend on various factors such as stress vs. strain of fibre and matrix, volume fraction, distribution and the direction in which load is applied (longitudinal or transverse) (Fig. 3.9). Also, the properties of fibre aligned composites are highly anisotropic. When fibres are aligned in the loaded longitudinal direction, the strength is normally taken as the maximum stress on the stress vs. strain curve and is considered as fibre fracture or onset of composite failure. These types of composite failure depend on fibre-matrix properties, nature and strength of the fibre-matrix interfacial bond. In such composites, fibre fracture strain (ε_f) is less than the matrix fracture strain (ε_m) because when the fibre is fractured, the majority of the load transferred is to the matrix. This being the case, the longitudinal strength of the composite (σ_{cl}) can be calculated using following equation.

$$(\sigma_{cl}) = \sigma_{\mathrm{m}}(1 - \mathrm{V}_{\mathrm{f}}) + \sigma_{\mathrm{f}}\mathrm{V}_{\mathrm{f}} \tag{3.12}$$

Here σ_m is the stress in the matrix at fibre failure and σ_f is the fibre tensile strength.

Unidirectional fibrous composites are normally designed to be loaded along high strength, longitudinal direction. However, during in-service applications transverse tensile loads may also be present. Transverse strength is extremely low and below the tensile strength of the matrix. Influence of various factors includes the properties of the fibre and the matrix, the fibre-matrix bond strength, and the presence of voids (Table 3.2).

Discontinuous aligned fibres composites are gaining interest in the commercial market such as chopped glass fibres, carbon fibres and aramid fibres. The reinforcement efficiency for such composite systems is less than for aligned continuous fibre composites. The longitudinal strength (σ_{cd}) of discontinuous aligned fibre composites with uniform distribution of fibres and in which fibre length (l) is greater than critical fibre length (l_c) can be calculated using following equation:

Table 3.2 Longitudinal and transverse tensile strengths of unidirectional fibre-reinforced composites

Material	Longitudinal tensile strength (MPa)	Transverse tensile strength (MPa)
Glass-polyester	700	20
Carbon-epoxy	1000	35
Kevlar™-epoxy	1200	20

Source: D. Hull and T.W. Clyne, An Introduction to composite Materials, 2nd edition, Cambridge University Press, 1996, p. 179

$$\sigma_{cd} = \sigma_f V_f (1-(l_c/2l) + \sigma_m (1-V_f) \qquad (3.13)$$

Where σ_f denotes fracture strength of the fibre and σ_m represents the and the stress in the matrix when the composite fails.

If $l_c < l$ then the longitudinal strength $\sigma_{cd'}$ is given by:

$$\sigma_{cd'} = (l\tau_c / d)Vf + \sigma'_m (1-V_f) \qquad (3.14)$$

Where d is the fibre diameter and τ_c is the smaller of either the fibre-matrix bond strength or the matrix shear yield strength.

3.17 Summary

Hydrogel based systems are currently viewed as the best candidates for tissue engineering applications, however due to the lack of appropriate mechanical properties, researchers have developed fibre reinforced hydrogels to improve the mechanical properties. Fibre reinforced hydrogel scaffolds not only improve the mechanical stability, but also mimic the morphological features of the natural ECM. The addition of nano-sized fibres embedded in the hydrogel increases biocompatibility, as a large surface area is beneficial for cell attachment. The use of bioreactor systems, which promotes tissue growth via nutrient provision and mechanical stimulation, helps to develop nanofibre composites for tissue engineering. As a growing research field, especially in the field of load-bearing tissue engineering, much care should be taken to characterisation and analysis of mechanical properties of the scaffold. Indeed there is need to optimise higher throughput production methods and analysis in realistic load-bearing conditions to evaluate their *in vivo* efficacy and expanding potential clinical application.

Fibre reinforced calcium phosphate cements are a new class of biomaterials that are attracting and growing attention as a replacement for conventional calcium phosphate cements. The main purpose of fibre reinforced calcium phosphate cements is to overcome the brittle nature of conventional calcium phosphate cements by improve their mechanical properties and biological activity. Inclusion of fibres not only improves the fracture toughness, but also creates porosity within the cement microstructure that can potentially help cell ingrowth for new native tissue formation. The success of fibre reinforced calcium phosphate cements mainly depends on the relationship between the different constituents that form the composite and the orientation of the fibres. To improve mechanical stability, fibre reinforced calcium phosphate cements have been developed containing resorbable fibres and fibrous constructs such as yarns, nonwoven, woven and knitted fabrics. Many publications have focused on the mechanical properties of these fibre reinforced cements – however there is still considerable work required to understand the dominant factors affecting the mechanical properties and fracture mechanism of fibre reinforced composites. As these fibres reinforced cements are still in the early stages of development – there is still significant scope to understand the mechanical and biological behaviour of these materials with respect to improving adhesion between fibre and matrix and the mechanical stability after implantation and during the bone remodeling process.

For tissue regeneration, scaffold reinforced with fibres have been growing due to their high mechanical properties. These properties are depend on various factors such as volume fraction, fibre diameter, direction of alignment, bond between matrix and fibres, method of loading either grafting or cross-link and chemical compatibility of these material. Therefore, while designing a composite for particular application, one should consider these properties. Also, the processing technique plays an important role, which determines the orientation of the fibres, improve the bond between fibre and matrix which ultimately produces homogeneous structure.

The addition of fibres to a scaffold not only improves the mechanical properties but also helps to improve the biocompatibility and biodegradability of the composite. For example, chitin fibres reinforced with PLA helps to maintain the pH during degradation, which helps for cell attachment and proliferation [132].

Tissue engineering is a very broad field that can be an effective tool for variety of composites that include fibre reinforced hydrogels and/or fibre reinforced calcium phosphate cements. As we have attempted to describe mechanical properties of fibres in some detail for tissue repair, as such there is no standard protocol developed for various composites. Innovation in the fabrication of fibre structures and characterisation continues to drive research for tissue repair and regeneration.

References

1. Atala A, Lanza R, Thomson J, Nerem R. Principles of regenerative medicine. 1st ed. Boston: Elsevier; 2008.
2. Liu SQ, Tay R, Khan M, Ee PLR, Hedrick JL, Yang YY. Synthetic hydrogels for controlled stem cell differentiation. Soft Matter. 2010;6:67–81.
3. Place ES, George JH, Williams CK, Stevens MM. Synthetic polymer scaffolds for tissue engineering. Chem Soc Rev. 2009;38:1139–51.
4. Zhu J. Bioactive modification of poly(ethylene glycol) hydrogels for tissue engineering. Biomaterials. 2010;31:4639–56.

5. Drury JL, Mooney DJ. Hydrogels for tissue engineering: scaffold design variables and applications. Biomaterials. 2003;24:4337–51.
6. Tuzlakoglu K, Reis RL. Biodegradable polymeric fibre structures in tissue engineering. Tissue Eng Part B Rev. 2009;15:17–29.
7. Teo WE, Ramakrishna S. Electrospun nanofibres as a platform for multifunctional, hierarchically organized Nanocomposite. Compos Sci Technol. 2009;69:1804–17.
8. Wintermantel E, Mayer J, Blum J, Eckert KL, Luscher P, Mathey M. Tissue engineering scaffolds with superstructures. Biomaterials. 1996;17:83–91.
9. Stevens MM, Marini RP, Schaefer D, Aronson J, Langer R, Shastri VP. In vivo engineering of organs: the bone bioreactor. Proc Natl Acad Sci. 2005;102:11450–5.
10. Moutos FT, Freed LE, Guilak F. A biomimetic three-dimensional wove composite scaffold for functional tissue engineering of cartilage. Nat Mater. 2007;6:162–7.
11. Moutos FT, Guilak F. Composite scaffolds for cartilage tissue engineering. Biorheology. 2008;45:501–12.
12. McCrum NG, Buckley CP, Bucknall CB. Principles of polymer engineering. 2nd ed. Oxford: Oxford Science Publications; 1997.
13. Boudriot U, Dersch R, Greiner A, Wendorff JH. Electrospinning approaches toward scaffold engineering – a brief overview. Artif Organs. 2006;30:785–92.
14. Merkle VM, Zeng L, Slepian MJ, Wu X. Core-shell nanofibres: integrating the bioactivity of gelatin and the mechanical property of polyvinyl alcohol. Biopolymers. 2014;101:336–46.
15. McManus MC, Boland ED, Koo HP, Barnes CP, Pawlowski KJ, Wnek GE, Simpson DG, Bowlin GL. Mechanical properties of electrospun fibrinogen structures. Acta Biomater. 2006;2:19–28.
16. Chawla KK. Composite materials: science and engineering. London: Springer Science Business Media; 2012.
17. Kingery WD, Bowen HK, Uhlmann DR. Introduction to ceramics. NewYork: Wiley-Blackwell; 1976.
18. Holloway JL, Lowman AM, Palmese GR. Mechanical evaluation of poly(vinyl alcohol)-based fibrous composites as biomaterials for meniscal tissue replacement. Acta Biomater. 2010;6:4716–24.
19. Holloway JL, Lowman AM, Van Landingham MR, Palmese GR. Chemical grafting for improved interfacial shear strength in UHMWPE/PVA-hydrogel fibre-based composites used as soft fibrous tissue replacements. Compos Sci Technol. 2013;85:118–25.
20. Gonzalez JS, Ludueña LN, Ponce A, Alvarez VA. Poly(vinyl alcohol)/cellulose nanowhiskers nanocomposite hydrogels for potential wound dressings. Mater Sci Eng C Mater Biol Appl. 2014;34:54–61.
21. Jonoobi M, Aitomäki Y, Mathew AP, Oksman K. Thermoplastic polymer impregnation of cellulose nanofibre networks: morphology, mechanical and optical properties. Compos Part A Appl Sci Manuf. 2014;58:30–5.
22. McMahon RE, Qu X, Jimenez-Vergara AC, Bashur CA, Guelcher SA, Goldstein AS, Hahn MS. Hydrogel-electrospun mesh composites for coronary artery bypass grafts. Tissue Eng Part C Methods. 2011;17:451–61.
23. Hong Y, Huber A, Takanari K, Amoroso NJ, Hashizume R, Badylak SF, Wagner WR. Mechanical properties and in vivo behaviour of a biodegradable synthetic polymer microfibre-extracellular matrix hydrogel biohybrid scaffold. Biomaterials. 2011;32:3387–94.
24. Tonsomboon K, Oyen ML. Composite electrospun gelatin fibre-alginate gel scaffolds for mechanically robust tissue engineered cornea. J Mech Behav Biomed Mater. 2013;21:185–94.
25. Strange DG, Tonsomboon K, Oyen ML. Mechanical behaviour of electrospun fibre-reinforced hydrogels. J Mater Sci Mater Med. 2014;25:681–90.
26. Nerurkar NL, Elliott DM, Mauck RL. Mechanics of oriented electrospun nanofibrous scaffolds for annulus fi- brosus tissue engineering. J Orthop Res. 2007;25:1018.
27. Courtney T, Sacks MS, Stankus J, Guan J, Wagner WR. Design and analysis of tissue engineering scaffolds that mimic soft tissue mechanical anisotropy. Biomaterials. 2006;27:3631–8.

28. Mirahmadi F, Tafazzoli-Shadpour M, Shokrgozar MA, Bonakdar S. Enhanced mechanical properties of thermosensitive chitosan hydrogel by silk fibres for cartilage tissue engineering. Mater Sci Eng C. 2013;33:4786–94.
29. Kai D, Prabhakaran MP, Stahl B, Eblenkamp M, Wintermantel E, Ramakrishna S. Mechanical properties and in vitro behaviour of nanofibre-hydrogel composites for tissue engineering applications. Nanotechnology. 2012;23:095705–14.
30. Moutos FT, Estes BT, Guilak F. Multifunctional hybrid three-dimensionally woven scaffolds for cartilage tissue engineering. Macromol Biosci. 2010;10:1355–64.
31. Coburn J, Gibson M, Bandalini P. Biomimetics of the extracellular matrix: an integrated three-dimensional fibre-hydrogel composite for cartilage tissue engineering. Smart Struct Syst. 2011;7:213–22.
32. Zhou C, Wu Q. A novel polyacrylamide nanocomposite hydrogel reinforced with natural chitosan nanofibres. Colloids Surf B: Biointerfaces. 2011;84:155–62.
33. Xu W, Ma J, Jabbari E. Material properties and osteogenic differentiation of marrow stromal cells on fibre-reinforced laminated hydrogel nanocomposites. Acta Biomater. 2010;6:1992–2002.
34. Koh C, Strange DGT, Tonsomboon K, Oyen ML. Failure mechanisms in fibrous scaffolds. Acta Biomater. 2013;9:7326–34.
35. Liu L, Wang B, Gao Y, Bai T-C. Chitosan fibres enhanced gellan gum hydrogels with superior mechanical properties and water-holding capacity. Carbohydr Polym. 2013;97:152–8.
36. Thayer PS, Dimling AF, Plessl DS, Hahn MR, Guelcher SA, Dahlgren LA, Goldstein AS. Cellularized cylindrical fibre/hydrogel composites for ligament tissue engineering. Biomacromolecules. 2014;15:75–83.
37. Tibbitt MW, Anseth KS. Hydrogels as extracellular matrix mimics for 3D cell culture. Biotechnol Bioeng. 2009;103:655–63.
38. Hoffman AS. Hydrogels for biomedical applications. Adv Drug Deliv Rev. 2002;54:3–12.
39. Shapiro JM, Oyen ML. Hydrogel composite materials for tissue engineering scaffolds. JOM. 2013;65:505–16.
40. Tan HP, Marra KG. Injectable, biodegradable hydrogels for tissue engineering applications. Materials. 2010;3:1746–67.
41. Nguyen MK, Lee DS. Injectable biodegradable hydrogels. Macromol Biosci. 2010;10:563–79.
42. Vanderhooft JL, Alcoutlabi M, Magda JJ, Prestwich GD. Rheological properties of cross-linked hyaluronan – gelatin hydrogels for tissue engineering. Macromol Biosci. 2009;9:20–8.
43. Rosenblatt J, Devereux B, Wallace DG. Injectable collagen as a pH-sensitive hydrogel. Biomaterials. 1994;15:985–95.
44. Qu X, Wirsen A, Albertsson AC. Novel pH-sensitive chitosan hydrogels: swelling behaviour and states of water. Polymer. 2000;41:4589–98.
45. Tan HP, Ramirez CM, Miljkovic N, Li H, Rubin JP, Marra KG. Thermosensitive injectable hyaluronic acid hydrogel for adipose tissue engineering. Biomaterials. 2009;30:6844–53.
46. Tan HP, Chu CR, Payne KA, Marra KG. Injectable in situ forming biodegradable chitosan-hyaluronic acid based hydrogels for cartilage tissue engineering. Biomaterials. 2009;30:2499–506.
47. Almany L, Seliktar D. Biosynthetic hydrogel scaffolds made from fibrinogen and polyethylene glycol for 3D cell cultures. Biomaterials. 2005;26:2467–77.
48. Landa N, Miller L, Feinberg MS, Holbova R, Shachar M, Freeman I, Cohen S, Leor J. Effect of injectable alginate implant on cardiac remodeling and function after recent and old infarcts in rat. Circulation. 2008;117:1388–96.
49. Eyrich D, Brandl F, Appel B, Wiese H, Maier G, Wenzel M, Staudenmaier R, Goepferich A, Blunk T. Long-term stable fibrin gels for cartilage engineering. Biomaterials. 2007;28:55–65.
50. Zhao HG, Ma L, Gong YH, Gao CY, Shen JC. A polylactide/fibrin gel composite scaffold for cartilage tissue engineering: fabrication and an in vitro evaluation. J Mater Sci Mater Med. 2009;20:135–43.

51. Bosworth LA, Turner LA, Cartmell SH. State of the art composites comprising electrospun fibres coupled with hydrogels: a review. Nanomed Nanotechnol, Biol Med. 2013;9:322–35.
52. Regev O, Reddy CS, Nseir N, Zussman E. Hydrogel reinforced by short albumin fibres: mechanical characterization and assessment of biocompatibility. Macromol Mater Eng. 2013;298:283–91.
53. Hsieh A, Zahir T, Lapitsky Y, Amsden B, Wan W, Shoichet MS. Hydrogel/electrospun fibre composites influence neural stem/progenitor cell fate. Soft Matter. 2010;6:2227–37.
54. Qin Y. The preparation and characterization of fibre reinforced alginate hydrogel. J Appl Polym Sci. 2008;108:2756–61.
55. Slivka MA, Leatherbury NC, Kieswetter K, Niederauer GG. Porous, resorbable, fibre-reinforced scaffolds tailored for articular cartilage repair. Tissue Eng. 2001;7:767–80.
56. Feingold-Leitman D, Zussman E, Seliktar D. Evaluation of a composite scaffold made from electrospun nanofibres and a hydrogel for tissue engineering. J Bionanosci. 2009;3:45–7.
57. Hayami JWS, Surrao DC, Waldman SD, Amsden BG. Design and charecterisation of a biodegradable composite scaffold for ligament tissue engineering. J Biomed Mater Res. 2010;92A:1407–20.
58. Ekaputra AK, Prestwich GD, Cool SM, Hutmacher DW. Combining electrospun scaffolds with electrosprayed hydrogels leads to three-dimensional cellularization of hybrid constructs. Biomacromolecules. 2008;9:2097–103.
59. Larsson S, Hannink G. Injectable bone-graft substitutes: current products, their characteristics and indications, and new developments. Injury. 2011;42:30–4.
60. Dorozhkin SV. Calcium orthophosphate cements for biomedical application. J Mater Sci. 2008;43:3028–57.
61. Dorozhkin SV. Calcium orthophosphate-based biocomposites and hybrid biomaterials. J Mater Sci. 2009;44:2343–87.
62. Canal C, Ginebra MP. Fibre-reinforced calcium phosphate cements: a review. J Mech Behav Biomed Mater. 2011;4:1658–71.
63. Callister WD Jr, Rethwisch DG. Materials science and engineering: an introduction. Hoboken: John Wiley Sons Inc; 2009.
64. Neumann M, Epple M. Composites of calcium phosphate and polymers as bone substitution materials. Eur J Trauma. 2006;32:125–31.
65. Low KL, Tan SH, Zein SHS, Roether JA, Mourino V, Boccaccini AR. Calcium phosphate based composites as injectable bone substitute materials: a review. J Biomed Mater Res B Appl Biomater. 2010;94B:273–86.
66. Hannant DH, Hughes DC, Kelly A. Toughening of cement and other brittle solids with fibres. Phil Trans R Soc Lond A. 1983;310:175–90.
67. Aveston J, Cooper GA, Kelly A. Single and multiple fracture. London: IPC Science and Technology Press Ltd.; 1971.
68. Li VC, Mishra DK, Wu HC. Matrix design for pseudo strain-hardening fibre reinforced cementitious composites. J Mater Struct. 1995;28:586–95.
69. Nelson PK, Li VC, Kamada T. Fracture toughness of microfibre reinforced cement composites. J Mater Civil,Eng. 2002;14:384–91.
70. Li VC, Maalej M. Toughening in cement based composites part II: fibre reinforced cementitious composites. Cem Concr Compos. 1996;18:239–49.
71. Li VC. On engineered cementitious composites (ECC) a review of the material and its applications. J Adv Concr Technol. 2003;1:215–30.
72. Gibson RF. Principles of composite material mechanics. 3rd ed. Boca Raton: CRC Press; 2011.
73. Rösler J, Harders H, Bäker M. Mechanisches verhalten der werkstoffe. Wiesbaden: Teubner Verlag; 2006.
74. Li VC, Wu HC. Condition for pseudo strain-hardening in fibre reinforced brittle matrix composites. Appl Mech Rev. 1992;45:390–8.

75. Pan ZH, Cai HP, Jiang PP, Fan QY. Properties of calcium phosphate cement synergistically reinforced by chitosan fibre and gelatin. J Polym Res. 2006;13:323–7.
76. Bao C, Chen W, Weir MD, Thein-Han W, Xu HHK. Effects of electrospun submicron fibres in calcium phosphate cement scaffold on mechanical properties and osteogenic differentiation of umbilical cord stem cells. Acta Biomater. 2011;7:4037–44.
77. Li VC, Stang H. Interface property characterization and strengthening mechanisms in fibre reinforced cement based composites. Adv Cem Based Mater. 1997;6:1–20.
78. Felekoglu B, Tosun K, Baradan B. A comparative study on the flexural performance of plasma treated polypropylene fibre reinforced cementitious composites. J Mater Process Technol. 2009;209:5133–44.
79. Li VC, Wu HC, Chang YW. Effect of plasma treatment of polyethylene fibres on interface and cementitious composite properties. J Am Ceram Soc. 1996;79:700–4.
80. Feldman D, Denes F, Zeng Z, Denes AR, Banu D. Polypropylene fibre matrix bonds in cementitious composites. J Adhes Sci Technol. 2000;14:1705–21.
81. Wilson AD, Nicholson JW. Acid-base cements. Cambridge: Cambridge University Press; 1993.
82. Dos Santos LA, de Oliveira LC, da Silva Rigo EC, Garcia Carrodéguas R, Ortega Boschi A, Fonseca de Arruda AC. Fibre reinforced calcium phosphate cement. Artif Organs. 2000;24:212–6.
83. Xu HHK, Quinn JB, Takagi S, Chow LC, Eichmiller FC. Strong and macroporous calcium phosphate cement: effects of porosity and fibre reinforcement. J Biomed Mater Res. 2001;57A:457–66.
84. Xu HHK, Eichmiller FC, Giuseppetti AA. Reinforcement of a self-setting calcium phosphate cement with different fibres. J Biomed Mater Res. 2000;52:107–14.
85. Low KL, Tan SH, Zein SHS, McPhail DS, Boccaccini AR. Optimization of the mechanical properties of calcium phosphate/multi-walled carbon nanotubes/bovine serum albumin composites using response surface methodology. J Mater Des. 2011;32:3312–9.
86. Chew KK, Low KL, Zein SHS, McPhail DS, Gerhardt LC, Roether JA, Boccaccini AR. Reinforcement of calcium phosphate cement with multi-walled carbon nanotubes and bovine serum albumin for injectable bone substitute applications. J Mech Behav Biomed Mater. 2011;4:331–9.
87. Zuo Y, Yang F, Wolke JGC, Li Y, Jansen JA. Incorporation of biodegradable electrospun fibres into calcium phosphate cement for bone regeneration. Acta Biomater. 2010;6:1238–47.
88. Xu HHK, Quinn JB. Calcium phosphate cement containing resorbable fibres for short-term reinforcement and macroporosity. Biomaterials. 2002;23:193–202.
89. Zhang Y, Xu HHK. Effects of synergistic reinforcement and absorbable fibre strength on hydroxyapatite bone cement. J Biomed Mater Res. 2005;75A:832–40.
90. Weir MD, Xu HKK, Simon CG Jr. Strong calcium phosphate cement-chitosan-mesh construct containing cell encapsulating hydrogel beads for bone tissue engineering. J Biomed Mater Res. 2006;77A:487–96.
91. Zhao L, Weir MD, Xu HHK. Human umbilical cord stem cell encapsulation in calcium phosphate scaffolds for bone engineering. Biomaterials. 2010b;31:3848–57.
92. Weir MD, Xu HHK. Human bone marrow stem cell encapsulating calcium phosphate scaffolds for bone repair. Acta Biomater. 2010;6:4118–26.
93. Akil HM, Omar MF, Mazuki AAM, Safiee S, Ishak ZAM, Abu Bakar A. Kenaf fibre reinforced composites: a review. Mater Des. 2011;32:4107–21.
94. Yan L, Chouw N, Jayaraman K. Flax fibre and its composites a review. Compos Part B-Eng. 2013;56:296–317.
95. Bajpai PK, Singh I, Madaan J. Comparative studies of mechanical and morphological properties of polylactic acid and polypropylene based natural fibre composites. J Reinf Plast Compos. 2012;31:1712–24.
96. Kumar S, Gupta DS, Singh I, Sharma A. Behavior of kevlar/epoxy composite plates under ballistic impact. J Reinf Plast Compos. 2010;29:2048–64.

97. Bogoeva-Gaceva G, Avella M, Malinconico M, Buzarovska A, Grozdanov A, Gen-tile G, et al. Natural fibre eco-composites. Polym Compos. 2007;28:98–107.
98. Singha AS, Thakur VK. Physical, chemical and mechanical properties of *Hibiscus sabdariffa* fibre/polymer composite. Int J Polym Mater. 2009d;58:217–28.
99. Azwa ZN, Yousif BF, Manalo AC, Karunasena W. A review on the degradability of polymeric composites based on natural fibres. Mater Des. 2013;47:424–42.
100. Eichhorn SJ, Dufresne A, Aranguren M, Marcovich NE, Capadona JR, Rowan SJ, et al. Review: current international research into cellulose nanofibres and nanocomposites. J Mater Sci. 2010;45:1–33.
101. George J, Sreekala MS, Thomas S. A review on interface modification and characterization of natural fibre reinforced plastic composites. Polym Eng Sci. 2001;41:1471–85.
102. Mukherjee T, Kao N. PLA based biopolymer reinforced with natural fibre: a review. J Polym Environ. 2011;19:714–25.
103. Singha AS, Thakur VK. Mechanical, morphological and thermal properties of pine needle-reinforced polymer composites. Int J Polym Mater. 2008a;58:21–31.
104. Singha AS, Thakur VK. Fabrication and characterization of *H. sabdariffa* fibre-reinforced green polymer composites. Polym-Plast Technol Eng. 2009b;48:482–7.
105. Thakur VK, Singha AS, Kaur I, Nagarajarao RP, Liping Y. Studies on analysis and characterization of phenolic composites fabricated from lignocellulosic fibres. Polym Polym Compos. 2011;19:505–11.
106. Megiatto JD, Silva CG, Rosa DS, Frollini E. Sisal chemically modified with lignins: correlation between fibres and phenolic composites properties. Polym Degrad Stab. 2008;93:1109–21.
107. Li X, Cai Z, Winandy JE, Basta AH. Selected properties of particleboard panels manufactured from rice straws of different geometries. Bio Res Technol. 2010;101:4662–6.
108. Oliveira FB, Gardrat C, Enjalbal C, Frollini E, Castellan A. Phenol-furfural resins to elaborate composites, reinforced with sisal fibres – molecular analysis of resin and properties of composites. J Appl Polym Sci. 2008;109:2291–303.
109. Mu Q, Wei C, Feng S. Studies on mechanical properties of sisal fibre/phenol formaldehyde resin in-situ composites. Polym Compos. 2009;30:131–7.
110. Singha AS, Thakur VK. Synthesis, characterization and study of pine needles reinforced polymer matrix based composites. J Reinf Plast Compos. 2009;29:700–9.
111. Joseph S, Sreekala MS, Thomas S. Effect of chemical modifications on the thermal stability and degradation of banana fibre and banana fibre-reinforced phenol formaldehyde composites. J Appl Polym Sci. 2008;110:2305–14.
112. Neelamana IK, Thomas S, Parameswaranpillai J. Characteristics of banana fibres and banana fibre reinforced phenol formaldehyde composites macro scale to nano scale. J Appl Polym Sci. 2013;130:1239–46.
113. Jawaid M, Khalil HPSA, Abu Bakar A. Mechanical performance of oil palm empty fruit bunches/jute fibres reinforced epoxy hybrid composites. Mater Sci Eng A. 2010;527:7944–9.
114. Jawaid M, Khalil HPSA, Abu Bakar A. Woven hybrid composites: tensile and flexural properties of oil palm-woven jute fibres based epoxy composites. Mater Sci Eng. 2011b;A528:5190–5.
115. Lu J, Askeland P, Drzal LT. Surface modification of microfibrillated cellulose for epoxy composite applications. Polymer. 2008;49:1285–96.
116. Sbiai A, Maazouz A, Fleury E, Sautereau H, Kaddami H. Short date palmtree fibres polyepoxy composites prepared using RTM process: effect of tempo mediated oxydation of the fibres. Bioresources. 2010;5:672–89.
117. Kaddami H, Dufresne A, Khelifi B, Bendahou A, Taourirte M, Raihane M, et al. Short palm tree fibres thermoset matrices composites. Compos Part-Appl Sci Manuf. 2006;37:1413–22.
118. Karaduman Y, Onal L. Water absorption behavior of carpet waste jute-reinforced polymer composites. J Compos Mater. 2011;45:1559–71.
119. Alamri H, Low IM. Mechanical properties and water absorption behaviour of recycled cellulose fibre reinforced epoxy composites. Polym Test. 2012b;31:620–8.

120. Boynard CA, Monteiro SN, d'Almeida JRM. Aspects of alkali treatment of sponge gourd (*Luffa cylindrica*) fibres on the flexural properties of polyester matrix composites. J Appl Polym Sci. 2003;87:1927–32.
121. Kushwaha PK, Kumar R. Influence of chemical treatments on the mechanical and water absorption properties of bamboo fibre composites. J Reinf Plast Compos. 2011;30:73–85.
122. Gao C, Wan Y, He F, Liang H, Luo H, Han J. Mechanical, moisture absorption, and photo degradation behaviors of bacterial cellulose nano fibre-reinforced unsaturated polyester composites. Adv Polym Technol. 2011;30:249–56.
123. Du Y, Wu T, Yan N, Kortschot MT, Farnood R. Pulp fibre-reinforced thermoset polymer composites: effects of the pulp fibres and polymer. Compos Part B-Eng. 2013;48:10–7.
124. Williams SF, Martin DP, Horowitz DM. PHA applications: addressing the price performance issue: I. Tissue engineering. Int J Biol Macromol. 1999;25:111–21.
125. Li XM, Feng QL, Jiao YF, Cui FZ. Collagen-based scaffolds reinforced by chitosan fibres for bone tissue engineering. Polym Int. 2005;54:1034–40.
126. Wang XN, Yan YD, Yost MJ, Fann SA, Dong S, Li XD. Nanomechanical characterization of micro/nanofibre reinforced type I collagens. J Biomed Mater Res Part A. 2007;83A:130–5.
127. Takamoto T, Hiraoka Y, Tabata Y. Enhanced proliferation and osteogenic differentiation of rat mesenchymal stem cells in collagen sponge reinforced with different poly(ethylene terephthalate) fibres. J Biomater Sci Polym Ed. 2007;18:865–81.
128. Gentleman E, Nauman EA, Dee KC, Livesay GA. Short collagen fibres provide control of contraction and permeability in fibroblast – seeded collagen gels. Tissue Eng. 2004;10:421–7.
129. Gentleman E, Nauman EA, Livesay GA, Dee KC. Collagen composite biomaterials resist contraction while allowing development of adipocytic soft tissue in vitro. Tissue Eng. 2006;12:1639–49.
130. Li XM, Feng QL. Porous poly-L-lactic acid scaffold reinforced by chitin fibres. Polym Bull. 2005;54:47–55.
131. Li XM, Huang Y, Zheng LS, Liu HF, Niu XF, Huang J, Zhao F, Fan YB. Effect of substrate stiffness on the functions of rat bone marrow and adipose tissue derived mesenchymal stem cells in vitro. J Biomed Mater Res A. 2014;102:1092–101.
132. Li XM, Liu XH, Dong W, Feng QL, Cui FZ, Uo M, Akasaka T, Watari F. In vitro evaluation of porous poly (L-lactic acid) scaffold reinforced by chitin fibres. J Biomed Mater Res B. 2009;90B:503–9.
133. Yildirim ED, Yin X, Nair K, Sun W. Fabrication, characterization, and biocompatibility of single-walled carbon nanotube-reinforced alginate composite scaffolds manufactured using freeform fabrication technique. J Biomed Mater Res B. 2008;87B:406–14.
134. Zhang J, Song H, Lin L, Zhuang J, Pang C, Liu S. Microfibrillated cellulose from bamboo pulp and its properties. Biomass Bioenergy. 2012;39:78–83.
135. Wang B, Sain M. Dispersion of soybean stock-based nanofibre in a plastic matrix. Polym Int. 2007a;56:538–46.
136. López-Rubio A, Lagaron JM, Ankerfors M, Lindström T, Nordqvist D, Mattozzi A, et al. Enhanced film forming and film properties of amylopectin using micro-fibrillated cellulose. Carbohydr Polym. 2007;68:718–27.
137. Alemdar A, Sain M. Biocomposites from wheat straw nanofibres: morphology, thermal and mechanical properties. Compos Sci Technol. 2008a;68:557–65.
138. Lima MMD, Borsali RD. Rodlike cellulose microcrystals: structure, properties, and applications. Macromol Rapid Commun. 2004;25:771–87.
139. Herrick FW, Casebier RL, Hamilton JK, Sandberg KR. Microfibrillated cellulose: morphology and accessibility. J Appl Polymer Sci: Appl Polymer Symp. 1983;37:797–813.
140. Zimmermann T, Bordeanu N, Strub E. Properties of nanofibrillated cellulose from different raw materials and its reinforcement potential. Carbohydr Polym. 2010;79:1086–93.
141. Nakagaito AN, Yano H. The effect of morphological changes from pulp fibre towards nanoscale fibrillated cellulose on the mechanical properties of high-strength plant fibre based composites. Appl Phys A – Mater Sci Process. 2004;78:547–52.

142. Stelte W, Sanadi AR. Preparation and characterization of cellulose nanofibers from two commercial hardwood and softwood pulps. Ind Eng Chem Res. 2009;48:11211–9.
143. Ferrer A, Filpponen I, Rodríguez A, Laine J, Rojas OJ. Valorization of residual Empty Palm Fruit Bunch Fibres (EPFBF) by microfluidization: production of nanofibrillated cellulose and EPFBF nanopaper. Bioresour Technol. 2012;125:249–55.
144. Lee SY, Chun SJ, Kang IA, Park JYD. Preparation of cellulose nanofibres by high-pressure homogenizer and cellulose-based composite films. J Ind Eng Chem. 2009;15:50–5.
145. Wang QQ, Zhu JY, Gleisner R, Kuster TA, Baxa U, McNeil SE. Morphological development of cellulose fibrils of a bleached eucalyptus pulp by mechanical fibrillation. Cellulose. 2012;19:1631–43.
146. Hassan ML, Mathew AP, Hassan EA, El-Wakil NA, Oksman K. Nanofibres from bagasse and rice straw: process optimization and properties. Wood Sci Technol. 2012;46:193–205.
147. Qing Y, Sabob R, Zhub JY, Cai Z, Wu Y. Comparative study of cellulose nanofibrils: disintegrated from different approaches. Bioresour Technol. 2013;130:783–8.
148. Spence KL, Venditti RA, Rojas OJ, Habibi Y, Pawlak JJ. A comparative study of energy consumption and physical properties of microfibrillated cellulose produced by different processing methods. Cellulose. 2011;18:1097–111.
149. Hui C-Y, Phoenix SL, Kogan L. Analysis of fragmentation in the single filament composite: roles of fibre strength distributions and exclusion zone models. J Mech Phys Solids. 1996;44:1715–37.
150. Drzal LT, Herrera-Franco PJ, Ho H. Fibre–matrix interface tests. In: Kelly A, Zweben C, editors. Comprehensive composite materials, vol. 5. New York: Pergamon Press; 2000.
151. Garkhail SK, Bertens P, George J, Klompen ETJ, Peijs T. Optimisation of flax-fibre-reinforced-polypropylene composites through interfacial adhesion. Proceedings of ECCM9, Brighton; 2000.
152. Zafeiropoulos NE, Baillie CA, Matthews FL. Modification and characterization of the interface in flax/polypropylene composite materials. Proceedings of ECCM9, Brighton; 2000.
153. Zafeiropoulos NE, Baillie CA, Matthews FL. A study of transcrystallinity and its effect on the interface in flax fibre reinforced composite materials. Compos Part A. 2001;32:525–43.
154. Stamboulis A, Baillie C, Schulz E. Interfacial characterization of flax fibre-thermoplastic polymer composites by the pull-out test. Angew Makromol Chem. 1999;272:117–20.

Chapter 4
The Biodegradability of Scaffolds Reinforced by Fibers or Tubes for Tissue Repair

Katerina E. Aifantis

Abstract The present chapter gives an overview of the fabrication, biocompatibility and biodegradability of flexible scaffold materials reinforced with stiffer nanofiber or nanotube inclusions. The need for such filler nanomaterials results from the necessity to increase the elastic modulus and fracture toughness of biopolymer materials, such as gelatin, chitosan, as well as synthetic biodegradable polymers, such as polylactic acid and polycaprolactone, that are to be used in large bone defect regeneration. For such biomedical applications it must be possible for the material to support the forces exerted by the skeletal network, until the new bone tissue is regenerated. Hence, the scaffold mechanical properties must not be altered much during the initial and intermediate degradation stages at which bone formation occurs. Furthermore, degradation of the scaffold must occur slowly not only so that sufficient tissue is formed, but so the degraded scaffold products can be at levels that can be tolerated by the cells without inducing cytotoxicity. The most widely examined fillers for reinforcing the scaffold strength are hydroxyapatite (HA) nanotubes, as HA is a natural mineral found in bone, while other promising candidates include silk, boron nitride, and magnesium alloy nanowires. Another common combination is the use of stiffer polymer fiber reinforcements, within a soft polymer matrix, or polymer fiber meshes embedded in hydrogels, to form laminates that are then stacked. The particular systems to be covered in this chapter are: (i) gelatin reinforced with hydroxyapatite nanofibers, (ii) chitosan reinforced with boron nitride nanotubes, (iii) polylactide–polycaprolactone reinforced with boron nitride nanotubes, (iv) nano-hydroxyapatite/collagen/PLLA reinforced by chitin fibers, (v) poly-lactic acid reinforced with magnesium alloy nanowires, (vi) polycaprolactone reinforced with poly-L-lactic acid fibers, (vii) polyactic acid reinforced with silk fibers, (viii) poly(lactide-co-ethylene oxide fumarate) gel reinforced with a mesh of poly-L-lactic acid fibers, and (ix) protein hydrogels reinforced with a nano-HAp/PHB fiber network. Standard mechanical testing (such tension and bending) has

K.E. Aifantis (✉)
Modern Functional Materials Lab, ITMO University, St. Petersburg, Russia

Lab of Mechanics and Materials, Aristotle University of Thessaloniki, Thessaloniki 54124, Greece
e-mail: k.aifantis@mom.gen.auth.gr

X. Li (ed.), *Tissue Repair*, DOI 10.1007/978-981-10-3554-8_4

shown that all these types of nanocomposite scaffolds exhibit preferred mechanical properties without reducing biocompatibility. Upon giving a brief overview of the fabrication, microstructure, strength, and biocompatibility for each scaffold, their biodegradability will be elaborated on as it is the focus of this chapter.

Keywords Reinforcements • Biodegradable • Fibers • Bone regeneration

4.1 Introduction

The main challenge in developing scaffolds for tissue regeneration is that it is not enough for them to have the necessary biodegradability and bioactivity that stimulates regeneration [1, 2], but their mechanical and structural properties must mimic that of the native tissue, in order to activate the surrounding cells (either *in vitro* or *in vivo*). Specifically, in order for materials to be used in tissue engineering they should fulfill the following criteria: (i) they must be biocompatible, meaning that the scaffold itself should not initiate inflammation or toxicity when placed in the body, (ii) their surface should promote cell adhesion, enhance cell growth and maintain differentiated cell functions, (iii) their mechanical properties must be such that will provide the necessary structural stability while retaining flexibility and avoiding fracture upon the development of internal stresses that occur *in vivo*, (iv) their porosity must mimic that of the extracellular matrix such that tissue formation occurs in a uniform manner, and the pore degree/size, geometry, interconnection, and orientation must be such so they allow for the appropriate cell–cell interactions, cellular density and the necessary transfer of oxygen, nutrients and waste products [3] (it has been shown that the optimum scaffold porosity should be over 80% [4–6]), (v) depending on the application their biodegradability must be tuned.

In order for a scaffold to combine all these properties it must be a composite of materials with complementary properties, whose combination at appropriate volume fractions will allow to meet a wider range of the necessary key characteristics. In doing so the most optimum structure is the reinforcement of a soft, biodegradable polymer material, with stiff secondary phases. Such structures are the most promising for meeting both the physiological and mechanical properties of the native tissue whose regeneration is targeted.

The main polymers that are used in biomedical engineering belong to the poly (a-hydroxyester) family (poly (lactic-acid), poly (glycolic-acid) and poly(3-caprolactone)), as they are biocompatible and it is easier to mold them into various shapes and sizes. Also it is easy to tune their porosity. Numerous fabrication techniques have been developed that allow for appropriate 3D configurations to be obtained, such as 3D printing, electrospinning, hydrogels via freeze-gelation, and fused deposition modeling. These polymers have predictable mechanical properties and degradation rates, which occur through ester hydrolysis and decarboxylation from the end chains; the kinetics of which depend on the particular intermolecular interactions [7]. Other natural biopolymers such as gelatin, which is derived from collagen, and chitosan, which is a derivative from chitin, are also good candidates. These can also be considered as protein based hydrogels, but they are not the only type of hydrogels that can be used in tissue engineering.

A straightforward approach that can be followed to increase the strength of polymers is to increase the concentration of the crosslinking agent or polymer, however, it has been shown that doing so affects the diffusion of the nutrients through the scaffold and hence reduces its biocompatibility [8, 9]. Hence, in order to increase the strength of these biodegradable polymers, which is particularly necessary when reconstructing large bone segments, the addition of fibers/tubes is becoming a common technique, resulting in polymer composite structures. The "filler" materials that will be elaborated in this chapter are: boron-nitride nanotubes, magnesium alloy fibers, hydroxyapatite (HA) nanofibers, and polymer fibers. It should be noted that HA on its own is a common material considered for scaffolds and implants since it has a very high compatibility and bioactivity as it is naturally present in bone, however, its practical applications are limited due its brittleness.

4.2 Gelatin Reinforced with Hydroxyapatite Nanofibers

Gelatin is commonly used in biomedical applications as it is derived from collagen, yet has lower costs and can retain information signals such as the Arginyl-glycyl-aspartic acid (RGD) tri-peptide which is known to promote cell adhesion [10, 11]. Furthermore, it has a low-toxicity, is non-immunogenic, and has good biodegradability and biocompatibility [12]. For applications, in which the extracellular matrix has a low elastic modulus, such as in the case of artificial skin, pure gelatin is appropriate. However, in mimicking tougher tissues such as bone, it is necessary to enhance the strength through composite structures that can withstand the stresses to which the human skeleton is subject to. This can be effectively done by enhancing gelatin with nanoscale hydroxyapatite (HA), which is found in natural bone and has a much higher elastic modulus. Hence, HA-gelatin composites appear to be a promising materials selection, as it is possible to increase the flexibility of gelatin to the desired stiffness by the appropriate concentration of HA. The optimum scaffold microstructure should be porous enough so as to allow bone ingrowth, however, a high porous microstructure results in low mechanical stability. The addition of HA nanorods can allow a retention of the porous microstructure of gelatin, while increasing the strength of the gelatin fibers. In addition, to the presence of HA, the cross-linking agent can further regulate the microstructure, surface roughness and compression strength of such nanocomposites. It should be noted that in addition to hydroxyapatite being used as a secondary phase reinforcement in scaffolds for tissue regeneration, one of its main applications is in bone implants where it can also be used as the main matrix material that is reinforced with nanofibers and nanotubers (such as carbon nanotubes) in order to increase its fracture toughness, and wear resistance.

In the following sub-sections a gelatin-HA nanocomposite as presented in [13] is summarized.

4.2.1 Fabrication and Characterization

In [13] HA nanorods that were ~28 nm in length and ~8 nm in diameter were added to a gelatin solution prepared and were extensively sonicated in order to achieve a homogeneous dispersion. After the pH was adjusted tanic acid (TA) powder was added and stirred in order for cross-linking to occur. In addition to TA being a natural phenolic compound, it reduces the solubility of gelatin in liquid environments. The mineralization process was then performed using the layer-by-layer deposition [14].

A first step in examining the applicability of such scaffolds in bone regeneration was to perform scanning electron microscopy to reveal their microstructure, so as to examine if the porosity of the composite was sufficient and the distribution of the nanoparticles uniform. Figure 4.1a, b depicts the gelatin-nanoparticle composite.

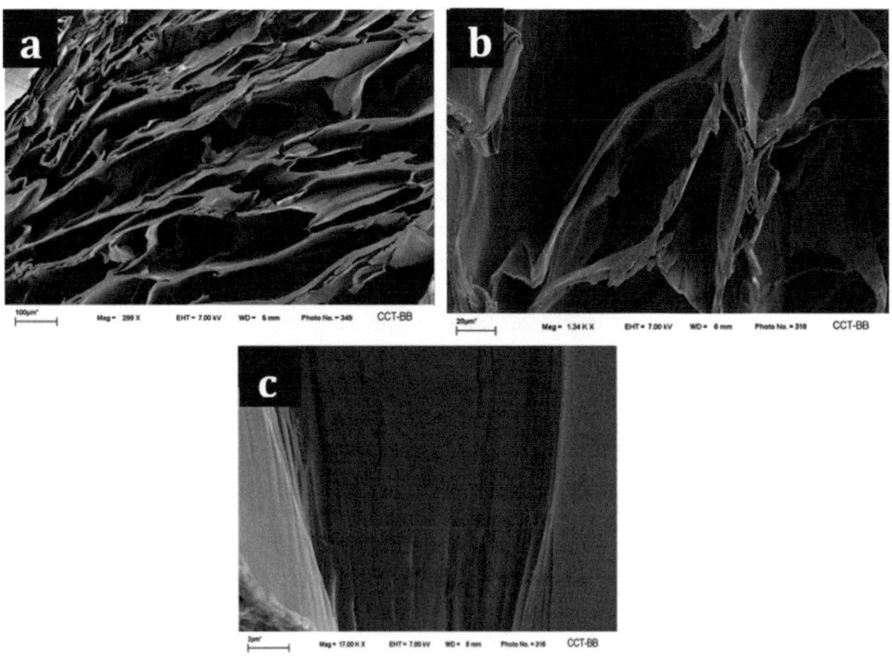

Fig. 4.1 Scanning electron microscopy images in the (**a**) longitudinal and (**b**) transverse direction of the lyophilized gelatin-hydroxyapatite; the scale bars in each image are: (**a**) 100 μm (**b**) 20 μm. (**c**) Pore wall morphology of lyophilized gelatin-HA scaffold, scale bar: 2 μm. These samples contained 0.80 g/mL gelatin and 0.85 mg/mL HA (Reproduced with permission from [13])

The scale bar in Fig. 4.1 is in the microscale, hence it was not expected to see the individual particles, and the absence of microscale aggregates indicates that the HA nanoparticles were indeed dispersed uniformly throughout the gelatin, which had a pore size between 35 and 288 μm, and was similar to the promising porosity of HA foams [15]. In Fig. 4.1 it can be seen that the HA particles were within the pore walls and aggregation did not occur. In order to examine the effect that the cross-linking concentration has on the graft morphology two different concentrations were used: 12.4 mg TA/g and 33.3 mg TA/g. Increasing the TA concentration resulted in a similar porous morphology, however, the fiber diameter decreased (~73 μm) while the interconnections increased, decreasing hence the overall porosity [13].

In addition to the internal structure of the scaffold, their surface structure, particularly their roughness also plays a significant role in determining how the native tissue will adhere to it. In Fig. 4.1c it is seen that the surfaces were rough as they consisted of aligned fibers [13]. When such rough implanted surfaces come in contact with the bone, which has a lower elastic modulus, stress concentrations occur at their interface [16], and it has been shown that high stress concentrations at the implant-bone interface increase bone resorption [17]. The magnitude of these stress concentrations is proportional to the sharpness of the implant [18], and therefore it is important to be able and control the scaffold roughness.

In addition to microstructure the strength of the scaffolds was examined, by uniaxial compression tests, in order to ensure that they could withstand the internal stresses exerted by the body after implantation. It was found that increasing the concentration of the HA nanoparticles and cross-linking agent resulted in an increase of the ultimate tensile strength and yield strength. However, for increasing bone ingrowth a high open porosity and pore size are preferred [3, 15], which dictates the use of low cross-linking concentrations in fabricating appropriate microstructures. Low contents of TA, such as 12.4 mg TA/g in GE, were therefore necessary in order to obtain highly porous microstructures, however, they had a poor mechanical strength, since only 650 kPa were enough to induce a 14% strain [13]. Addition of HA nanorods, however, in this polymer network required twice the stress level (~1.2 MPa) in order to reach such strains; it is interesting to note that similar stress-strain curves were obtained for a TA concentration of 33.3 mg/g in GE without addition of HA. Hence, by reinforcing gelatin biopolymers with HA allowed to retain an open porous structure while increasing their strength. It follows that adding HA-nanorods to the 33.3 mg TA/g GE resulted in brittle biopolymer fibers that would collapse easily. Hence low TA concentrations in GE reinforced with HA resulted in the preferred properties for scaffold engineering.

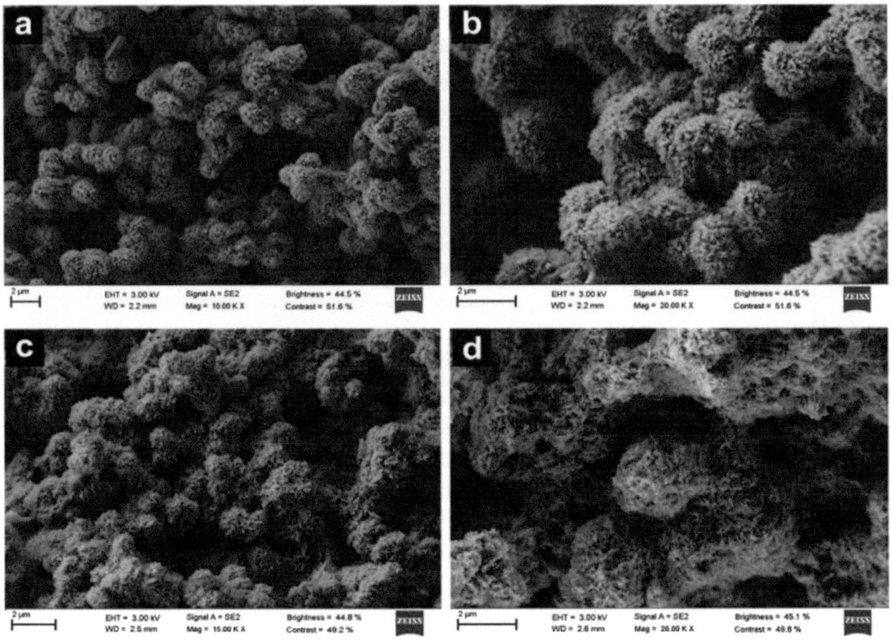

Fig. 4.2 Formation of the calcium-phosphate layer on the surface of gelatin-HA scaffolds to examine their biocompatibility and biodegradability. (**a**–**b**) after 7 days of immersion in SBF; (**c**–**d**) after 14 days of immersion in SBF (Reproduced with permission from [13])

4.2.2 Biodegradability

Having determined that the 12.4 mg TA/g GE-HA scaffold had the preferred physical, microstructural and mechanical properties it was chosen for biodegradability and bioactivity evaluation studies. The effectiveness of HA based implants relies on the ability of the scaffold to form on its surface a calcium-phosphate layer [19], as it can initiate bone formation through the activation of signaling proteins [20]. In order to examine the ability of such a layer to form on these scaffolds *in vitro* studies were performed by immersing them into a simulated body fluid (SBF) solution. It was shown that after 7 days of immersion such a bone-like apatite film formed throughout the scaffold surface. The microstructure depicted in Fig. 4.2a, b illustrates that this film was comprised of spherical nano-plate aggregates with a 150 nm length. After 14 days the microstructure of the film was slightly altered as the spherical aggregates were elongated and eroded (Fig. 4.2c, d). This change in shape was attributed to the initiation of degradation of the scaffold, since as the HA-gelatin material dissolved stresses evolved at the calcium-phosphate/scaffold interface, which resulted in the deformation of the film crystals. Examination of the scaffold 21 days after immersion in the SBF indicated that the bone-like apatite coating was not continuous, since the scaffold supporting it had further disintegrated and therefore the attached Ca/P exfoliated in the solution. It should be noted that throughout the 21 days of immersion the Ca/P ratio that remained attached on the scaffold was at the values present in trabecular and cortical bone and the observed microstructure of the bone-like apatite crystals was similar to that reported for promoting osteo-integration [19, 20].

In addition to examining the effect that SBF fluid had on the scaffold microstructure, their biodegradability was also examined by immersing them in three additional solutions and monitoring their weight loss after 3, 12, 16 and 25 days following immersion. The first solution was comprised of PBS with a pH of 7.4 and corresponded to degradation in physiological fluids, the second solution included acetic acid/sodium acetate buffer solution with a pH of 4.2 and corresponded to the resorption process of incharge osteoblasts and the third solution included lysozyme (4 mg/mL) into the PBS in order to simulate decomposition by enzymatic catalysis. After 25 days the weight loss was 15% for the physiological fluid conditions, 75% for the 4.2 pH solution and 95% for the enzymatic degradation [13]. The ability of the scaffold to degrade under a 7.2 pH is necessary such that the deposition of biogenic hydroxyapatite crystals can occur as described earlier and shown in Fig. 4.2.

4.3 Boron Nitride Nanotube Reinforced Polymers

4.3.1 Chitosan Reinforced with Boron Nitride Nanotubes

Another biopolymer that is commonly used in tissue engineering is the polysaccharide chitosan as it is non-toxic, has good biocompatibility, enzymatic degradation and antimicrobial activity against viruses, bacteria and fungi. It is produced by deacetylation of chitin, and its main applications are in wound patches for healing skin traumas [21–23]. Drawbacks that limit its applications in scaffold engineering are its low mechanical strength (like gelatin) but also its high swelling ratio and most importantly uncontrolled degradation. In order to address these issues the reinforcement with boron nitride nanotubes (BNNTs) has been examined.

BNNTs are biocompatible and have a high elastic modulus ~1.22 TPa [24], which allows them to increase the low strength of chitosan networks. Despite their high modulus, they have a high ductility and therefore their addition does not affect the ductility of the host polymer they reinforce [25]. Since BNNTs are not biodegradable cytoxicity measurements have been performed and it was shown that they did not affect cell death [26].

In fabricating BNNT-polymer composites it is necessary to functionalize the BNNTs in order to increase their hydrophyllicity and hence dispersion within the aqueous solution during the fabrication process. Use of surfactants is not preferred for biomedical applications as they can increase toxicity and therefore chemical modification with –OH is commonly performed. The following subsections summarize the work by [27].

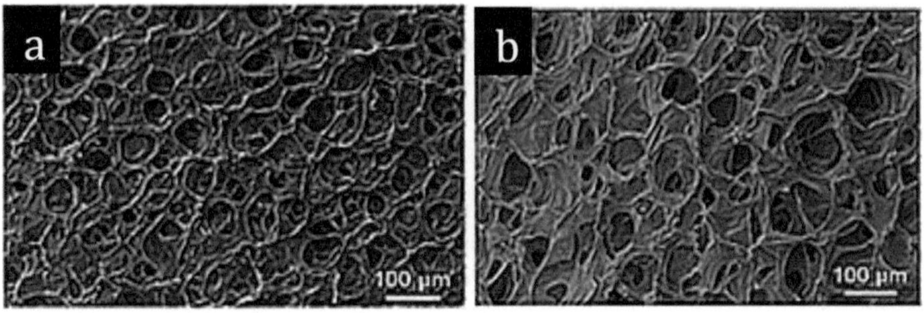

Fig. 4.3 Scanning electron microscopy images of (**a**) pure chitosan scaffold surface; (**b**) chitosan reinforced with BNNT-OH (Reproduced with permission from Ref. [27])

4.3.1.1 Fabrication and Characterization

Hydroxylated BNNTs (BNNT-OH) chitosan scaffolds were prepared by mixing 2 mL of chitosan with 1 mL BNNT-OH suspension in H_2O as described in [27]. In order to examine the effect of the BNNTs a control group of pure chitosan scaffolds was also prepared.

Scanning electron microscopy images, such as those depicted in Fig. 4.3a, b, illustrated a porous microstructure for both chitosan and chitosan/BNNT-OH scaffolds. However, the addition of BNNT-OH resulted in an increased pore size, which is preferred for tissue ingrowth. This was due to the hydrophobicity of BNNTs. The average pore sizes were determined from the randomly chosen pores on the SEM images and as seen, the pore sizes of the chitosan/BNNT-OH scaffold were larger than the pure chitosan scaffold. Although the BNNTs became more hydrophilic with the introduction of hydroxyl groups into their structures, they still exhibited some hydrophobicity. Hence, when they were added into the hydrophobic chitosan solution, the water present formed larger droplets than in pure chitosan in order to avoid the BNNTs. As a result when H_2O sublimated larger pores formed in the nanocomposite scaffolds [28].

In addition to increasing the pore size, the hydrophobicity of BNNTs also resulted in a lower swelling ratio as it repelled water molecules and reduced the amount that could enter the composites. Particularly, the swelling ratio was found to be 60 and 100 for the chitosan/BNNT-OH and pure chitosan scaffolds, respectively.

Finally, as expected, the BNNT-OH reinforced composites also exhibited an increased strength, despite their larger pore size. Performing compression tests indicated that for pure chitosan 35 kPa were necessary to induce a 70% strain, while the chitosan/BNNT-OH required a 55 kPa stress [27]. A drawback here is that up to a ~35% strain the stress-strain curves for both types of materials coincided; hence a higher content of BNNT should most likely be considered in future studies, to attain better mechanical properties at lower strains.

4.3.1.2 Biodegradability

In-vitro biodegradability of the pure chitosan and chitosan/BNNT-OH scaffolds was examined by incubating both types of materials in an enzymatic solution that contained a concentration of lysozyme that was similar to that present during cellular digestion [27]. Particularly the solution contained 2 mL of DMEM and 1 mg of lysozyme and the weight of the scaffolds was measured after 1, 3 and 7 days. Afterwards they were washed in distilled water and dried. The degradation ratio after 7 days of incubation was similar in both cases, as it was 5.2% for the chitosan/BNNT-OH and 6.2% for the pure chitosan. However, for the latter the biodegradation process was irregular, and at a higher rate, and therefore could not be controlled.

Although the degradation rates do not differ significantly between the two scaffolds it is clear that the addition of BNNT-OH does result in a slower and more controlled biodegradability. A low biodegradation rate is preferred since it allows the accommodation of the degradation products by the body, minimizing hence possible toxic effects they may have on cells, and allowing for ingrown tissue within the scaffold. In order to verify that the BNNT-OH reinforced materials did not have cytotoxic effects, cell cultures with human dermal fibroblasts (HDF) were performed using a WST-1 cell viability assay. Prior to the toxicity tests, the HDF cells was seeded and incubated for 24 h. Then the scaffolds were incubated for 1, 3 and 7 days (at 37 °C) in a solution containing 2 mL of DMEM. After each incubation time the solution which contained the scaffold degradation products was used to replace the cell culture medium and incubate the cells for 24 h. Then their viability was tested.

It was shown that the cell viability decreased by 80% on the pure chitosan scaffolds, while it increased to 115% for the ones that included BNNT-OH. This highly enhanced cell viability of HDF cells is not only due to the lower degradation rate, but also to the ability of chitosan/BNNT-OH to retain a higher strength during immersion in lysozyme. Particularly after allowing the scaffolds to degrade in lysozyme the chitosan/BNNT-OH scaffolds required 35 kPa to reach a strain of 70% while the pure chitosan scaffold needed 18 kPa [27].

In addition to viability tests, it is important to examine how the scaffold type affects the morphology of the adhered cells. Fluorescent and scanning electron microscopy was therefore performed 1, 3 and 7 days after incubation. After the first day the HDF cells in both cases had a round shape without their natural extensions. Such extensions were visible at the end of the third day for both scaffolds and they had an expanded/relaxed morphology which is their natural state. After the seventh day, however, this preferred state was lost for the cells on the chitosan scaffold (due to degradation of the chitosan and loss of adhesion with the cells), but retained for the chitosan/BNNT-OH, which degraded at lower rates. Representative SEM and fluorescent microscopy images from the 1st and 7th culture days are depicted in Fig. 4.4a–h [27].

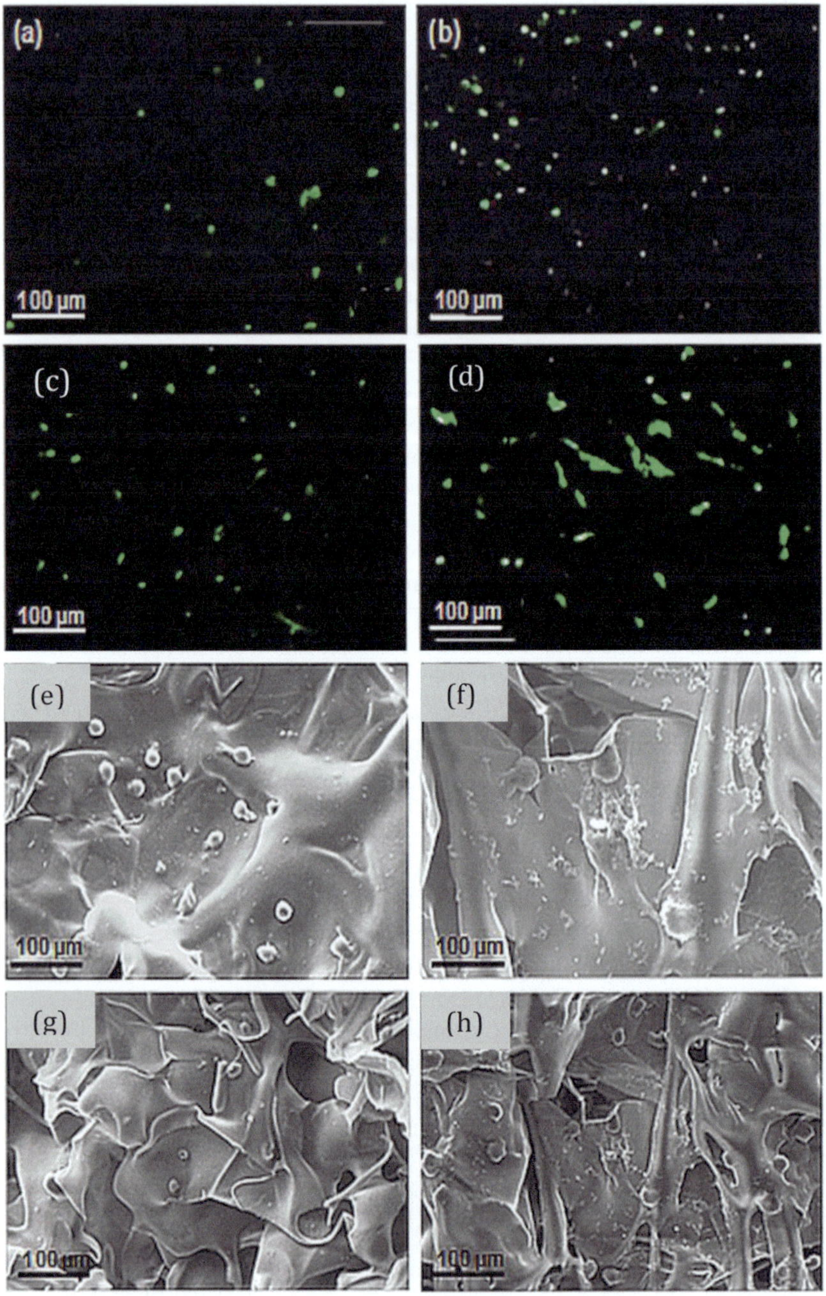

Fig. 4.4 Fluorescence and scanning electron microscopy images of cell cultures on the pure chitosan and chitosan/BNNT-OH scaffolds. (**a**) & (**c**) Correspond to the first day of culture on pure chitosan; (**b**) & (**d**) correspond to the first day of culture on chitosan/BNNT-OH scaffolds; (**e**) & (**g**) correspond to the seventh day of culture on pure chitosan; (**f**) & (**h**) correspond to the seventh day of culture on chitosan/BNNT-OH scaffolds (Reproduced with permission from Ref. [27])

4.3.2 Polylactide–Polycaprolactone Reinforced with Boron Nitride Nanotubes

In addition to chitin being reinforced with boron nitride nanotubes, another matrix material used is polylactide–polycaprolactone (PLA-PCL). PLA and PCL are both promising biodegradable polymers, which have significant differences in mechanical strength and biodegradability, and therefore their blends allow to tune their properties to those desired for tissue engineering applications. PLA has a crystalline structure, and is brittle, with a high strength and low elongation and fracture. It degrades at fast rates by enzymatic or alkali hydrolysis [29]. PCL has a semicrystalline structure and its mechanical properties have characteristics of elastomers. It has a low degradation rate as it is hydrophobic [30].

In order to increase the strength of PLA-PCL copolymers BNNTs are proposed as reinforcements, but at low contents as they do not biodegrade. Particularly, 2 wt% and 5 wt% BNNTs were considered in the study to be summarized below [26]. The length of these nanotubes ranged between 0.43 and 5.8 μm, and their diameter between 32 and 145 μm, while their aspect ratio was 15–84.

4.3.2.1 Fabrication and Characterization

The copolymer PLA-PCL blend contained a 70/30 PLLA/PCL ratio, and was termed PLC for abbreviation. First a colloidal solution was prepared by mixing PLC with acetone. BNNTs were also mixed separately with acetone and the solution was added to PLC colloid, which was appropriately mixed and finally cast to form films. For comparison purposes pure PLC films were also prepared using a similar process. The thickness of the resulting films was 500 μm, 250 μm and 200 μm for the PLC, 2 wt% BNNT/PLC and 5 wt% BNNT/PLC, respectively [26].

Figure 4.5a, b reveals a top view of the microstructure of the pure PLC and 5 wt% BNNT/PLC composite. It is seen that the addition of BNNTs decreased the porosity and also resulted in a smoother surface. The polymer coated BNNTs are shown Fig. 4.5c which is a cross-sectional view of a composite film.

Performing uniaxial tensile tests indicated that the addition of BNNTs increased the strength of the PLC. The samples were loaded to a maximum strain of 2.4%, and upon unloading they all resumed their initial shape, i.e. neither plastic deformation nor fracture was observed. The stress values recorded at this maximum strain were 2.67 MPa, 4.98 MPa and 5.59 MPa for the PCL, 2 wt% BNNT/PLC and 5 wt% BNNT/PLC, respectively [26]. The ability of composites to retain the flexibility of pure PLC, despite the addition of BNNTs, was due to the high flexibility of BNNTs [31] and the ability of acetone to allow for a good incorporation of this second phase into the polymer matrix. The higher strength of the composite materials was due to short fiber strengthening which resulted from the addition of stiffer fibers, of an appropriate aspect ratio, in a matrix.

Fig. 4.5 Scanning electron microscopy images of the cross-section of (**a**) PLC and (**b**) PLC–5BNNT composite showing the difference in porosity. Inset shows high magnification SEM images. (**c**) Scanning electron microscopy image of the fracture surface of PLC–5%BNNT composite, illustrating polymer coated BNNTs of varying diameters (Reproduced with permission from Ref. [26])

In order to test the applicability of these scaffolds for bone regeneration human osteoblasts (ATCC CRL-11372) were seeded in culture medium (at 34 °C) and after 60 h they were stained with a solution containing fluorescein di-acetate (FDA) for observing live-dead cells with a fluorescent microscope. Such fluorescent microscopy indicated that the viability of osteoblasts significantly increased for the BNNT/PLC composites, as it was 60% for PLC, while for both BNNT composites it was ~90% [26]. This suggests that the cell viability did not differ between the 2 wt% and 5 wt% BNNT content composites, but the addition of BNNTs increased the biocompatibility of such scaffolds. This could be partly due to the fact that rough surfaces have been shown to obstruct the proliferation of osteoblasts [32] and since the BNNT composites had smoother surfaces they could promote osteoblast viability.

4.3.2.2 Biodegradability

As biodegradability of polylactide–polycaprolactone materials has been examined it is important to examine the cytotoxicity effects of the BNNTs. This is of significant importance, as upon disintegration of the scaffold in the body the BNNTs may not be internalized completely by the macrophages (phagocytosis). Experiments have shown that maintaining a fiber length that is below the macrophage diameter [33] can allow for complete phagocytosis and prevent cytotoxocity to occur. In examining the cytotoxicity of BNNTs upon the polymer scaffold degradation murine macrophages (with a diameter of 20 μm) macrophages and human osteoblasts (ATCC CRL-11372) were seeded in culture medium and allowed to attach on the plastic well surface for 24 h. Then new solution was added which contained a concentration of 1 μg ml^{-1} of pure BNNTs (which would correspond to the case of complete PLA-PCL degradation). For comparison purposes control cultures were also performed without the addition of BNNTs. After 60 h cytotoxicity tests were performed on all cell cultures (with and without BNNTs for both macrophages and osteoblasts) using the Cyto-Tox 96 Non-Radioactive Cytotoxicity Assay kit. It was found that the cultures containing the BNNTs did not show an increase in dead cells, either for the osteoblast or the macrophage cultures. Hence, upon complete degradation of polymer scaffolds containing BNNT this study [26] suggested that there will be no toxic effects from the free standing BNNTs that will be released. This could be due to the structural stability and chemical inactivity of these types of fiber reinforcements.

4.4 Nano-Hydroxyapatite/Collagen/PLLA Reinforced by Chitin Fibers

In addition to using chitin as the scaffold matrix, it is also possible to use chitin fibers for reinforcing other biocompatible matrix materials. One such example are nano-hydroxyapatite/collagen (nHAC)/Poly-l-lactic acid (PLLA) scaffolds. PLLA is promising for biomedical applications as it is biocompatible, biodegradable and non-toxic. However, it has poor mechanical properties, it is hydrophobic and the products it is decomposed to during degradation by hydrolization are acidic. The existence of such acidic products increases the hydrolization rate and therefore degradation occurs faster. PLLA is therefore commonly combined with other materials, such as chitin fibers since they are alkalescent and can neutralize the acidity, decreasing hence the degradation rate. Furthermore, such fibers increase its mechanical strength. In this section the properties for porous nHAC/PLLA materials reinforced chitin fibers are described, as presented in [34]. To increase the strength of the interfacial bonding the chitin fibers were first linked with PLLA and then added to the nHAC/PLLA solution. The chitin fibers had a diameter of 12.5 μm and tensile strength of 550 MPa. To examine the effect of fiber content samples were prepared at separate chitin volume % of 10, 20, 30, 40 and 50.

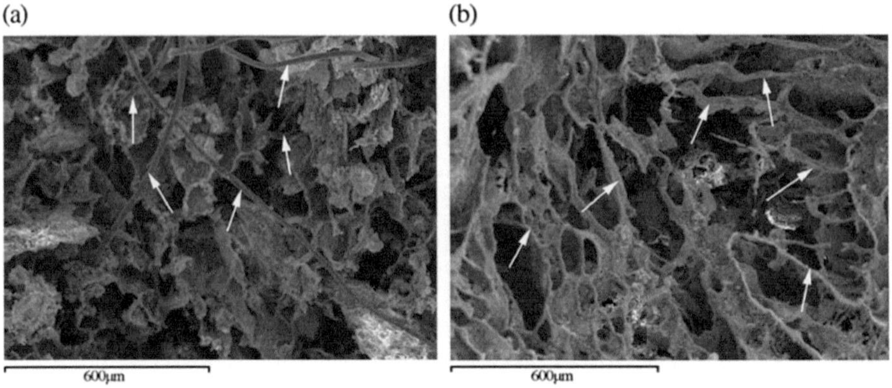

Fig. 4.6 Scanning electron micrograph of: (**a**) the samples in which chitin was not linked with PLLA; (**b**) the samples in which chitin was linked with PLLA (Reproduced with permission from Ref. [34])

4.4.1 Fabrication and Characterization

In [34] the chitin fibers were linked with PLLA, by dissolving chitin, PLLA dicyclohexylcarbodimide into dichloromethane. The collagen nano hydroxyapatite solution was prepared by mixing an appropriate concentration, lyophilizing and then grinding into a fine powder according to the method described in [35]. Then the PLLA was dissolved in dioxane, which allowed for pore formation, and the nHAC was added to the solution. The treated chitin fibers were added to the solution and after appropriate processing the liquid was lyophilized for 12 h.

The SEM images Fig. 4.6 illustrate that when the chitin fibers were not linked with PLLA, free-standing fibers could be clearly distinguished within the composite as they did not bind well with the matrix [34]. In the case, however, that the chitin was linked with PLLA, the fibers were distributed very well within the matrix such that it was difficult to distinguish them. Furthermore in the chitin treated case the pore size was larger and more uniform than in the untreated case. A large size is preferred for scaffold engineering since the tissue can grow within the pores, while a uniform microstructure allows for uniform degradation. Furthermore, the porosity of the samples decreased by increasing the fiber content as would be expected, however, the decrease was small, since the 10% chitin samples had a porosity of ~90% while the 50% samples had a porosity of ~80% [34]. The porosity between the treated and untreated scaffolds was similar for the 10% and 50% fiber content composites but it was slightly lower for the treated case for the 30% and 40% chitin composites.

The compression strength of the nHCA/PLLA scaffolds was 2 MPa, while mechanical mixing with the chitin fibers increased it to ~3 MPa. Treating the chitin fibers resulted in this strength to double as it was measured to be ~6.2 MPa [34]. This significant increase was due to the strong interface bonding between the fiber and matrix, which allowed for the stresses to be transferred to the fiber and therefore support the nHCA/PLLA matrix mechanical stability.

4.4.2 *Biodegradability*

The degradation of pure nHCA/PLLA, as well as treated and untreated chitin scaffolds was measured by incubating the samples in phosphate buffer saline (PBS) solution for 1, 2, 3 and 4 weeks at 37 °C. At the completion of each incubation time the scaffolds where rinsed three times with distilled water and dried at 45 °C. The solution was changed every week. For all samples the compressive strength decreased with increasing degradation time, however, the rate and total reduction depended significantly on the scaffold composition. The nHCA/PLLA exhibited the highest continuous decrease throughout the 4 week immersion reaching a value of 0 MPa [34]. The composite with untreated fibers reached a compressive strength of ~2.3 MPa after 4 weeks, hence a ~0.8 MPa decrease from the initial value took place. The treated chitin composite exhibited almost a constant compressive strength during PBS treatment, as only a 0.3 MPa decrease took place after the total immersion period. Treating the fibers resulted in strong chemical bonds between the fiber and matrix, which could withstand the degradation effects. Even after such interface bonds are destroyed by degradation of the high molecular weight PLLA in the matrix to lower molecular weight PLLA, the –OOCR groups in the PLLA can again bind to the treated fibers, rebuilding hence the interface. If the fibers are untreated the bonding between the fiber and matrix relies on molecular forces and hydrogen bonds, which are destroyed by the water molecules in the PBS [34]. Enhancing hence the interface strength allows for the scaffold to retain its mechanical strength for a longer time during degradation.

4.5 Poly-Lactic Acid Reinforced Magnesium Alloy Nanowires

Poly-lactic acid (PLA) is another common biomaterial used as it is naturally biodegradable since it is derived from plants, such as starch and sugarcane. In using it for orthopedic applications several drawbacks must be overcome, such as poor mechanical strength, heterogeneous degradation [36, 37], and formation of acidic hydrogen ions during degradation, which lower the pH giving rise to bone resorption. Various materials have been used to create PLA composites with enhanced properties, including hydroxyapatite [36, 38], titania nanoparticles [39], and magnesium alloy fibers. In this section a summary will be given of PLA-Mg alloy composites.

Mg based fillers are biodegradable and have a very high biocompatibility. They are preferred for orthopedic applications since the density and elastic modulus of Mg are similar to those of cortical bone. Mg, however, cannot be used in its pure form as it is highly corrosive, and therefore it is either alloyed with another metal [40, 41] or coatings are deposited on its surface [42, 43]. A recent promising alloy that has been used to reinforce PLA is AZ31, which is comprised of 96 wt% Mg, 3 wt% Al and 1 wt% Zn, and will be described in the sequel [44].

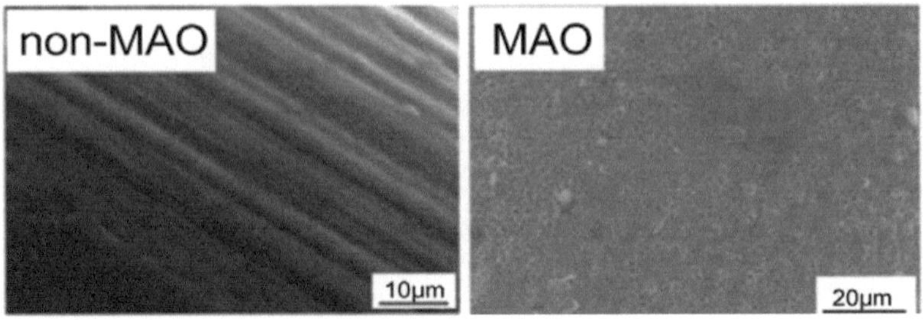

Fig. 4.7 Scanning electron microscopy images illustrating the Mg-alloy reinforced scaffolds, with either treated or non-treated alloy surface (Reproduced with permission from Ref. [44])

4.5.1 Fabrication and Characterization

The Mg alloy reinforcements were in the form of fibers/wires that were 300 μm in diameter. Once they were arranged directionally a solution of PLA dissolved in chloroform was poured on them and left to dry in air for 12 h. Then the lamina stacking method was employed as described in [44]. To further increase the mechanical strength of the alloy fibers micro arc oxidization (MAO) treatment was performed on a batch of fibers, which were used to fabricate a separate set of PLA/Mg alloy treated scaffolds. The volume fraction of the fibers in the composites was 5%, 10%, 20% and 40%. Finally, pure PLA scaffolds, again using the lamina stacking method were fabricated to use as the control group and better understand the effect of the Mg alloy and Mg alloy treated reinforcements.

Scanning electron microscopy, shown in Fig. 4.7 revealed that the PLA had a uniform structure into which the fibers were directionally distributed in a uniform manner. The difference between the MAO treated fibers is that their surface was coated with a porous structure which allowed for better adhesion with the PLA as it could penetrate into the pores and result in a stronger interface.

Inclusion of the untreated fibers in the PLA resulted in a significant increase in the tensile strength of PLA. Performing tension tests indicated that the pure PLA scaffold had a tensile strength of 46 MPa and increasing the fiber content continually increased it to 50 MPa (5% fibers), 65 MPa (10% fibers), 80 MPa (20% fibers), and 108 MPa (40% fibers). Three point bending was further used to determine the bending strength which was found to be 88 MPa for PLA and increased with fiber addition to 99 MPa (5% fibers), 119 MPa (10% fibers), 136 MPa (20% fibers), and 190 MPa (40% fibers). This increase in strength was further enhanced for the composites fabricated with the MAO treated reinforcements. Particularly, for the 20% MAO treated composite the tensile strength was 100 MPa, i.e. 14% higher than without surface modification and the bending strength was 150 MPa, i.e. 11% higher than the untreated case [44]. As mentioned MAO treatment resulted in better PLA/fiber bonding and therefore the interface strength was increased which enhances the overall composite strength.

4.5.2 Biodegradability

The biodegradability of the scaffolds was examined by immersing them in Sorensen's phosphate-buffered solution and monitoring the percentage mass loss, the viscosity average molecular weight, and bending strength. In one case rapid degradation was simulated by immersing for 3 weeks at 50 °C, while in the other regular degradation was simulated for 56 days at 37 °C. Upon immersion the untreated fiber composites underwent MAO and therefore rapid corrosion was avoided.

For both rapid and normal degradation a low mass loss was observed that was always below 1%. It did increase with an increase of fiber concentration which suggests that the MAO fibers were prone to some degradation, which was however negligible. The viscosity average molecular weight (M_v) decreased faster for both PLA and PLA-fiber composites at rapid degradation conditions. However, the degradation rate of both types of scaffolds was at comparable values for both degradation conditions until the last incubation day (i.e. 21st and 56th) in which the M_v was significantly higher for the fiber reinforced scaffolds. The bending stiffness for all types of scaffolds examined under the normal degradation conditions decreased between 60 and 75% over the 56 day period. However, at rapid degradation conditions the pure PLA and low content fiber composites (up to 10%) exhibited a ~ 90% decrease in bending stiffness. Increasing, however, the fiber content to 20% resulted in a 65% retention of the initial value [44]. This suggests that at the initial stages of implantation the role of Mg-based fibers in the mechanical stability of the scaffold is not that crucial, however, as bone regeneration requires more than a few weeks it is important to use such fiber reinforcement to assure that the implant will be able to withstand the skeletal forces until sufficient osteogenessis has occurred. The morphology of the scaffold after degradation at normal and accelerated rates is shown in Fig. 4.8.

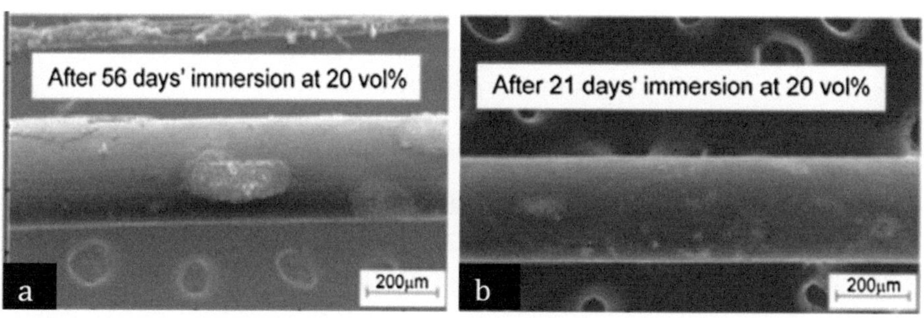

Fig. 4.8 Changes in the morphology of the Mg-based alloy fibers after immersion in Sorensen's phosphate-buffered solution for (**a**) 56 days at 37 °C and (**b**) 21 days at 50 °C (Reproduced with permission from Ref. [44])

The degradation process proposed for the PLA fiber composites is as follows: first the PLA reacts with water which initiates the hydrolysis of the PLA and forms acidic carboxyl end groups [44]. As the PLA begins to degrade the water reacts with the surface of the fibers (i.e. the MAO coating) fracturing it and penetrating further into the Mg-alloy fiber. During degradation of the coating, OH^- and Mg^{2+} are dissolved and released into the immersion fluid, giving rise to a higher mass loss with an increase in % fiber in the composites. The release of OH^- is desired since they diffuse into the PLA and react with the acidic hydrogen ions from the acidic carboxyl end groups that form during degradation of the PLA. This allows for a steady pH to be attained throughout implantation, which allows for a significantly lower and controlled degradation rate of PLA but also minimizes bone resorption, which can result from decreasing pH values [45, 46]. Therefore, the present fiber PLA composites have a long term stability that is significantly enhanced over pure PLA scaffolds.

4.6 Polycaprolactone Reinforced with Poly-L-Lactic Acid Fibers

Polycaprolactone (PCL) is a biodegradable polyester with good biocompatibility and high hydrophobicity. Its degradation rate is very slow as it occurs by random hydrolytic ester cleavage and weight loss through the diffusion of oligomeric species. PCL is promising for implants requiring a longer life time, as it can provide the necessary mechanical and structural support until the necessary tissue regeneration has taken place [47].

In order, however, to enhance regeneration studies have shown that it is useful to reinforce the scaffold with fibers so as to mimic the structure of the extracellular matrix, which contains collagen fibers. Such a fiber network can guide the growth of the cells and, if they have a faster degradation rate than PCL, pathways for cell migration can be created within the polymer matrix. Poly-L-lactic acid (PLLA) has a higher degradation rate then PCL, and it is also hydrophilic. In the sequel an overview will be given of PCL reinforced with PLLA fibers [48].

4.6.1 Fabrication and Characterization

PCL solution was produced by dissolving PCL pellets in N,N-dimethylacetamide according to standard procedures [48]. Then PLLA semicrystalline fibers, which had been prepared by the filament winding technique, were added to the solution in order to obtain the composite structures.

Scanning electron microscopy revealed a bi-modal porosity as there were two pore ranges that could be distinctly observed, as shown in Fig. 4.9; micropores were

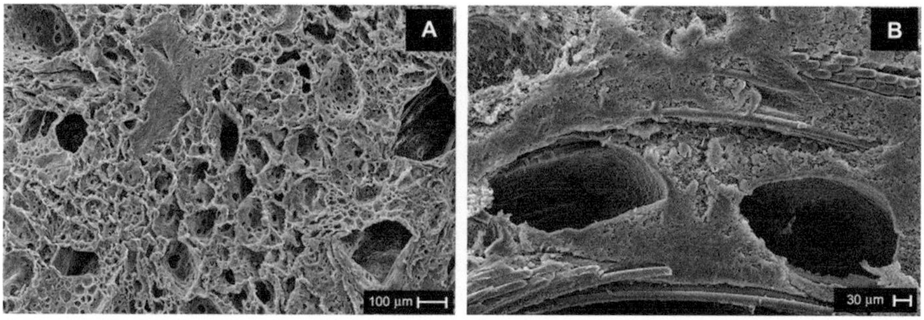

Fig. 4.9 Scanning electron microscopy analysis of PLLA fiber-reinforced PCL scaffolds: (**a**) bimodal pore size distribution and (**b**) organization of PLLA fibers inside porous PCL matrix (Reproduced with permission from Ref. [48])

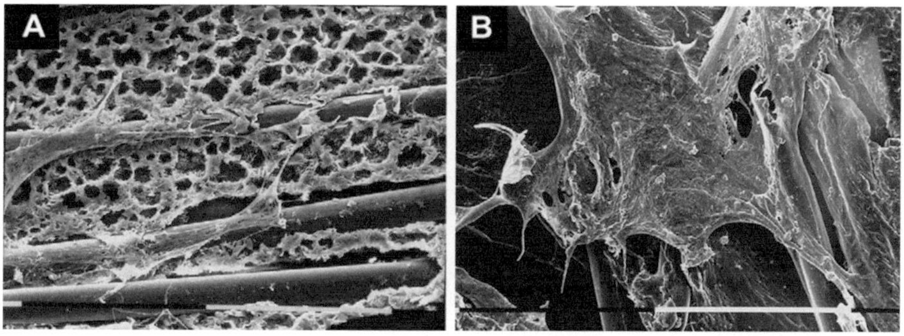

Fig. 4.10 SEM image of cells seeded on PCL/PLLA for 5 weeks: (**a**) marrow stromal cells aligned along PLLA fibres (scale bar was 100 µm) and (**b**) human osteoblasts formed clusters on the matrix and PLLA fibres (scale bar was 100 µm) (Reproduced with permission from Ref. [48])

1–10 µm, while macropores were 100–400 µm. The PLLA fibers were ~10 µm in diameter and their orientation was parallel to the porous structure. In order to determine the wt% composition for each polymer, thermogravimmetry analysis (TGA) was performed. The PCL and PLLA contents were estimated from the 330–420 °C and 200–330 °C regions, respectively, to be 85 wt% PCL and 15 wt% PLLA. Only this type of composition was considered.

To test the biocompatibility of these scaffolds cell cultures were performed using human osteoblasts (HOB) and marrow stromal cells (MSC) that were obtained from patients who underwent hip-joint replacement surgery. The SEM images, Fig. 4.10, revealed that after 5 weeks of culture the MSC cells had their long axis aligned along the PLLA fibers and had developed their filopodia, while the HOB formed clusters on both the PCL matrix and PLLA fibers [48]. Although the HOB were able to attach and grow on the scaffold, confluence was not reached within 5 weeks.

4.6.2 Biodegradability

In order to assess their biodegradability the materials were immersed for 35 days at 37 °C in three different types of solutions: (i) phosphate-buffered saline (PBS) at a 7.5 pH, (ii) simulated body fluid buffered (SBF) at 7.5 pH, and (iii) 0.01 M NaOH solution. The solution was changed once a week [48]. The weight of the samples was recorded prior to initiation of the tests and thereafter they were weighed once a week, after they were first washed with distilled water and vacuum dried at room temperature. For the tests carried out in PBS and SBF negligible weight loss was observed indicating a low degradation rate. However, in the case where NaOH was used as the immersion fluid a linear decrease in weight was observed throughout the 35 days, indicating a 14% decrease in the initial sample weight. This high degradation rate in NaOH was expected since OH^- plays a key role in the degradation of aliphatic polyesters [49, 50]. TGA revealed that after immersion in NaOH the PCL/PLLA ratio was higher than in the as prepared materials, indicating that PLLA degraded at a faster rate. Raman spectroscopy revealed that during degradation the crystallinity of PLLA increased, while its polymer chains shortened. This is due to the ability of water to penetrate easier into the amorphous parts, which hence degrade first. Also the water attacks the ester back bone, which results in a smaller chain length and the formation of short mobile oligomers that can recrystallize [49, 50].

HOB cells were less active and did not induce morphological changes on the scaffolds during the *in vitro* tests. The MSC cells however, did have an effect on the surface of the scaffold after 5 weeks of culture. The ability of human cells to induce degradation changes on such polymer composites suggests their possible use for *in vivo* applications where degradability is desired.

4.7 Polylactic Acid Reinforced with Silk Fibers

Another promising type of reinforcements used in degradable polymers are silk fibers produced from silkworms. Silk fibers have a hierarchical microstructure: 15 nm fibrils align to form a 1 μm fiber; several fibers form a fibroin. The fibroins are encased within a sericin coating, which is mainly comprised from protein. Such silk fibers can be embedded within various polymers and examined for biomedical applications since they are biocompatible, bioresorable and have good mechanical properties. It is interesting to note that silk in medicine has been used in wound dressing for centuries [51]. In this section, therefore, PLA reinforced with silk fibers will be elaborated on.

PLA is hydrophobic, while silk is hydrophilic since it contains hydroxyl groups, and therefore interface bonding in the composite is a major issue. A significant difference between the PLA-silk composites and the reinforced scaffolds of the previous sections is that the silk fibers were not chemically treated [52] to increase their bond strength with the PLA. Only the sericin layer was removed using the simple degumming process. Other chemical treatments were avoided as the target application of the proposed scaffolds was for bone implants. The connectivity that existed at the interface was weak and occurred only due to mechanical interlocking resulting from the fiber roughness.

It should be noted that the fabrication method employed in [52] was injection molding, and therefore the fibers had to be ~5 μm in order to prevent them from getting tangled with screws, or deforming plastically during fabrication. As this was a first study towards examining in detail the mechanical and degradable properties of silk reinforced composites only a 5 wt% fiber content was considered in [52] and described below.

4.7.1 Fabrication and Characterization

In order to release the fibroin fibers and have them react directly with the polymer matrix [53], the sericin layer was removed (although not completely) and the silk fibers were in the form of a reeling. Then [52] the fibers were cut into segments of 5 μm and placed them in an oven in order for them to dry and minimize their moisture. Finally, the fibers were mixed with PLA and the composites were fabricated using a Hakke MiniLab twin-screw micro-extruder. The melting temperature used during fabrication was kept constant at 180 °C, which is below the degradation temperature for the silk fibers. The content of the fibers in all samples was 5 wt%.

In Fig. 4.11a, b it is seen that the fibers in the PLA were aligned in the same direction, while in Fig. 4.11c,d it is seen that during molding some of the 1 μm fibers were separated into the 15 nm fibrils which comprise them, and dispersed into the PLA.

The tensile and flexural properties of the samples were tested with an MTS machine (Alliance RT/50). The elastic modulus of pure PLA samples was ~3.2 GPA, while for the PLA/fiber composites it increased by 27% to ~4.1 GPa [52]. The flexural modulus of pure PLA (~4 GPa) remained practically unaffected by the addition of the silk fibers. Furthermore, the tensile strength was ~70.7 GPa with and without the silk fibers, while the strain at fracture was 31% lower for the PLA/fiber composite than the pure PLA [52]. This decrease in fracture strain could be due to various factors, since the mechanical properties of fiber composites depend on the fiber concentration, length, diameter, and fiber-matrix interface strength. The morphology of a cross section of the deformed composite structures upon fracture is shown in Fig. 4.12. A gap was present between the fiber-PLA interface, indicating that the interface bonding was not strong, and therefore the stress from the PLA was not transmitted efficiently to the fibers. This could be due to residual sericin left on the fiber surface, which inhibits good chemical bonding with the PLA [53]. The large gaps between the fiber and the matrix act as flaws (voids) in the material, which reduce its strength. It should be noted that the apparent increase in elastic modulus is due to the fact that the modulus is measured when the material undergoes small loads, at which the stress transfer from the matrix to the fiber can occur,

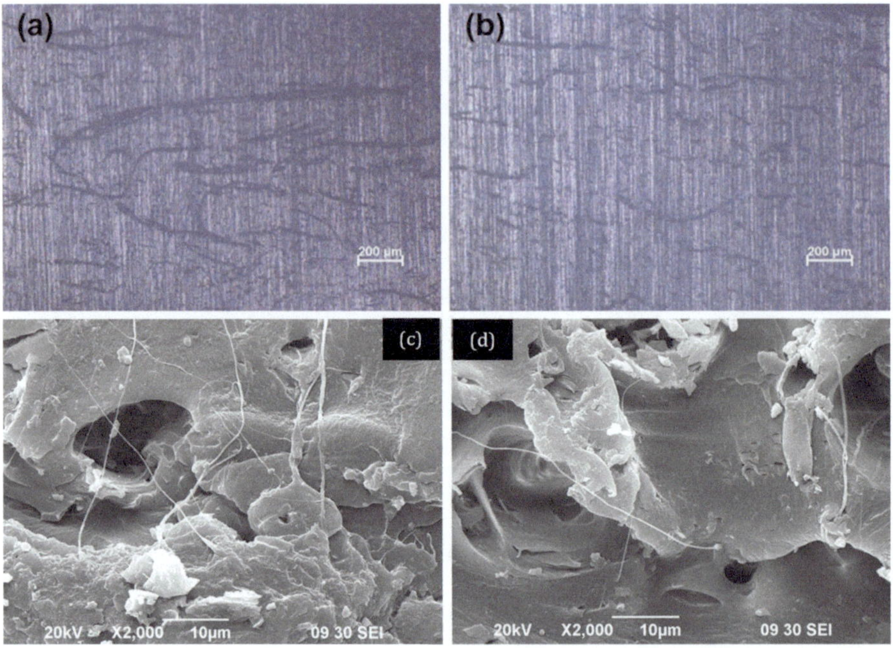

Fig. 4.11 (**a**) & (**b**): Optical micrographs of cut-off view (along the longitudinal direction of the sample) of the 5% silk fiber/PLA composite; and scanning electron microscopy of the micro fibrils inside the PLA (**c**) and (**d**) (Reproduced with permission from Ref. [52])

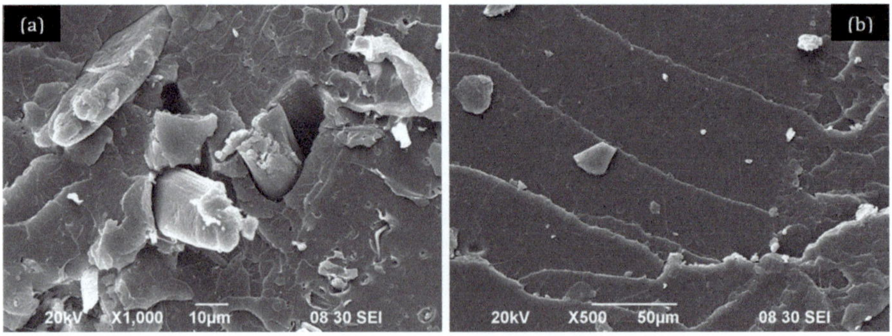

Fig. 4.12 Scanning electron microscopy images showing the fractured surface of (**a**) silk/PLA composites, in which the fiber pull out is seen and (**b**) pure PLA (Reproduced with permission from Ref. [52])

even when the interface strength is weak. With increased loading, however, the stress transfer cannot occur through weak interfaces and de-bonding takes place, creating voids, which reduce the tensile strength and fracture strain. In addition to poor interface contact, the low content of the silk fibers can also affect the lack of increase in tensile strength, over pure PLA.

4.7.2 *Biodegradability*

In order to test the biodegradability of the PLA/silk composites the samples were immersed into phosphate buffered saline solution and kept in an incubator at 37 °C that was shaking at 100 rpm. PLA has a slow degradation rate and no changes in mass were observed for the pure PLA scaffolds during a 4 month period in the PBS, however, the composite materials exhibited a slight increase of ~2.3%. This was due to the silk fiber hydrophilicity, which attracted water molecules that could be stored between the gaps at the silk-PLA interface, and therefore higher moisture absorption was present for these samples, which was reflected in a weight increase. Furthermore the interface between the fibers and PLA, which contained voids and flaws could enhance capillary transports, and also the fabrication process could result in cracks within the PLA allowing for better diffusion of the water molecules to the voids.

The mechanical properties of the samples were tested after 1, 2, 3, and 4 months after immersion. The tensile strength was reduced to ~55 MPa and ~38 MPa for PLA and PLA/silk materials after 2 months and the elastic modulus was reduced to ~3 GPa for both samples after 2 months. Measuring these properties throughout the 4 month period did not show any further decrease in their value. The flexural modulus, however, continuously decreased throughout 4 months, dropping to ~3.2 and ~2.7 GPa, for both samples after 2 and 4 months, respectively. Hence, although the PLA/silk samples had a higher water uptake, only their tensile strength decreased significantly faster than that of pure PLA. It should be noted that the 5% silk fiber content of these materials is rather low and results in a localized strain to occur at the matrix regions surrounding the fiber. Hence the interface can easily fracture and affect the overall strength of the composite. It would therefore be important to examine the mechanical properties of higher content composites, as it is anticipated they would be able to maintain their strength for a longer period of time. However, care must be made in selecting the optimum fiber content, as increasing above certain levels can result in composites that are unsuitable for bioapplications, due to high rigidity.

4.8 Poly (Lactide-co-Ethylene Oxide Fumarate) gel Reinforced with a Poly-L-Lactic Acid Fibers

A more recent study [54] tried to mimic the structure of bone to a greater extent by fabricating laminates of electrospun PLLA fibers coated with hydrogel solution that contained hydroxyapatite nanoparticles, and then stacking the laminates together. The reasons this microstructure is more representative of bone are: (i) the fibers within the extracellular matrix of bone form a random network as opposed to the aligned one produced in Sect. 4.6, (ii) bone contains apatite crystals in the extracellular matrix that allow for an increased mechanical stability.

The hydrogel chosen in [54] was based on a poly(lactide-co-ethylene oxide fumarate) (PLEOF) precursor, which was crosslinked in an aqueous environment with redox or ultraviolet initiators. Its biodegradability was tuned by adjusting the molecular weight of the lactide chains and also the ratio between the lactide and ethylene glycol [55, 56]. It should be noted that in order to increase the elastic modulus of the PLEOF/HA precursor the HA crystals were treated with an acrylate-functionalized glutamic acid sequence, while conjugating bioactive peptides to the PLEOF can increase its biocompatibility. Hence, an acrylamide-terminated integrin-binding Arg-Gly-Asp (RGD) peptide (Ac-RGD) was used in the PLEOF/HA precursor.

It should be noted that other research groups have used electrospun fibers to reinforce hydrogels, such as polyethylene glycol (PEG) [57] and chitosan–hyaluronic acid [28] hydrogels reinforced with PCL fiber mats. However, those constructs did not have a multilayer structure to the extent of the system that will be described in the sequel and they also did not consider the addition of HA in order to further increase the biocompatibility and mechanical properties of the possible scaffolds. Below the work of [54] is summarized.

4.8.1 Fabrication and Characterization

First a homogeneous solution of HA nanowhiskers (80 nm long) was prepared, then the (Ac-RGD) was dissolved into the solution, followed by addition of the PLEOF. The PLEOF had a PEG/LMW-PLA ratio of 70:30 in order to ensure hydrophilicity and water solubility. The PLLA fiber mats were prepared by electrospinning 9 wt% of low molecular weight PLLA in a 1,1,1,3,3,3-hexafluoro-2-propanol (HFIP) solvent. A fiber mat was placed on a Teflon plate and coated with the PLEOF/HA solution, then another fiber mat was placed on top and the process was repeated until 4 mats were stacked, which were then compressed by applying a pressure of 4.7 kPa. The thickness of the final laminate structure was ~85 μm [54]. The structure of the resulting scaffolds is shown in Fig. 4.13a–c, which includes images of the electrospun fiber mesh, a separate laminate and the four laminate composite.

Scanning electron microscopy, on the separate fiber mesh (Fig. 4.13a), indicated that the average diameter of the PLLA fibers was 610 ± 320 nm. The elastic properties were measured separately for the fiber mesh, the hydrogel with and without HA and the 4-layer laminates, using a rheometrics dynamic mechanical analyzer. The PLLA fibers do not absorb water and therefore their elastic modulus was the same under dry and wet conditions, at ~140 MPa [54]. The hydrogels, however, do absorb water and therefore a significant difference in their strength was recorded prior to immersion (~139 MPa) and after water immersion (0.5 MPa). The PLLA-hydrogel composite laminate had a significantly higher elastic modulus than both of its components (570 ± 130 MPa). For the case where the HA nanoparticles and Ac-RGD

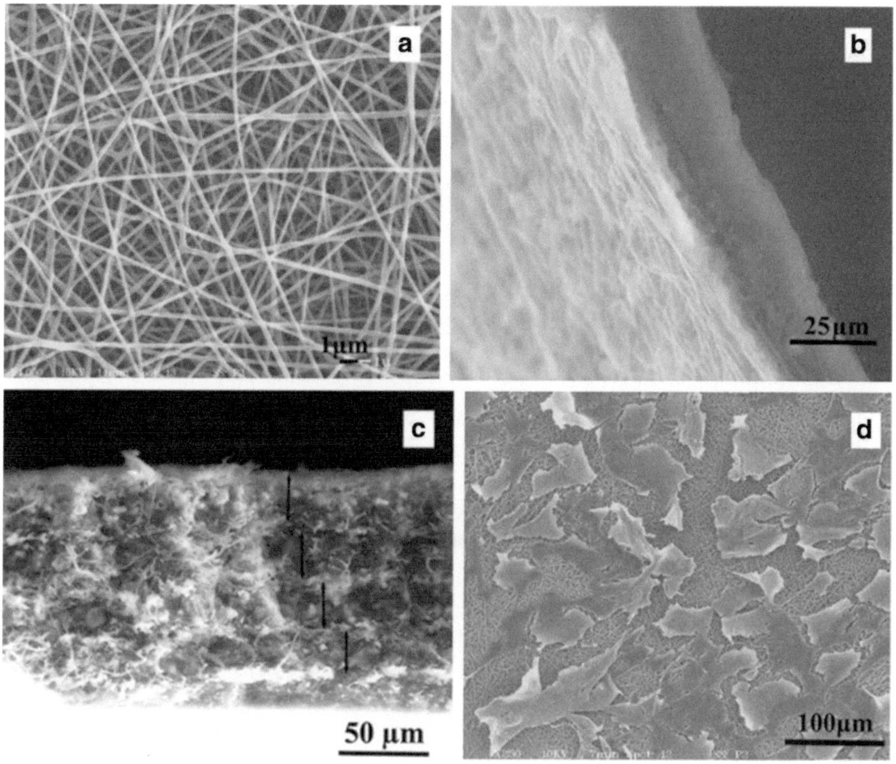

Fig. 4.13 Scanning electron microscopy images of (**a**) fiber mesh, (**b**) one-layer laminate, (**c**) and four-layer laminate; the arrows in (**c**) show each layer in the multi-layer laminate, (**d**) a typical image of the bone marrow stromal cells on the laminate with 10% HA (Lam-HA) (Reproduced with permission from Ref. [54])

were added to the hydrogel solution the elastic modulus was 470 ± 25 under dry conditions and 380 ± 47 MPa in the liquid. These mechanical characteristics can be further tuned by various approaches, such as changing the PLLA wt% in the solution used to obtain the electrospun fiber meshes, altering the number of layers used to create the laminate, vary the HA content, increasing or decreasing the water uptake by the hydrogel.

Bone marrow stromal (BMS) cells were cultured on the laminated composites in osteogenic medium for 21 days. The SEM image of Fig. 4.13d depicts the cuboidal to polygonal morphology of the BMS cells on the laminates. It was noted that when the PLEOF was not reinforced with HA or conjugated with Ac-RGD a lower area of the laminate was covered with cells. Furthermore, when the laminate contained HA and Ac-RGD a higher ALPase activity, calcium content and expression of osteogenic markers was noted during the cell culture, as compared to the laminates without HA and Ac-RGD.

4.8.2 Biodegradability

In order to assess biodegradability each component comprising the laminate was examined separately first and then the laminate as a whole. The samples were immersed in a solution comprised of: 5 ml primary medium (DMEM supplemented with 10% FBS, 100 Uml^{-1} penicillin (PEN), 100 μg ml^{-1} streptomycin (SP), 50 μg ml^{-1} gentamicin sulfate (GS) and 250 ng ml^{-1} fungizone (FZ)) and left at 37 °C. Once the desired degradation time had passed they were washed with DDI water and left to dry in a vacuum and their weight was measured in order to determine the weight loss in the degradation medium. The hydrogel-RGD material degraded completely after 3 weeks; the PLLA fiber mat lost less than 10% of its mass after 4 weeks; the laminate without HA lost 18%, 29%, 34% and 44% of its mass after the 1st, 2nd, 3rd, and 4th week, respectively; while the laminate containing HA had a mass loss of 15%, 17%, 34% and 34% after the 1st, 2nd, 3rd, and 4th week [54]. It is seen that including HA in the hydrogel resulted in a slower degradation rate, and particularly during the last week of immersion (week 4) no mass loss was observed.

Similar to the mechanical properties, the degradable properties of each laminate component (and hence overall degradation of the laminate) can be tuned through their fabrication process. Particularly increasing the amount of lactide in the PLEOF hydrogel increases the density of the ester bonds and hence its degradation. It has been observed that PLEOF containing lactide fractions of 10%, 20%, 30% and 40% resulted in a mass loss of 19%, 23%, 70% and 86% over a four-week period. The ability of tuning the hydrogel degradation rate to such extents can allow for the fabrication of laminates with a bimodal degradation rate, since the hydrogel could degrade much faster than the PLLA fiber network. As a result cell migration can occur in place of the hydrogel, while the PLLA fibers provide the necessary structural and mechanical support until sufficient tissue has been regenerated.

4.9 Protein Hydrogels Reinforced with a Nano-HAp/PHB Fiber Network

Another type of hybrid scaffold similar to the one described previously is that of using an electrospun fiber mesh containing nano-HAp but with the biodegradable polymer polyhydroxybutyrate (PHB), and a gelatin-based hydrogel [58]. Such a hybrid scaffold can increase the functionality of the system, over other studies which use synthetic hydrogels.

The gelatin was treated with methacrylate making it photocrosslinkable, while PHB was chosen as it is the simplest member of microbial polyhydroxyalkanoates (PHAs), which are recently being considered for tissue engineering. The HA nanoparticles allowed to further enhance the mechanical properties of the scaffold, as well as its osteoconductivity.

4.9.1 Fabrication and Characterization

The HA nanoparticles and methacrylated gelatin were prepared according to standard procedures. In order to distinguish the gelatin from the PHB nanofibers, fluorescein isothiocyanate (FITC) was added to it during fabrication. The hydrogel consisted of a photoinitiator, 0.1 mg/mL HA and 5% (w/v) of the as prepared gelatin [58].

To prepare the fiber mats, 15 wt% HA nanoparticles were dispersed in chloroform and then Poly(3-hydroxybutyric) acid was added to the solution as described in [58], after which the solution was placed in the electrospinning syringe to produce the desired fiber network.

In fabricating the 3D layered scaffolds, the following procedure was followed: (i) a 5 μL droplet was placed on a glass slide; (ii) the 7 mm diameter electrospun fiber mats were soaked in the hydrogel solution and then placed on the hydrogel droplet of step (i); (iii) a 5 μL droplet was placed on the soaked fiber mat and then a glass slide on the top. The composite structure was then exposed to UV light and incubated in PBS for 2 h. While in the PBS the glass was detached and therefore the gelatin/PHB/HA scaffolds were obtained. For comparison purposes a similar process was followed to make control scaffolds, without the fiber mat reinforcement, using only gelatin and HA.

Scanning electron microscopy images of the fiber mats indicated that they had a random microstructure and the fiber diameter was ~2 μm. By performing electron-dispersive spectroscopy (EDS) it was possible to obtain a map of the HA distribution, which indicated that it was evenly distributed throughout. The nanoparticles were ~25 nm along the (002) plane and their crystallinity is 60%. The SEM and EDS mapping indicated that the HA did not attach on the surface of the fibers but were fully encapsulated within the fiber. Such encapsulation is preferred from a mechanical point of view, although surface adhesion increases bioactivity [59].

Furthermore, performing fluorescent microscopy on the cross section of a fiber indicated that the gel solution diffused to the center of the fibers, as they also appeared to be fluorescent. In sectioning no deformation took place at the fiber-gel interface, suggesting that the interface between the fibers and hydrogels was very strong. SEM was also performed on freeze-dried samples, in order to get a higher resolution of the resulting microstructure and it was observed that the surface and material volume had a high porosity. Particularly, the layers of the gelatin consisted of pores between 30 and 70 μm, which can allow for preferred adhesion and proliferation of osteoblasts. The use of electron dispersive spectroscopy indicated a uniform distribution of HA nanoparticles within the hydrogel. In order to determine the thermal stability, thermogravimmetry analysis was performed, indicating that the electrospun fibers began to degrade at 250 °C, due to thermal decomposition of PHB. The hydrogel layers began to decompose at ~325 °C [58].

Tension tests were performed to determine the mechanical strength of the hybrid PHB/hydrogel/HA scaffolds. Their elastic modulus was calculated as 7.0 ± 1.2 MPa, while their tensile strength as 329 ± 18 kPa. Such tests could not be performed on the control hydrogel/HA materials as they were very soft; the modulus and strength of such pure hydrogels has been estimated to be ~0.1 MPa and 100 kPa [60], respectively, hence the addition of the fiber mat allowed for a significant increase in mechanical properties.

4.9.2 Biodegradability

Cell viability was examined by encapsulating mouse pre-osteoblasts (MC3T3-E1) inside the hydrogel/HA layers of the PHB/hydrogel/HA scaffold, and it was shown that the same number of cells were viable throughout a course of 7 days. Cell proliferation was monitored by measuring the cell metabolic activity and DNA content throughout a 14 day period. Both were found to significantly increase during the first 7 days, illustrating that high proliferation took place, however after that, the number of cells remained the same, with an increase noted again on the 14th day [58]. Performing similar viability and proliferation tests for the hydrogel/HA scaffolds without the electrospun fiber mats, indicated similar cell behavior. Hence, the nanofiber reinforcement did not affect the cellular response of the MC3T3-E1 cells to the hydrogel.

As expected the cells displayed a different migration and spreading behavior in the 3D scaffold configurations than in 2D cultures. Staining the actin microfilament cytoskeleton and nucleus after the 1st, 4th, 7th and 14th day of culture indicated that the encapsulated cells had a round shape until day 4, but had elongated by the 7th day and had spread out throughout the scaffold by the 14th day. To examine the interaction of the cells with the fiber reinforcements, cross sections of the scaffolds were cut and stained [58]. Fluorescent microscopy images illustrated that after the 4th day of culture the MC3T3-E1 cells were located mostly in the hydrogel/HA matrix and the interface between the fibers and the matrix. However, on the last day of culture (14th) the cells had migrated inside the fibers and had aligned along its center. The resulting cell migration allowed for the cells to be spread throughout all three layers of the scaffold in both the vertical and horizontal directions, indicating tissue in-growth.

The ability of the cells to penetrate the fibers is due to the biodegradability of the PHB, which became significant after the 7th day, as PHB degraded pores began to form, which made it softer and also gave it the ability to host the cells. These 3D constructs hence are promising scaffolds for tissue regeneration since the cells can be initially encapsulated in the hydrogel/HA matrix and in the initial stages of proliferation, the fiber reinforcement can provide the necessary structural and mechanical support required for tissue regeneration in bone defects. By the time the fibers begin to biodegrade, enough cells have formed that can migrate into the fibers and form an interconnected network throughout the complete scaffold volume, and therefore bone-tissue can then provide the necessary support to the skeletal system.

4.10 Conclusions

From this overview it can be seen that nanofiber or nanotube reinforced polymers, biopolymers, and hydrogels posses promising biodegradable properties, while providing the necessary mechanical strength and biocompatibility required for bone regeneration. The biodegradable properties for each composite system can be tuned by the content of the fiber inclusions, as well as the material properties of both the inclusion and matrix, while hydrophobicity also plays a crucial role. The interface strength between the fiber and matrix is a key parameter as it controls the response of the composite system to all external parameters, and it can be modified by chemically treating the fibers, or mechanically altering their surface to increase their roughness. Hence, it is difficult to draw a conclusion on which of the presented reinforced scaffolds is the most promising for inducing bone growth, as all show promise, with the appropriate tuning of parameters.

Acknowledgments The European Research Council Starting Grant MINATRAN 211166 is acknowledged.

References

1. Bose S, Mangal R, Bandyopadhyay A. Recent advances in bone tissue engineering scaffolds. Trends Biotechnol. 2012;30:497–554.
2. Hutmacher DW. Scaffolds in tissue engineering bone and cartilage. Orthop Polym Biomaterials: Appl Biodegradables. 2000;21:2529–43.
3. Velema J, Kaplan D. Biopolymer-based biomaterials as scaffolds for tissue engineering. In: Lee K, editor. Tissue engineering 1, Advances in biochemical engineering/biotechnology, vol. 102. Heidelberg: Springer; 2006. p. 187–238.
4. Attawia MA, Herbert KM, Uhrich KE, Langer R, Laurencin CT. Proliferation, morphology, and protein expression by osteoblasts cultured on poly (anhydride-*co*-imides). J Biomed Mater Res. 1999;48:322–7.
5. Holland SJ, Yasin M, Tighe B. Polymers for biodegradable medical devices: VII. Hydroxybutyrate-hydroxyvalerate copolymers: degradation of copolymers and their blends with polysaccharides under in vitro physiological conditions. Biomaterials. 1990;11:206–15.
6. Pollok JM, Vacanti JP. Tissue engineering. Semin Pediatr Surg. 1996;5:191–6.
7. Freed LE, Vunjak-Novakovic G, Biron RJ, Eagles DB, Lesnoy DC, Barlow SK, et al. Biodegradable polymer scaffolds for tissue engineering. Nat Biotech. 1994;12:689–93.
8. Fedorovich NE, Alblas J, de Wijn JR, Hennink WE, Verbout AJ, Dhert WJ. Hydrogels as extracellular matrices for skeletal tissue engineering: state-of-the-art and novel application in organ printing. Tissue Eng. 2007;13:1905–25.
9. Nichol JW, Koshy ST, Bae H, Hwang CM, Yamanlar S, Khademhosseini A. Cell-laden microengineered gelatin methacrylate hydrogels. Biomaterials. 2010;31:5536–44.
10. Lien SM, Ko LY, Huang TJ. Effect of crosslinking temperature oncompression strength of gelatin scaffold for articular cartilage tissueengineering. Mater Sci Eng C Mater Biol Appl. 2010;30:631–5.
11. Zhang F, He C, Cao L, Feng W, Wang H, Mo X, Wang J. Fabrication of gelatin–hyaluronic acid hybrid scaffolds with tunable porous structures for soft tissue engineering. Int J Biol Macromol. 2011;48:474–81.

12. Nezu T, Winnik FM. Interaction of water-soluble collagen with poly (acrylic acid). Biomaterials. 2000;21:415–9.
13. Sartuqui J, Gravina AN, Rial R, Benedini LA, Yahia L, Ruso JM, et al. Biomimetic fiber mesh scaffolds based on gelatin and hydroxyapatite nano-rods: designing intrinsic skills to attain bone reparation abilities. Colloids Surf B-Biointerfaces. 2016;145:382–91.
14. Taguchi T, Kishida A, Akashi M. Apatite formation on/in hydrogel matricesusing an alternate soaking process: II. Effect of swelling ratios of poly (vinylalcohol) hydrogel matrices on apatite formation. J Biomater Sci Polym Ed. 1999;10:331–9.
15. Sepulveda P, Bressiani AH, Bressiani JC, Meseguer L, König B. In vivoevaluation of hydroxyapatite foams. J Biomed Mater Res. 2002;62:587–92.
16. Timoshenko S, Goodier JN. Theory of elasticity, vol. 5. NewYork: McGraw-Hill; 1951. p. 500.
17. Isidor F. Loss of osseointegration caused by occlusal load of oral implants. A clinical and radiographic study in monkeys. Clin Oral Implants Res. 1996;7:143–52.
18. Hansson S, Werke M. The implant thread as a retention element in corticalbone: the effect of thread size and thread profile: a finite element study. J Biomech. 2003;36:1247–58.
19. Kokubo T, Kushitani H, Sakka S, Kisugi T, Yamamuro T. Solutions able to reproduce in vivo surface-structure changes in bioactive glass-ceramic A-W3. J Biomed Mater Res. 1990;24:721–34.
20. Vanzillotta PS, Sader MS, Bastos IN, de Almeida SG. Improvement of in vitro titanium bioactivity by three different surface treatments. Dent Mater. 2006;22:275–82.
21. Cho Y, Shi R, Borgens RB. Chitosan produces potent neuroprotection and physiological recovery following traumatic spinalcord injury. J Exp Biol. 2010;213:1513–20.
22. Hirose K, Onishi H, Sasatsu M, Takeshita K, Kouzuma K, Isowa K, Machida Y. *In vivo* evaluation of kumazasa extract and chitosan films containing the extract against deep skin ulcer model in rats. Biol Pharm Bull. 2007;30:2406–11.
23. Mohd Hilmi AB, Halim AS, Hassan A, Lim CK, Noorsal K, Zainol I. *In vitro* characterization of a chitosan skin regenerating template as a scaffold for cells cultivation. Springer Plus. 2013;2:79.
24. Li L, Chen Y, Stachurski ZH. Boron nitride nanotube reinforced polyurethane composites. Prog Nat Sci: Mater Int. 2013;23:170–3.
25. Lahiri D, Hadjikhani A, Zhang C, Xing T, Li LH, Chen Y, Agarwal A. Boron nitride nanotubes reinforced aluminum composites prepared by spark plasma sintering:microstructure, mechanical properties and deformation behavior. Mater Sci Eng. 2013;574:149–56.
26. Lahiri D, Rouzaud F, Richard T, Keshri A, Bakshi S, Kos L, Agarwal A. Boron nitride nanotube reinforced polylactide–polycaprolactone copolymer composite: mechanical properties and cytocompatibility with osteoblasts and macrophages in vitro. Acta Biomater. 2010;6(9):3524–33.
27. Emanet M, Kazanc E, Cobandede Z, Culha M. Boron nitride nanotubes enhance properties of chitosan-based scaffolds. Carbohydr Polym. 2016;151:313–20.
28. Mallick KK, Cox SC. Biomaterial scaffolds for tissue engineering. Front Biosci. 2013;1:341–60.
29. Zhang J, Xu J, Wang H, Jin W, Li J. Synthesis of multiblock thermoplastic elastomers based on biodegradable poly (lactic acid) and polycaprolactone. Mater Sci Eng C. 2008;29:889–93.
30. Lee BT, Quang DV, Youn MH, Song HY. Fabrication of biphasic calcium phosphates/polycaprolactone composites by melt infiltration process. J Mater Sci Mater Med. 2008;19:2223–9.
31. Golberg D, Bai XD, Mitome M, Tang CC, Zhi CH, Bando Y. Structural peculiarities of in situ deformation of a multi-walled BN nanotube inside a high-resolution analytical transmission electron microscope. Acta Mater. 2007;55:1293–8.
32. Khosroshahi ME, Mahmoodi M, Saeedinasab H. In vitro and in vivo studies of osteoblast cell response to a titanium-6 aluminium-4 vanadium surface modified by neodymium:yttrium–aluminium–garnet laser and silicon carbide paper. Lasers Med Sci. 2008;24(96):925–39.
33. Blake T, Castranova V, Schwegler-Berry S, Baron P, Deye GJ, Ci L, et al. Effect of fiber length on glass microfiber cytotoxicity. J Toxicol Environ Health A. 1998;54:243–59.

34. Li X, Feng Q, Cui F. In vitro degradation of porous nano-hydroxyapatite/collagen/PLLA scaffold reinforced by chitin fibres. Mater Sci Eng. 2006;26(4):716–20.
35. Du C, Cui FZ, Feng QL, et al. Tissue response to nano-hydroxyapatite/collagen composite implants in marrow cavity. J Biomed Mater Res. 1998;42:540.
36. Böstman O, Viljanen J, Salminen S, Pihlajamäki H. Response of articular cartilage and subchondral bone to internal fixation devices made of poly-llactide: a histomorphometric and microradiographic study on rabbits. Biomaterials. 2000;21:2553–60.
37. Nordström P, Pohjonen T, Törmälä P, Rokkanen P. Shear-load carrying capacity of cancellous bone after implantation of self-reinforced polyglycolic acid and poly-l-lactic acid pins: experimental study on rats. Biomaterials. 2001;22:2557–61.
38. Niiranen H, Pyhältö T, Rokkanen P, Kellomäki M, Törmälä P. In vitro and in vivo behavior of self-reinforced bioabsorbable polymer and self-reinforced bioabsorbable polymer/bioactive glass composites. J Biomed Mater Res A. 2004;69A:699–708.
39. Liu H, Slamovich EB, Webster TJ. Less harmful acidic degradation of poly (lactic-co-glycolic acid) bone tissue engineering scaffolds through titania nanoparticle addition. Int J Nanomedicine. 2006;1:541–5.
40. Song G, Atrens A. Understanding magnesium corrosionda framework for improved alloy performance. Adv Eng Mater. 2003;5:837–58.
41. Zong Y, Yuan GY, Zhang XB, Mao L, Lin NJ, Ding WJ. Comparison of biodegradable behaviors of AZ31 and MgeNdeZneZr alloys in Hank's physiological solution. Mater Sci Eng B. 2012;177:395–401.
42. Gray JE, Luan B. Protective coatings on magnesium and its alloys - a critical review. J Alloys Compd. 2002;336:88–113.
43. Rudd AL, Breslin CB, Mansfeld F. The corrosion protection afforded by rare earth conversion coatings applied to magnesium. Corros Sci. 2000;42:275–88.
44. Li X, Chu CL, Liu L, Liu XK, Bai J, Guo C, Xue F, Lin PH, Chu KP. Biodegradable poly-lactic acid based-composite reinforced unidirectionally with high-strength magnesium alloy wires. Biomaterials. 2015;49:135–44.
45. Suganuma J, Alexander H. Biological response of intramedullary bone to poly- L-lactic acid. J Appl Biomater. 1993;4:13–27.
46. Taylor MS, Daniels AU, Andriano KP, Heller J. Six bioabsorbable polymers: in vitro acute toxicity of accumulated degradation products. J Appl Biomater. 1994;5:151–7.
47. Griffith L. Polymeric biomaterials. Acta Mater. 2000;48:263–77.
48. Guarino V, Causa F, Taddei P, Foggia Di M, Ciapetti G, Martini D, Fagnano C, Baldini N, Ambrosio L. Polylactic acid fibre-reinforced polycaprolactone scaffolds for bone tissue engineering. Biomaterials. 2008;29(27):3662–70.
49. Taddei P, Di Foggia M, Causa F, Ambrosio L, Fagnano C. In vitro bioactivity of poly (epsilon-caprolactone)-apatite (PCL-AP) scaffolds for bone tissue engineering: the influence of the PCL/AP ratio. Int J Artif Organs. 2006;13:719–25.
50. Taddei P, Tinti A, Fini G. Vibrational spectroscopy of polymeric biomaterials. J Raman Spectrosc. 2001;32:619–29.
51. Wang YZ. In vivo degradation of three-dimensional silk fibroin scaffolds. Biomaterials. 2008;29:3415–28.
52. Ho MP, Lau KT, Wang H, Bhattacharyya D. Characteristics of a silk fibre reinforced biodegradable plastic. Compos Part B. 2011;42(2):117–22.
53. Cheung HY, Ho MP, Lau KT. Natural fibre reinforced composites for bioengineering and environmental engineering applications. Compos Part B Eng. 2009;40:655–63.
54. Xu W, Ma J, Jabbari E. Material properties and osteogenic differentiation of marrow stromal cells on fiber-reinforced laminated hydrogel nanocomposites. Acta Biomater. 2010;6:1992–2002.
55. He X, Jabbari E. Material properties and cytocompatibility of injectable MMP degradable poly(lactide ethylene oxide fumarate) hydrogel as a carrier for marrow stromal cells. Biomacromolecules. 2007;8(3):780–92.

56. Sarvestani AS, He XZ, Jabbari E. Viscoelastic characterization and modeling of gelation kinetics of injectable in situ cross-linkable poly(lactide-co-ethylene oxide-co-fumarate) hydrogels. Biomacromolecules. 2007;8(2):406–15.
57. Tseng H, Puperi DS, Kim EJ, Ayoub S, Shah JV, Cuchiara ML, West JL, Grande-Allen KJ. Anisotropic poly (ethylene glycol)/polycaprolactone hydrogelfiber composites for heart valve tissue engineering. Tissue Eng A. 2014;20:2634–45.
58. Sadat-Shojai M, Khorasani M, Jamshidi A. A new strategy for fabrication of bone scaffolds using electrospun nano-hap/phb fibers and protein hydrogels. Chem Eng J. 2016;289:38–47.
59. Bianco A, Di Federico E, Moscatelli I, Camaioni A, Armentano I, Campagnolo L, Dottori M, Kenny JM, Siracusa G, Gusmano G. Electrospun poly (e-caprolactone)/Ca-deficient hydroxyapatite nanohybrids: microstructure, mechanical properties and cell response by murine embryonic stem cells. Mater Sci Eng C. 2009;29:2063–71.
60. Drury JL, Dennis RG, Mooney DJ. The tensile properties of alginate hydrogels. Biomaterials. 2004;25:3187–99.

Chapter 5
The Biocompatibility of the Scaffolds Reinforced by Fibers or Tubes for Tissue Repair

Yanfeng Luo, Juan Xing, and Manping Lin

Abstract Protein adsorption on scaffolds is believed as the initial event happened before cell adhesion. The physiochemical properties of scaffolds regulate protein adsorption which further modulates cell adhesion and the subsequent cell responses, i.e. protein-cell interactions. In addition, scaffolds provide a temporary physical support for cells so that their mechanical properties determine the mechanical compatibility of scaffolds with the surrounding mechanical environment and the adhered cells. Scaffolds-protein interactions, protein-cell interactions and the mechanical compatibility of scaffolds dynamically interplay to control the scaffolds' biocompatibility. Protein adsorption and mechanical compatibility are essentially determined by scaffolds themselves, which, however, have received insufficient attentions in 3D levels. Understanding the fundamentals on protein adsorption and mechanical compatibility is potentially helpful to scaffolds design. Fiber- or tube-reinforced scaffolds are featured with improved mechanical properties and altered architecture and chemical compositions when compared to their unreinforced counterparts. Moreover, fibers or tubes themselves provide parameters such as fiber or tube size and alignment to affect the protein adsorption and thus the biocompatibility of the reinforced scaffolds. Therefore, in this chapter, the effects of scaffold chemistries and pore structure, and fiber/tube size and alignment on protein adsorption were detailed, and the regulation of cell/tissue responses by mechanical properties of scaffolds was understood. In addition, the effects of scaffold architecture and biochemical cues on cell/tissue responses were included as well.

Keywords Scaffold • Nanofiber • Nanotube • Biocompatibility • Protein adsorption • Mechanical compatibility

Y. Luo (✉) • J. Xing • M. Lin
Lab for Smart/Bioinspired Materials, College of Bioengineering, Chongqing University, Chongqing 400030, China
e-mail: yfluo@cqu.edu.cn

X. Li (ed.), *Tissue Repair*, DOI 10.1007/978-981-10-3554-8_5

5.1 Introduction

The term "biocompatibility" has been defined as "the ability of a material to perform with an appropriate host response in a specific application" [1]. According to the definition, biocompatibility depends not only on material itself but also on the material's application. The latter determines the environment with which the material interacts, including biological and mechanical ones. Accordingly, being satisfied by both the biological and mechanical environments is the precondition for a material to promise sound biocompatibility.

In tissue repair, scaffolds function as a template to allow new tissue growth and provide a temporary physical support as well. Cell adhesion to scaffolds is fundamental to multiple processes such as cell migration, proliferation, and tissue morphogenesis. It is commonly believed that cell adhesion to scaffold surfaces begins with the adsorption of proteins such as fibronectin, fibrinogen, vitronectin and immunoglobulins onto the surfaces. Subsequently, the layer of adsorbed proteins can be perceived by the molecules such as intergrin on cell surface and then initiates cell adhesion (Fig. 5.1). Therefore, proteins and cells are two key components in the biological environment of a specific application. Appropriate scaffold-protein interaction and the resulting protein-cell interaction are required for sound cell adhesion and good biocompatibility.

As a temporary physical support, the mechanical properties of scaffolds determine their mechanical compatibility with the surrounding mechanical environment, which further affects their biocompatibility. On one hand, the scaffolds should be strong enough to keep their mechanical integrity under various mechanical loadings, such as compression, stretch, shear, twist, etc. On the other hand, the scaffolds should not too strong to shield the loadings and thus prevent the embedded cells from sensing and responding to the loading. Moreover, the embedded cells take the mechanical properties of the scaffolds as a mechanical stimulation and may perceive and respond to them as well. How do the cells perceive and respond to the mechanical stimulations? One important mechanism is through the focal adhesions and cytoskeletons, which are highly controlled by the scaffold-protein interactions. The scaffold-protein interactions are markedly affected by the physiochemical

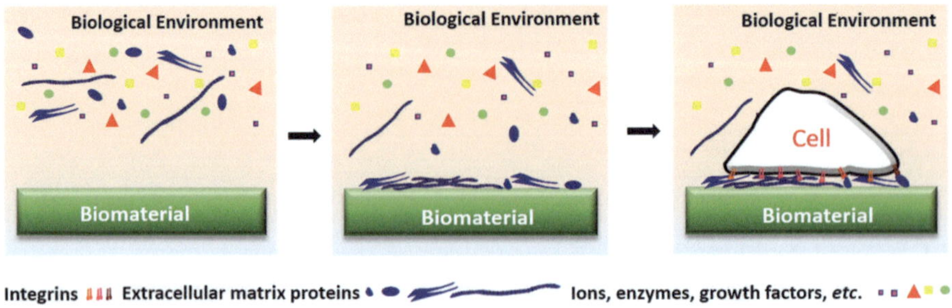

Fig. 5.1 Protein initially adsorbed onto biomaterial surface before cell adhesion

properties of the scaffolds, including the chemical compositions, surface energy, wettability, topography, fibers/tubes size and orientation, pore structures, etc. So do the resulting protein-cell interactions.

According to the above information, it would be reasonably deduced that the physiochemical properties of the scaffolds regulate the protein-scaffolds interactions and thus the protein-cell interactions, which further regulate the mechanical compatibility of the scaffolds. Scaffolds-protein interactions, protein-cell interactions and the mechanical compatibility of scaffolds dynamically interplay to control the scaffolds' biocompatibility. This chapter will focus on the interactions between the fibers-or tubes-reinforced scaffolds with proteins and their effects on the biocompatibility of the scaffolds. Another emphasis is the effects of scaffolds mechanical properties on biocompatibility. Up to date, these two aspects have not been widely and extensively explored on a 3D level. Nevertheless, we believe that they are key factors to be considered when designing and fabricating scaffolds with sound biocompatibility.

5.2 Fundamentals on Scaffold-Protein-Cells Interactions

5.2.1 Ubiquity of Protein Adsorption in vitro *and* in vivo

Protein adsorption is essentially important for cell fate both *in vitro* and *in vivo*. For *in vitro* cell culture, serum-containing medium is generally indispensable for desirable cell adhesion, morphology and proliferation. Serum was initially intended to provide cells with nutrients such as amino acids or growth factors. In fact, however, the proteins in serum are equally or even more important. It was found that proteins in serum such as albumin and fibronectin preferentially adsorb onto the surface of a material before a cell attaches to the material. The adsorbed proteins act as a bridge between the cell and the material and the status of the "bridge" determines the interactions between the cells and the material. In other word, whether the information from the material could be transmitted to cells or what information the cells could perceive is determined by the adsorbed protein layer. In cases of serum-free culture, pre-adsorbed proteins are commonly required for appropriate cellular responses, though it is believed that the cells may rapidly secret sufficient extracellular matrix proteins to coat the material surface. *In vivo*, proteins are also present in blood or interstitial fluid and may immediately coat the foreign material introduced into the body. In brief, the actual surface perceived by cells, no matter *in vitro* or *in vivo*, is the protein-modified surface.

5.2.2 *Material-Protein-Cell Interactions*

After proteins are adsorbed onto a material, cells will adhere onto the protein layer and subsequently initiate various cellular responses such as cell migration, cell proliferation and differentiation, and tissue formation. Therefore, cell adhesion, as the initial and basic protein-cell interaction, influences the subsequent cellular responses. Generally, cell adhesion is realized via the specific interactions between the transmembrane receptors of cells and the ligands in the proteins. A typical transmembrane receptor is integrin superfamily. Integrins recognize specific peptide sequences in the proteins to form cell adhesion. The RGD sequence found in fibronectin, fibrinogen, vitronectin and type I collagen is one of the best studied sequences. A large number of researches have proved that the quality of cell adhesion is dependent on the density and distribution of RGD on material surfaces, which is basically controlled by status, i.e. the amount, type and conformation of the adsorbed proteins.

The status of adsorbed proteins are modulated by the physicochemical properties of material surfaces including surface chemistry [2], wettability [3, 4] and topography [5], which will further regulate cell adhesion, morphology and subsequent cellular responses and functions [2, 6–8]. This cell-material interaction originating from materials is defined as a "bottom-up" or "outside-in" mode. After cells become specifically adhered onto the surface of biomaterials, the complex between integrins on cell membranes and adhesive ligands on adsorbed proteins may assemble to form focal adhesions (FAs) and fibrillar adhesions (FBs), during which cell traction force is generated *via* contraction of actin cytoskeleton [9, 10]. This traction force can be transmitted to the adsorbed protein layer *via* FAs and FBs (i.e. cell-protein interface) and then to the underlying biomaterial *via* protein-biomaterial interface. This kind of cell-material interaction originating from cells to biomaterials is called a "top-down" or "inside-out" mode. These two modes of cell-materials interactions are believed to be cooperative and competitive. Only when the cell traction force (top-down) can be resisted by the underlying protein layer, protein-material interface and materials, FAs will form and grow up. Correspondingly, cell traction force is increased since the number and size of FAs are positively proportional to cell traction force [11, 12]. Otherwise, FAs will be degraded and cell traction force is reduced [13–15]. By this way, a dynamic equilibrium is achieved between these two kinds of biomaterial-cell interactions. Since protein-material interface is an essential node for the transfer of cell traction force, its interfacial force, or called protein adsorption force, should play an important role in cell adhesion and cell traction force.

Gonzalez-Garcia et al. [5] and Altankov et al. [16] noticed the importance of the interaction strength between protein and the underlying material for biocompatibility. Lin et al. [17] quantitatively measured the adsorption force of fibronectin on various surface chemistries and found the significant dependence of the adsorption force on surface chemistry and the regulation of the adsorption force to late cell adhesion. Specifically, the fibronectin adsorption forces on OH, NH_2 and CH_3

surfaces followed an order of OH (1.5 ± 0.4 pN/FN) < < CH_3 (14.1 ± 4.1 pN/FN) < NH_2 (25.6 ± 7.5 pN/FN) whereas the accessibility of cell binding domains such as RGD adhesive peptide followed a divergent chemistry-dependence as OH > NH_2 ≈ CH_3. The initial cell traction force (2 h of culture), denoted by expression and organization of vinculin and tensin, expectedly matched well with the accessibility of cell binding domains, i.e. OH > NH_2 ≈ CH_3 (Fig. 5.2 a_1, b_1). However, after 12 h of culture, the fibronectin layer adsorbed on OH surfaces could not resist the cell traction so as to partially detach from the surface (Fig. 5.3-OH), while on CH_3 surfaces the protein layer was almost intact due to the weak cell traction and relatively strong protein adsorption force (Fig. 5.3-CH_3). An exception was found on NH_2 surfaces, where the fibronectin layer was reorganized to take an orientation parallel to the cytoskeleton (Fig. 5.3-NH_2), which was regarded as the synergistic consequence of cell traction, protein adsorption force and protein layer rigidity. As a result, the late cell adhesion (12 h of culture) took a chemistry-dependence of NH_2 > > OH ≈ CH_3 (Fig. 5.2 a_2, b_2).

Though the importance of protein adsorption force in cell adhesion has been preliminarily confirmed, its thorough roles in various cell responses and tissue formation are waiting for further verification. This chapter will focus more on the amount, type and conformation of the adsorbed proteins.

5.3 Protein Adsorption on Fibers- or Tubes-Reinforced Scaffolds

Given that adsorbed protein layer acts as a bridge between the cell and the material, it is important to investigate the protein adsorption on materials and uncover the underlying mechanism. It is widely proved that the status of adsorbed proteins in a 2D system can be regulated by the surface chemistry [7], wettability [3, 4], and topography [5]. In terms of nanofiber and nanotube scaffolds, protein adsorption was more easily impacted by the physical properties of the scaffolds, such as the wettability, the size and alignment of the incorporated fibers and tubes, and the pore structure.

It is worth noting that the status of the adsorbed proteins, especially for protein conformation, depends not only on the physiochemical properties of scaffolds but also on the categories of protein [18, 19]. For instance, α-chymotrypsin showed unfolded conformation and significantly lost its activity when adsorbed onto single-walled carbon nanotubes due to the hydrophobic surface exposure, while soybean peroxidase retained most of its native structure and activity after adsorbed [20]. Vlasova et al. [21] reported that PEGylated single-walled carbon nanotubes displayed 8.2 and 18.8 times higher adsorption capacity for immunoglobulin G and fibrinogen respectively compared to albumin. Even so, the following sections will not take into account of this aspect.

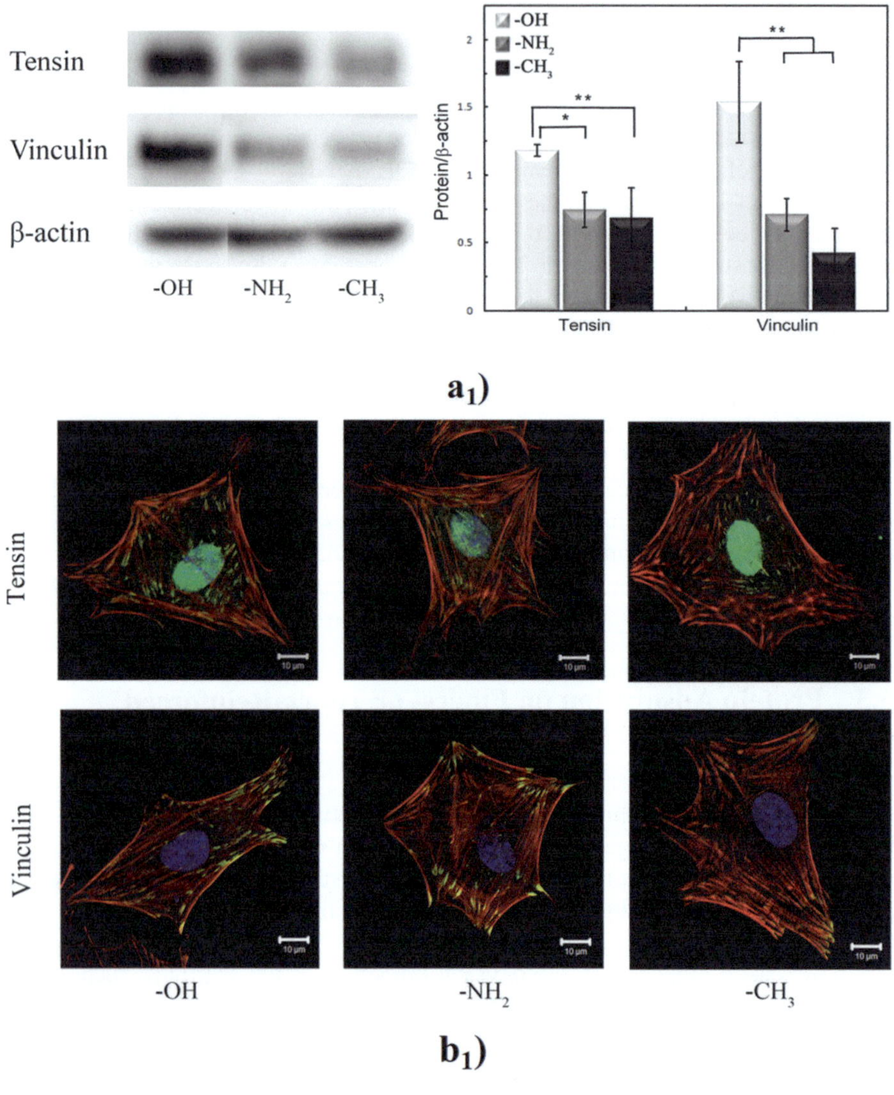

Fig. 5.2 Expression and organization of vinculin and tensin after osteoblasts were seeded in serum free medium for 2 h (**a**$_1$ and **b**$_1$) and 12 h (**a**$_2$ and **b**$_2$) on various fibronectin-adsorbed surfaces. (**a**) Western blotting images and quantitative analysis of tensin and vinculin. (**b**) Immunofluorescence staining images of tensin (*green*) and vinculin (*green*), together with F-actin (*red*) and nuclei (*blue*). Scale bar indicates 10 μm. * denoted $p < 0.05$, ** denoted $p < 0.01$ [17]

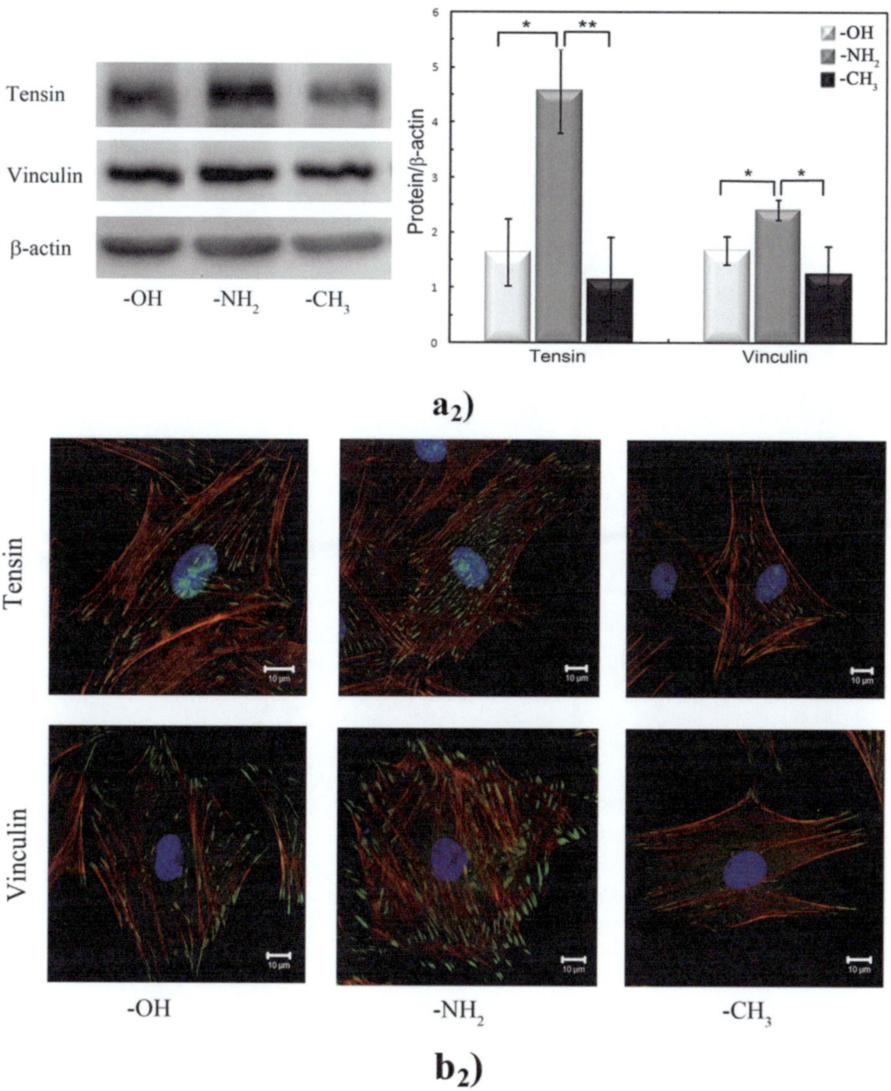

Fig. 5.2 (continued)

5.3.1 *Effects of Wettability*

A large number of materials are deficient in proper protein adsorption and cell adhesion, which may result from the high hydrophobicity. Though it is widely accepted that proteins adsorb stronger on hydrophobic than hydrophilic surfaces [22–24], hydrophobic surface is easier to induce changes in protein structure and activity than hydrophilic one [25]. Accordingly, much attention has been paid to the scaffold

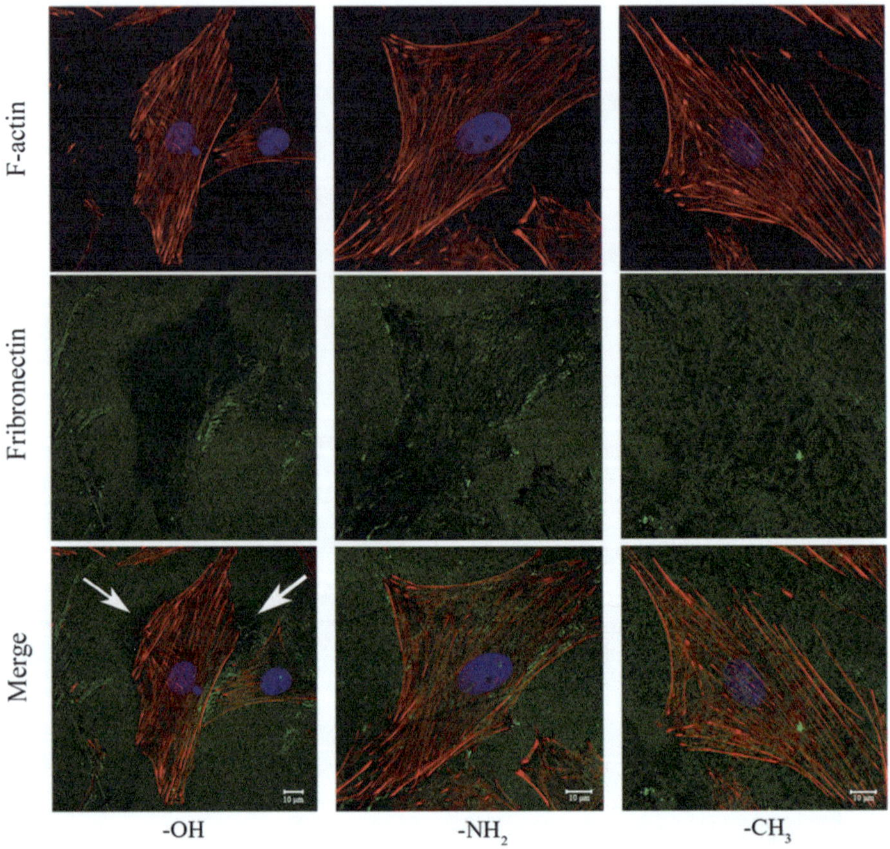

Fig. 5.3 Reorganization of adsorbed fibronectin on various chemistries after osteoblasts were seeded in serum free medium for 12 h. Immunofluorescence staining images of F-actin (*red*), FN (*green*) and nuclei (*blue*). Scale bar indicates 10 μm [17]

modification. For example, by treating of nitrogen plasma, highly hydrophobic poly(caprolactone) (PCL) nanofibers showed more significant reduction in water contact angle than microfibers and micro/nano fibers, which enhanced their serum protein adsorption and increased cell attachment and proliferation [26]. Moreover, the wettability of scaffolds can be improved by compositing with other hydrophilic components like chitosan and hyaluronan. Li et al. [27] reported that the incorporation of hyaluronan to polycaprolactone (PCL)/silk fibroin (SF) nanofibers significantly reduced the hydrophobicity of the scaffolds and subsequent non-specific protein adsorption, which may account for the reduction of fibrosis formation after implantation *in vivo* (Fig. 5.4) The presence of chitosan in PCL nanofibrous also aided a prominent improvement in the hydrophilicity of the scaffold and enhanced the serum protein adsorption on the scaffold [28]. Jose et al. showed that adding collagen to hydroxyapatite/poly[(D,L-lactide)-co-glycolide] nanofibrous scaffolds

Fig. 5.4 (**a**) Water contact angle decreased with the increasing of hyaluronan (HA) ratio. (**b**) Fluorescence images of fibers after incubating with 2.5 mg/ml FITC-BSA for 2 h [27]

helped to enhance fibronectin and vitronectin adsorption [29]. Furthermore, the wettability of scaffolds can be altered by reinforcing with nanotubes. Although pure carbon nanotube surface is extremely hydrophobic (contact angle of more than 160°), CNT-poly(carbonate urethane) became more hydrophilic and enhanced protein adsorption after compositing with high ratios of CNT because of higher nano surface roughness [30].

5.3.2 *Effects of the Fiber and Tube Size*

The protein adsorption on the reinforced scaffolds should be influenced by the size of the incorporated fibers and tubes as well. It has been proved that the amount of the adsorbed proteins on nanofibers, including total serum protein, fibronectin and vitronectin is much larger than that on microfibers because of the high surface area to volume ratio [31]. In addition to the protein amount, the conformation of the adsorbed proteins is also affected by fiber diameter. Vertegel et al. [32] found that the structure and enzymatic activity of the adsorbed lysozyme was strongly influenced by the size of silica nanoparticles. Particularly, greater loss of α helicity and thus greater loss of enzyme activity were observed for the lysozyme adsorbed onto larger nanoparticles. This might be attributed to the divergent curvature for nanoparticles with different diameter. It has been proved that the protein-material interaction depends on the curvature of the material surface [33]. Specifically, a small globular protein, namely albumin, demonstrated a native-like conformation on surfaces with high curvature yet became denatured upon binding to surfaces with low curvature; on the other hand, a rod-like shape protein, namely fibrinogen, was denatured on nanospheres with higher curvature (Fig. 5.5) [19].

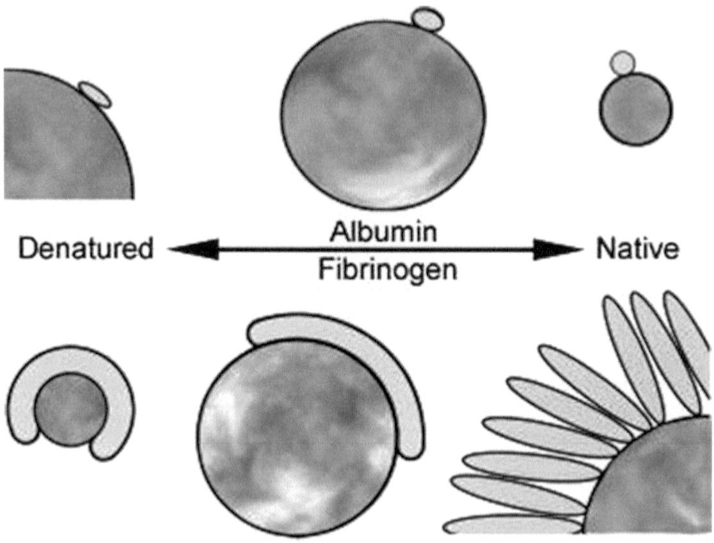

Fig. 5.5 Schematic demonstrating control of protein conformation and orientation by surface curvature. Global albumin took native conformation on surfaces with lower curvature yet became denatured on surfaces with larger curvature. In contrast, rod-like fibrogen took native conformation on surfaces with larger curvature yet became denatured on surfaces with lower curvature [19]

Similarly, nanofibers of different diameter are different in surface curvature. More complex surface curvatures are encountered by nanotubes. In addition to the divergent curvature caused by diameters, their inner surface and outer surface may have totally divergent curvature even for the same tubes. Therefore, it is reasonably claimed that the incorporated fibers and tubes in the fiber/tube reinforced scaffolds should regulate the protein adsorption via their fiber or tube size. Li et al. [34] employed CNT compacts to study the protein adsorption in 10% fetal bovine serum and found that CNT could adsorb more proteins including bone-inducing proteins. Enhanced protein adsorption was also observed by other groups [35]. The suggested mechanism is that the dimensionality of CNT is comparable with that of natural ECM proteins in addition to the increased specific area [36, 37]. Despite the increased protein amount, the conformational changes in both secondary and tertiary structures of the adsorbed proteins might be rendered on carbon nanotubes surface [38, 39], which may be advantageous or disadvantageous to cell responses. To more clearly uncover the controlling mechanism for protein adsorption by nanotube size, Shen et al. [40] investigated the interaction between human serum albumin and carbon nanotubes by using dynamics simulations. They found that the interaction between albumin and carbon nanotubes becomes strengthened with the increase of tube diameter. Furthermore, the atoms near the tube wall rearrange to fit the curving surfaces of carbon nanotubes; thereby, the final conformation and orientation of albumin are altered on the tubes and exhibit curvature-dependence (Fig. 5.6) [40]. There is an interesting observation about the effect of fiber diameter

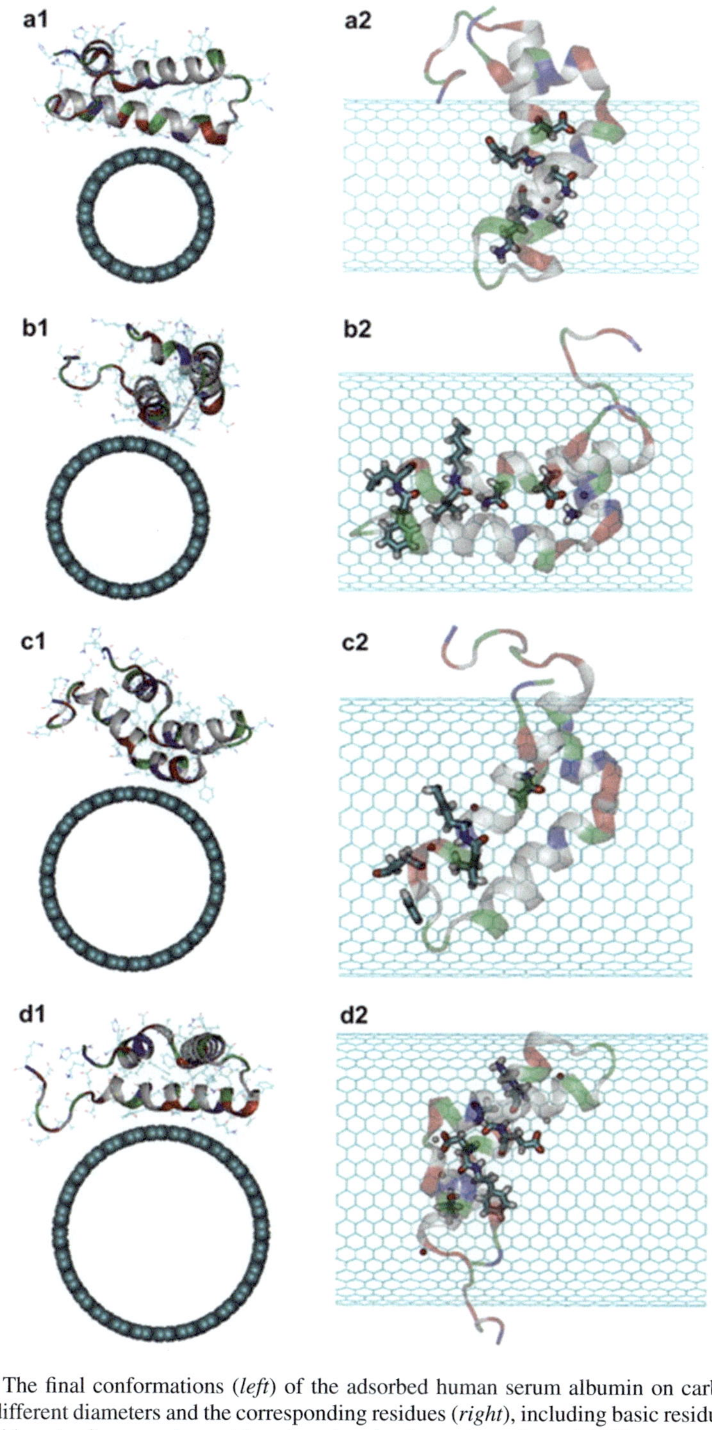

Fig. 5.6 The final conformations (*left*) of the adsorbed human serum albumin on carbon nanotubes of different diameters and the corresponding residues (*right*), including basic residues (*blue*), acidic residues (*red*), non-polar residues (*gray*) and polar residues (*green*) [40]

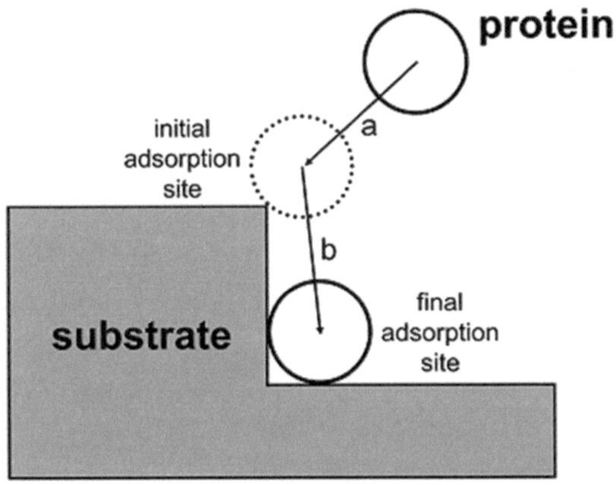

Fig. 5.7 Schematic of the protein adsorption mechanism [44]

on protein adsorption. It is found that the protein amount adsorbed on random nanofibrous scaffolds increased with the decrease of fiber diameter [41], whereas on homogenously aligned nanofibrous scaffolds, similar protein amount was observed regardless of the fiber diameter [42]. This suggests that protein adsorption is regulated by fiber alignment as well.

5.3.3 Effects of the Fiber and Tube Alignment

Aligned nanofibers enhanced fibronectin adsorption compared to smooth films [43]. Moreover, compared with random fibers, aligned fibers adsorb higher amount of proteins as well [42]. The mechanism may be obtained by assuming the aligned nanofibers as regular convex-concave nanostructures. By combining Brownian dynamics with numerical field calculation methods, Elter et al. [44] found that proteins initially adsorb on surfaces near the convex edges because of the strengthened electric field; subsequently, the adsorbed proteins migrate into the grooves via surface diffusion (Fig. 5.7), leading to increased protein amount per surface area by concave nanostructures. Aligned nanofibers render obvious convex-concave nanostructures. Therefore, the protein amount on nanofibers increases with the degree of alignment. Mukhatyar et al. [43] further demonstrated that aligned poly-acrylonitrile methylacrylate fibers promoted the formation of fibronectin fibrils network and subsequent ECM organization compared to smooth films where fibronectin had globular morphology (Fig. 5.8).

5.3.4 Effects of Pore Structure

Scaffolds for tissue engineering should be porous and permeable to permit the ingress of cells, nutrients and oxygen diffusion, as well as waste removal [45, 46]. In nanofibers/nanotubes-reinforced scaffolds, the incorporated fibers and tubes may

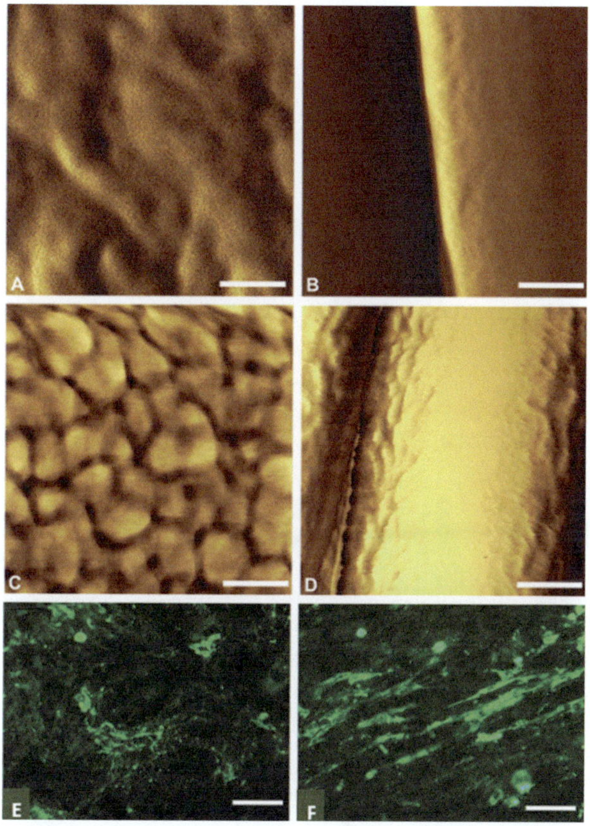

Fig. 5.8 AFM images of poly-acrylonitrile methylacrylate (**a–d**). Smooth film without (**a**) and with fibronectin (**c**). Aligned fibers without (**b**) and with fibronectin (**d**). Fluorescence images of ECM organization of glial cells on smooth film (**e**) and aligned fibers (**f**). Scale bar indicates 100 nm [43]

alter the pore structure, which plays vital roles in regulating cell/tissue responses (see Sect. 5.5 for details). In fact, this regulation role by pore structures is realized through protein adsorption. Ma et al. [47] demonstrated that the pore size of CNT reinforced poly(L-lactide) scaffolds was reduced with the increasing of the CNT incorporation (Fig. 5.9a–d). More importantly, the capacity of protein adsorption was dramatically improved when the percentage of CNT was higher than 1.0 wt% (Fig. 5.9e). Wei et al. [48] fabricated nano-hydroxyapatite/poly(L-lactic acid) (NHAP/PLLA) composite scaffolds with fibrous pore (Fig. 5.10a,b) and smooth pores (Fig. 5.10c,d) and investigated the effects of pore structures on protein adsorption. It was found that the amount of adsorbed serum proteins on scaffolds with fibrous pores was four times higher than that with smooth pores (Fig. 5.10e). Leong et al. [49] further fabricated nanopores on poly(DL-lactic acid) (PDLLA) nanofibrous scaffolds by using vapor-induced phase separation method. After incubated in a dilute solution of fetal bovine serum, the nanoporous PDLLA fiber scaffolds were found to have ~ 80% more proteins than the solid fiber scaffolds, though the former demonstrated much weaker mechanical properties. The role of pore structure to protein adsorption is generally attributed to the higher surface area to volume ratio.

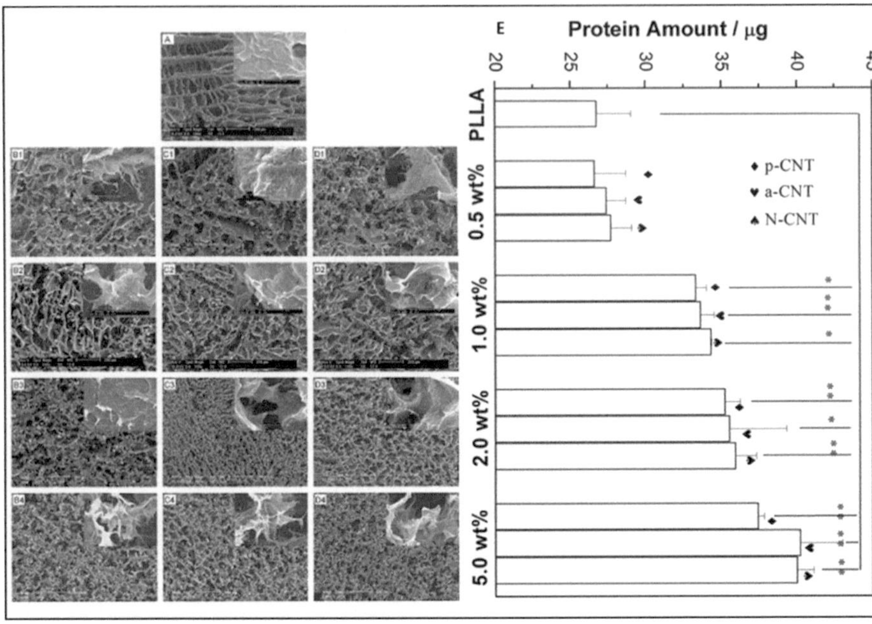

Fig. 5.9 (**a**–**d**) Micro-morphologies of CNT reinforced poly(L-lactide) scaffolds with different functionalized CNT and loadings (wt%) and corresponding total adsorption proteins (**e**) [47]

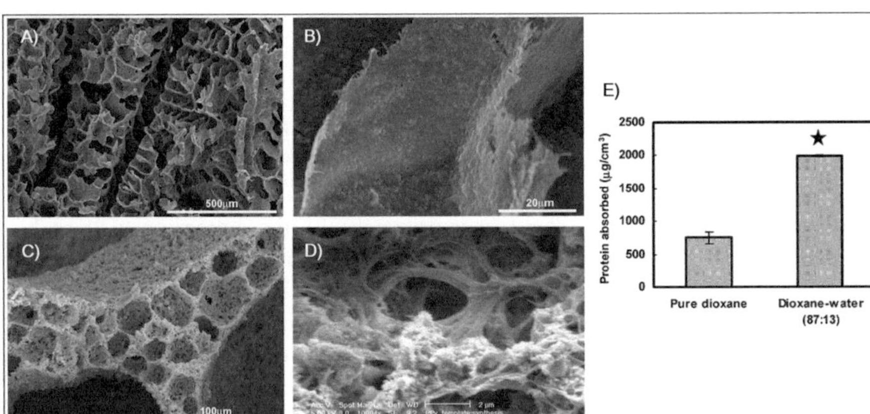

Fig. 5.10 Micro-micrographs of NHAP/PLLA scaffolds fabricated from pure dioxane (**a**, **b**) and dioxane: water = 87: 13 (**c**, **d**), and corresponding protein adsorption (**e**) [48]

5.4 Regulation of Cell/Tissue Responses by Mechanical Properties of Scaffolds

An ideal reinforcement material for medical applications should impart mechanical integrity to the composite without offsetting the biocompatibility of the bulk material [50]. Preferably, the addition of reinforcement material could improve the biocompatibility of the composite. Nanofibers and nanotubes are both featured with small dimensions and high aspect ratio (length to diameter). Their reinforcement role primarily comes from their high strength and stiffness, and high orientation [51–56]. The cell/tissue responses on nanofibers- or nanotubes-reinforced scaffolds are strongly affected by the wettability, architecture (fiber/tube size and alignment, pore structures) and mechanical properties of the scaffolds. The effects of wettability, fiber/tube size and alignment, and pore structure on protein adsorption and further on cell/tissue responses have been thoroughly discussed in Sect. 5.3. This section will pay more attentions to the effects of the mechanical properties. More particularly, an in-depth understanding of the potential mechanisms will be detailed regarding how the mechanical properties of scaffolds regulate the biocompatibility.

5.4.1 Potential Regulation Mechanisms by the Mechanical Properties of Scaffolds

5.4.1.1 Regulating the Transmission of External Mechanical Stimulations to Cells

It has been unanimously recognized that sufficient mechanical properties of scaffolds lay the foundation of the successful implant surgery for a specific application. Only if they possess proper mechanical properties, can the scaffolds keep their shape and characters after being suffered from complex mechanical environments in the body.

To construct functional 3D tissues such as blood vessels or bones, cells anchoring to the scaffold require appropriate mechanical stimulations, which are generally applied to the cells either through scaffolds stretch [57], compression [58], or fluid flow through scaffolds [59]. Accordingly, scaffolds, as the physical support of cells, are the medium to transmit mechanical stimulations. In a scaffold-cell system, the mechanical stimulation such as compression or shear stress could be directly applied to the cell membrane of the embedded or attached cells, which may deform the cell membrane and further activate mechanosensitive ion channels, heterotrimeric G proteins, protein kinases and other membrane-associated signal-transduction molecules [60–62]. In this case, the mechanical properties of scaffolds, to a certain extent, determine how much of the applied loading could be transferred to the cells. More frequently, however, the mechanical loading is applied to the cells through a certain mechanotransduction mechanics relating to focal adhesions and cytoskeletons [63–67], which is strongly affected by the material chemistry [68–71], and fiber/tube size and orientation of the scaffolds [72–74]. These scaffold-based parameters realize their regulation to focal adhesions and cytoskeletons through affecting protein adsorption (see Sect. 5.3).

Extremely soft scaffolds might collapse and lose their mechanical integrity under a mechanical loading. On the other hand, extremely rigid scaffolds might demonstrate negligible strain while maintaining the mechanical integrity of scaffolds. In the latter case, the loading is shielded by the scaffold and the cells embedded cannot sense the mechanical loading and thus cannot secret and deposit appropriate matrix, especially for those mechanosensitive cell lineages such as osteoblasts and endothelial cells. Common clinical phenomena by load shielding include the formation of bone tissues incapable of load-bearing in bone repair and delayed interbody fusion or even intervertebral non-union in intervertebral fusion surgeries [75]. One critical shortcoming by metal scaffolds or metal fixation devices is the loading shielding or stress shielding due to their excessively high mechanical properties compared with their contact tissues.

5.4.1.2 Acting as a Mechanical Stimulation to Regulate Cell Responses

In addition to the mechanical stimulations externally applied through scaffolds, the mechanical properties of scaffolds themselves, such as elasticity or stiffness, act as a mechanical stimulation as well that the attached cells can directly feel or sense [76, 77]. In order for a cell to sense the mechanical properties of the underlying substrate, the substrate should be strong enough to resist the traction force exerted by the adhered cell. Otherwise, for an extremely soft scaffold, even when the mechanical integrity is maintained under external loading, the soft scaffold cannot allow a balance to be established between the cell traction forces and the resistance of scaffolds to these forces, which is necessary for the assembly of functional cell-matrix adhesion complexes, actin cytoskeleton and cell spreading [67]. Consequently, cells are not able to form detectable cell-material adhesion complexes as well as actin cytoskeleton, and thus remain rounded, non-spread and become non-viable [78]. The cell traction force to be balanced by the scaffolds is dependent on cell lineage. Osteoblasts require higher while neurons require much lower scaffold stiffness to resist their traction force [77]. This determines the dependence of cell responses on the mechanical properties of scaffolds. Chatterjee et al. [79] investigated the effects of scaffold modulus on the mineralization of osteoblasts by fabricating a poly(ethylene glycol) hydrogel with gradient compressive modulus spanning from $\approx$10 kPa to $\approx$300 kPa. It was found that gel moduli of $\approx$225 kPa and higher enhanced osteogenesis compared to those softer gels and gel stiffness of scaffolds could be leveraged to induce cell differentiation in 3D culture as an alternative to biochemical cues such as soluble supplements, immobilized biomolecules and vectors.

5.4.1.3 Guaranteeing Larger Pore Sizes and Higher Porosity for a Specific Application

It is well-known that larger pore size and higher porosity are good for nutrients delivery and wastes removal and thus for cell migration and tissue ingrowth [80, 81]. For instance, pore sizes greater than 500 μm are required for rapid vascularization and for the survival of transplanted cells in fibrovascular tissues [82]. Macroporous scaffolds with 150–350 μm pore sizes were seen with optimum osteoblast adhesion, proliferation and bone formation [83, 84]. On the other hand, nanofiber scaffolds, which have been developed to mimic the topographical structure of extracellular matrix, have nanosized interfibrillar pores, which disable efficient ingrowth of cells into deeper layers of the nanofibrillar support [85]. More researches involving the contributions of pore sizes on cell responses have been recently reviewed by Perez and Mestres [46]. Despite the positive contribution of larger pore size and higher porosity, they are commonly accompanied by weakened mechanical properties. Balancing the mechanical properties and pore structures have been being the topic of scaffolds design and fabrication. Nanofibers and nanotubes can reinforce the mechanical properties of the composite scaffolds to a large extent when even added with a small amount [86, 87]. The strong mechanical properties of the reinforced scaffolds provide more possibility to guarantee that the scaffolds meet the mechanical requirement for a specific application even with larger pore sizes and higher porosity. Correspondingly, tissue ingrowth and regeneration may be accelerated. This might be another mechanism for the improved biocompatibility of nanofibers- or nanotubes-reinforced scaffolds by enhanced mechanical properties. More description about the effect of pore structures on cell/tissue responses will be detailed in Sect. 5.5.

5.4.2 Experimental Evidences

There have no researches directly investigating the relationship between the mechanical properties of the nanofibers/nanotubes-reinforced scaffolds and cell/tissue responses. However, many studies examined both the mechanical properties and the cell/tissue responses of these scaffolds. A careful classification of these studies may find some relationship between the mechanical properties and the cell/tissue responses.

Carbon nanotubes (CNTs) are featured with high mechanical properties so that they have been widely employed as reinforcement fillers in scaffolds fabrication, including both ceramics and polymers scaffolds. Several researches consistently found that CNTs reinforced scaffolds can improve both the mechanical properties and proliferation or differentiation capabilities of cells.

Ogihara et al. [88] used CNTs to reinforce an alumina composite for bone tissue replacement. The CNTs/alumina composite demonstrated a 120% increase in fracture toughness. Meanwhile, the CNTs/alumina composite showed superior proliferative ability for osteoblasts, chondrocytes, smooth muscle cells and fibroblasts when compared to noncomposite alumina. Further in vivo implantation into 15 week-old Japanese white rabbits found that the CNTs/alumina composite demonstrated direct integration into the bone and a comparable or lessened inflammatory response to that of the alumina control. Chen et al. [86] found that the elastic modulus and compressive strength of chitosan-multiwalled carbon nanotubes/hydroxyapatite (CS-MWNTs/HA) nanocomposites increased sharply from 509.9 to 1089.1 MPa and from 33.2 to 105.5 MPa with an increase of multiwalled carbon/chitosan weight ratios from 0 to 5%. As a result, preosteoblast MC3T3-E1 cells cultured on the CS-MWNTs/HA composites spread out well and the cell proliferation on the composites was observed to be higher than that on the CS-HA composites. Similar positive correlation between the mechanical properties and the cell/tissue responses were observed for CNTs reinforced polymeric scaffolds. Jell et al. [89] added CNTs into polyurethane (PU) scaffolds and found that both the compressive strength and the osteoblasts proliferative ability of the resulting CNT/PU scaffolds were increased. Lin et al. [90] produced MWCNTs-reinforced poly(lactic acid-co-glycolic acid) (PLGA) scaffolds and found that the inclusion of MWCNTs into PLGA scaffolds yielded a 500% increase in the stiffness and a 270% increase in the tensile strength. The proliferative potential was superior to that of pure PLGA scaffolds as well. Similarly, MWCNTs reinforced PLLA scaffolds (PLLA/MWCNTs) were observed to be more beneficial to the viability of osteoblasts than pure PLLA scaffolds [91]. In vivo test by Sitharaman et al. [92] demonstrated that the inclusion of short (20–80 nm) SWNTs into poly(propylene fumarate) (PPF) scaffolds resulted in 300% increase in bone area as observed in micro-CT analysis and histological analysis. The histological scoring at 12 weeks was also significantly higher than pure PPF scaffolds, indicating higher tissue quality, greater tissue ingrowth, reduced inflammatory cell density, increased connective tissue organization and a greater bone volume. CNTs are also widely employed to reinforce natural polymer scaffolds. Collagen has excellent biocompatibility due to its natural origin. Hirata et al. [93] found that, when coated collagen scaffolds with MWCNT, the resulting composite scaffolds demonstrated earlier osteoblastic differentiation than pure collagen scaffolds based on data of alkaline phosphatase activity and calcium and osteopontin contents. Moreover, the CNT/collagen scaffolds showed a significant increase in bone formation surrounding the scaffolds at both 4 and 8 weeks. Chitosan, as another widely employed natural polymer in biomedical fields, was reinforced with MWCNT by Venkatesan et al. via freezing and lyophilization method [94]. The obtained MWCNT/ chitosan scaffolds showed higher osteoblasts proliferation, alkaline phosphatase and mineralization than pure chitosan scaffolds.

In addition to carbon-based nanotubes, boron nitride nanotubes (BNNTs), as an important noncarbonic nanotubes, can be employed as reinforcement fillers of biomedical scaffolds as well due to their nontoxicity to various cells including osteoblasts, macrophages, human embryonic kidney cells and human neuroblastoma cells [45]. Lahiri et al. [95] prepared BNNTs reinforced polylactide-polycaprolactone (PLC) copolymer composite and investigated the cytocompatibility with osteoblasts and macrophages *in vitro*. The results show that the BNNTs addition to PLC enhanced the tensile strength and also resulted in an increase in osteoblasts viability and differentiation. BNNTs were employed to reinforce hydroxyapatite (HA) as well. Addition of 4 wt% BNNTs resulted in 120% increment in elastic modulus, 129% higher hardness, and 86% more fracture toughness as compared to HA [96]. Correspondingly, the proliferation and viability of osteoblast cells of BNNTs/HA scaffolds are better than pure HA ones.

Scaffolds reinforced by fibers were observed with enhanced mechanical properties and improved biocompatibility as well. Li et al. [97] reinforced poly(L-lactic acid) (PLLA) porous scaffolds by chitin fibers of 12 μm in diameter, and found that the compressive strength of the resulting composite scaffolds was 4 times higher than the pure PLLA scaffolds. Correspondingly, the reinforced scaffold demonstrated significantly better attachment, proliferation, differentiation, and mineralization of osteoblasts than pure PLLA scaffolds [98]. Being similar to CNTs, carbon nanofibers (CNFs) possess unique mechanical properties as well. Compared with the CNTs, some special characteristics are featured by CNFs such as larger diameter (60–150 nm) and longer length (30–100 μm), and different surface morphology [99], which render them more suitable as a reinforcement agent. Rajzer et al. [100] produced CNFs-reinforced HA fibrous scaffolds for bone tissue engineering. It was demonstrated that the composite scaffolds could establish direct chemical bonds with bone tissue after implantation and the mineralization activity was drastically increased.

5.4.3 Brief Comments

All together, the mechanical properties of scaffolds play crucial roles in scaffolds biocompatibility. The selection of scaffold mechanical properties should take consideration of both the local mechanical properties which is directly sensed by cells and the bulk mechanical properties which is responsible for the transmission of the surrounding mechanical loadings to cells. For scaffolds with identical chemical compositions, the architecture of scaffolds affects both the mechanical properties of scaffolds and the cell/tissue responses. Therefore, the exemplified cell/tissue responses in Sect. 5.4.2 should not be considered as the sole contributions from the mechanical properties of the fibers/tubes-reinforced scaffolds. The core objective of Sect. 5.4 is to provoke concerns for the regulation roles of mechanical properties in 3D scaffold systems to cell/tissue responses and for purposefully tailoring the mechanical properties of scaffolds to induce sound biocompatibility for specific applications.

5.5 Regulation of Cell/Tissue Responses by Architecture of Scaffolds

Scaffolds for tissue engineering should be porous and permeable to permit the ingress of cells, nutrients and oxygen diffusion, as well as waste removal. Porosity and pore size of scaffolds play critical roles in tissue formation, which have been extensively discussed elsewhere [74, 101–104]. Generally, the scaffolds should possess certain characteristic including high porosity with an interconnected pore structures for efficient cell ingrowth. In addition, a minimum pore size is required, which is correlated with cell size. For example, the minimum pore size is considered to be about 100 μm for the migration of osteogenetic cells into the pores [102], whereas pore sizes greater than 500 μm are beneficial for rapid vascularization and survival of transplanted cells in fibrovascular tissues [82]. Satisfaction of the minimum pore size together with mechanical integrity requires sufficient mechanical properties of the scaffolds.

As mentioned in Sect. 5.4.1.3, the high mechanical properties of the reinforced scaffolds may augment the porosity and thus provide microstructure with interconnected pores and extremely large surface-area-to-volume ratio. This is advantageous to cell ingrowth, proliferation, and tissue regeneration [98, 105–107]. Moreover, the high mechanical properties may balance the mechanical function with cell ingrowth, thus providing a sequential transition in which the regenerated tissue assumes function as the scaffold degrades [103].

In addition to the pore structure, the nanoscale characteristics of the incorporated nanofibers or nanotubes positively contribute to the cell adhesion and proliferation [108]. Several studies have revealed that the incorporation of nanofibers or nanotubes into scaffolds facilitates cell attachment and cell proliferation [94, 106, 109]. For instance, Woo et al. [110] found that the nano-fibrous architecture built in three-dimensional scaffolds increased cell adhesion by two-fold. It is widely accepted that macro-, micro- and nano-sized topographical factors stimulate behavioral changes in both cells and tissues [78]. However, the nanostructures elicit a higher degree of biological plasticity compared with micro- or macro- structures. For example, osteoblasts cultured on a nanograined Al_2O_3 and TiO_2 substrate (grain sizes less than 100 nm) were found with obvious improvement in adhesion, proliferation and mineralization compared with those cultured on a micrograined surface [111, 112]. The fate of hMSCs could be controlled only by the nanotopography of culture substrates [113]. Specifically, small TiO_2 nanotubes ($\approx$30 nm diameter) promoted hMSC adhesion without noticeable differentiation, whereas larger ($\approx$ 70 to 100 nm diameter) nanotubes elicited a dramatic stem cell elongation ($\approx$ 10-fold increased), which induced cytoskeletal stress and selective differentiation into osteoblast-like cells. One important factor responsible for the increased adhesion, growth and maturation of cells in the reinforced scaffolds is the surface-area-to-volume ratio. Incorporation of nanofibers/nanotubes may increase the surface-to-volume ratio of scaffolds [45], which provides more room for cell adhesion and survival. Similarly, more room is provided for protein adsorption and ECM deposition, resulting in

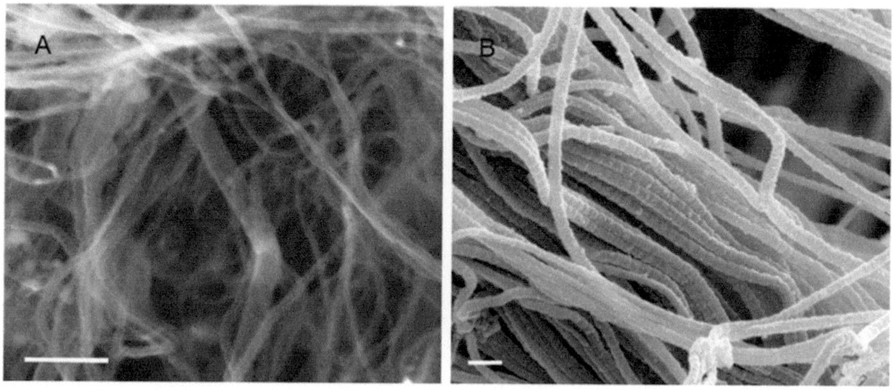

Fig. 5.11 Scanning electron micrographs demonstrating the morphological similarities between (**a**) nitrogen-doped-MWNTs and (**b**) collagen, both structures are long and rod-like with a periodic structure along their long axis, scale bars: 200 nm [37]

more ligands for cell adhesion (see Sect. 5.3). In addition, the organization of endogeneous ECM protein may be affected. Lee et al. [114] once found that the endogenous ECM produced by human ACL fibroblast organized along the fibers. Another factor for the improved cell adhesion and growth is the morphological similarities of nanofibers and nanotubes to the natural ECM, thus providing cells a more *in vivo*-like environment [115]. For instance, CNTs can be synthesized with dimensions comparable to the proteins of ECM (Fig. 5.11), forming a fibrillar/ECM-protein mimicking morphology, thus stimulating and strengthening integrin mediated adhesions between CNTs and cells [37]. The embedded fibers in nanofiber-reinforced hydrogel may form ECM-like microstructures and thus provide more binding sites for cell adhesion and proliferation, which was observed to significantly accelerate the proliferation of hMSCs [106].

In addition to facilitating cell adhesion and proliferation, nanofibers/nanotubes were observed to regulate cell morphology and orientation [116]. For example, fibroblast cells on gelatin/PCL nanofiber composites containing unidirectional nanofibers clearly orientated along the direction of the aligned nanofibers; on the other hand, cells on composites containing bidirectional nanofibers took an orientation in the criss-crossed pattern of the fibers (Fig. 5.12) [117]. The same phenomenon was observed on the case of Keratinocytes grown for 48 h in the presence of parallel or rectangularly arranged polyvinyl alcohol (PVA) nanofibers [74]. Previous study has also found that cell elongation resulted from the nanofibers in scaffolds could force hBMSCs into an osteogenic morphology [73]. Specially, hBMSCs cultured on PCL nanofibers had smaller cell area, higher aspect ratio, lower roundness and more branching than hBMSCs on PCL films. This spindly, branched morphology was similar to hBMSCs cultured on PCL films in the presence of osteogenic supplements, suggesting that the nanofibers could drive osteogenic differentiation of hBMSCs via controlling cell shape (Fig. 5.13). Linear cell orientation on nanofibers should be contributed to contact-cell guidance. According to Albuschies's

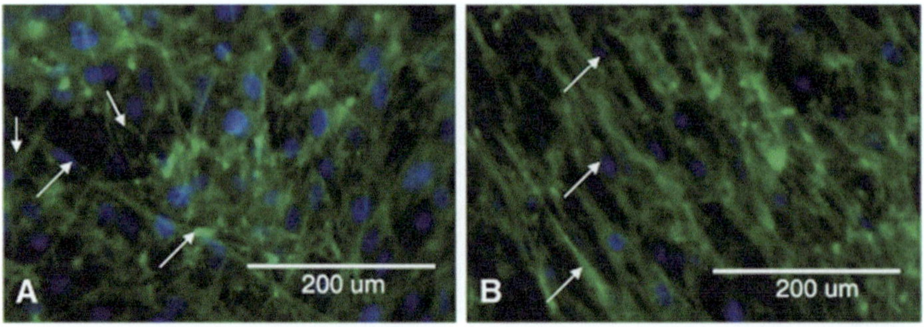

Fig. 5.12 3 T3 fibroblasts grown on gelatin/PCL nanofiber composites with (**a**) bi-directional and (**b**) unidirectional fiber orientations (*white arrows* show orientation of cells) [117]

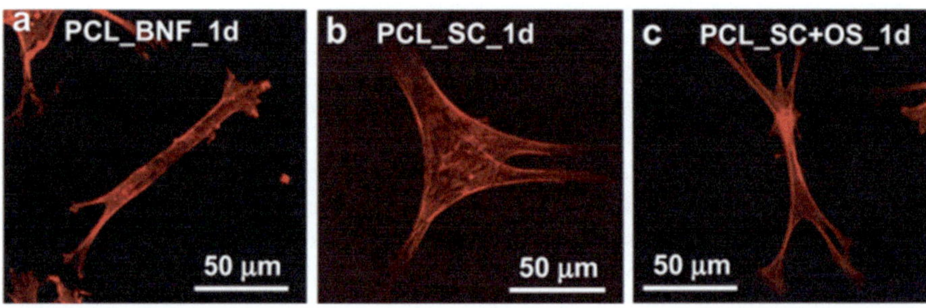

Fig. 5.13 hBMSC morphology (400×) after 1d culture for (**a**) PCL _BNF (PCL nanofibers), (**b**) PCL_SC (PCL films) and (**c**) PCL_SC + OS (PCL films in the presence of osteogenic supplements). Actin is *red* (AlexaFluor 546 phalloidin) and projections of confocal z-stacks are shown [73]

research, when cells cultured on flat versus nanofibrillar interfaces, transient filopodia are initially ($t < 1$ min) in contact with both flat surfaces and nanowires; however, as time passes ($t \approx 5$ min), the transient filopodia adhere and align with individual nanowires and quickly direct cell spreading towards nanowire adhesion sites, while most filopodia retract from the flat glass surfaces (Fig. 5.14) [118]. The retraction of filopodia on flat surface and the formation of aligned filopodia on nanowires indicate that the interaction between cells and substrates is influenced by nanostructures on substrates. The aligned filopodia formation can explain why cells adopt an orientated morphology when they grow on aligned nanofibers. It should be noted that cell alignment due to fiber orientation is dependent on the fiber diameter and cell type. For example, Liu et al. [119] found that a critical minimum diameter of 0.97 μm exists for cell orientation to occur. At lower fiber diameter, no big difference in aspect ratio was observed relative to the control samples on film. On the contrary, for endothelial cells, aligned fiber diameter over 1.2 μm seems to have less influence on cell orientation. Using electrospun aligned polycaprolactone (PCL)/collagen fibers with different fiber diameters (100 nm, 300 nm and 1200 nm), Whited and Rylander [120] found that the fiber-directed cell orientation was signifi-

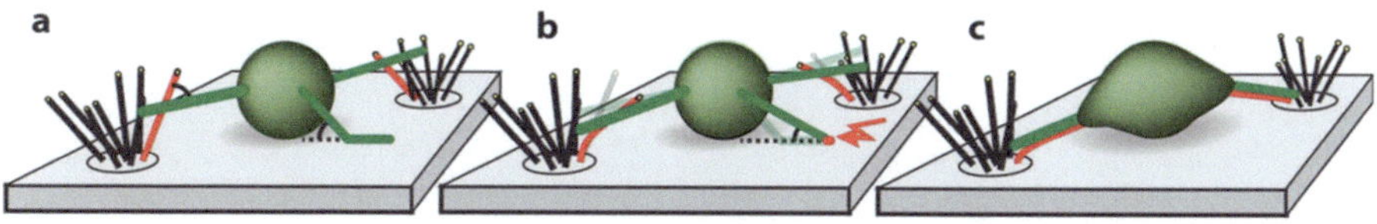

Fig. 5.14 Schematics of the topography recognizing function of filopodia on flat surface and nanowires (NMs). (**a**) filopodia initiate the very first substrate contacts to both, NWs and flat surfaces at a time where no ruffles and no lamellipodia were present; (**b–c**) as time progresses, most of the initial filopodia peeled off from flat surfaces, while the filopodia were able to pull the NWs into alignment thereby increasing the NW-filopodia contact areas. [118]

cantly greater for primary human umbilical vein endothelial cells (HUVEC) on 100- and 300-nm scaffolds as compared to 1200 nm scaffolds.

Not only cell orientation but also cell elongation was observed on nanofibers, especially on aligned fibers. For example, the length of keratinocytes grown on a flat glass surface is around 30 mm, but in the case of aligned nanofibers the cells reach 80 mm or more in length. Besides the whole cell morphology, the cell nuclei also adopted a distinctly elongated shape [74]. Similar linear cell orientation and elongation guided by nanofibers were observed in the case of stem cells [73]. The morphological similarity between nanofibers and ECM, as well as the organized ECM along the fibers should be responsible for cell elongation on the aligned nanofibers. The phenomena of cell orientation and elongation along nanofibers indicate the potential applications of nanofibers reinforced scaffolds in promoting the formation of various artificial tissues or organs, such as blood vessel, tendon, nerve, etc., in which cell orientation and elongation are crucial for their performance.

As indicated in Sect. 5.4.2, nanofibers and nanotubes reinforced cell proliferation and growth are better than their unreinforced counterparts, one contributor of which is the enhanced mechanical properties. Beside, the elongated cell morphology on nanofibers might be another contributor. Cell elongation is accompanied by a high degree of cytoskeleton organization parallel to the direction of fiber alignment, which results in greater cytoskeletal stiffness [85, 120, 121]. Thereby, the cell-matrix interface force (ligand-receptor binding) increases, which further promotes the assembly of focal adhesions (FAs). It has been found for a long time that FAs function as a node on which a complex signaling network is assembled and regulated [122]. Specifically, FA assembly and signaling result in the phosphorylation of FA kinase (FAK). FAK appears to be a key component of the mechanotransduction apparatus, and both FAK and its interacting proteins are candidate transducers (Integrins in mechanotransduction), which can activate pathways known to promote cell proliferation, such as Shc pathway and MAPK pathway [123, 124]. It should be reminded that FAs and cytoskeletal stiffness are the mechanotransduction mechanics by which the external mechanical loadings are transmitted to cells and the mechanical properties of the scaffolds are perceived by cells (Fig. 5.15) (see Sect. 5.4). Besides, the mechanical properties of the scaffolds are influenced by the scaffold architecture as well. Therefore, the observed cell/tissue responses on nanofibers/nanotubes reinforced scaffolds are more often the combinational or syner-

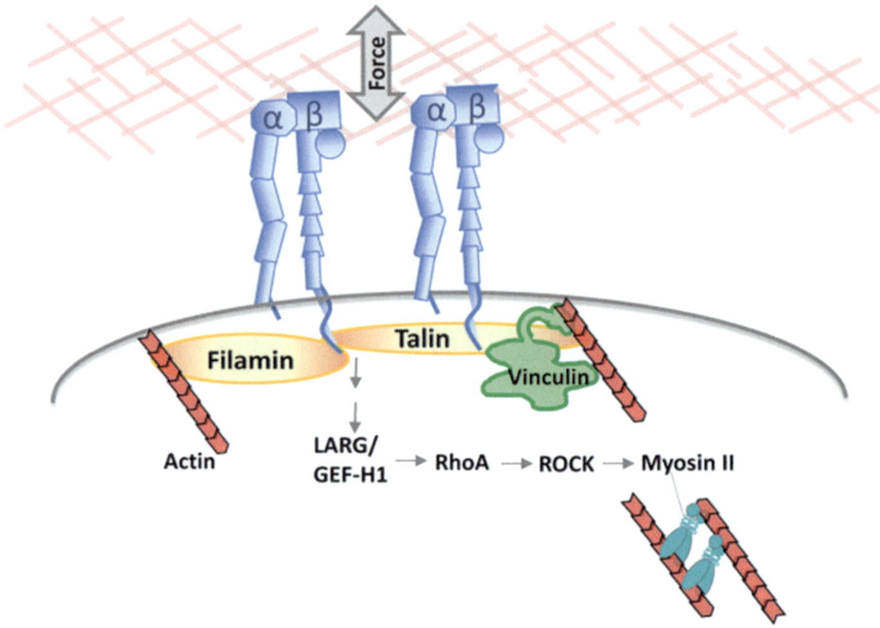

Fig. 5.15 Intergrin-mediated force transmission. Integrins (*blue*) are heterodimers of α and β subunits that form bonds with the extracellular matrix on the outside of the cell. In response to force, several proteins are recruited to the integrin cytoplasmic β domain. Critical proteins among these are talin and filamin. Talin, in turn, recruits vinculin, a prominent mechanotransducing protein [125]

getic results from the mechanical and architectural and even chemical properties of the scaffolds. Understanding of their interplays helps to take advantage of their positive contributes to cell/tissue responses and further to design desirable scaffolds for specific application.

5.6 Regulation of Cell/Tissue Responses by Biochemical Cues of Scaffolds

Since the reinforcing nanofibers or nanotubes are different with the bulk scaffolds in chemical compositions, incorporation of nanofibers or nanotubes into the scaffolds may change the chemical cues provided to cells on scaffolds. As described in Sect. 5.3, the chemical composition of a material will affect the type, quantity and activity of the adsorbed proteins, which may subsequently regulate the cell-material interactions. For instance, Allen et al. showed differential gene expression for several cell types on various chemistries [126]. Our previous studies indicated that material chemistry regulates the sensitivity of osteoblasts to various magnitudes of fluid shear stress and the potential mechanism was that the surface chemistry influences the cell adhesion and morphology which further regulate the perceiving and responding of osteoblasts to the fluid shear stress [68, 70]. For 3D scaffolds such as poly(lactic acid) (PLA), its strong hydrophobicity due to the existence of CH_3 side chains and ester backbone may inhibit cell adhesion and proliferation. Incorporation of hydrophilic cellulose nanofibers into PLA scaffolds was found to facilitate proteins adsorption in an appropriate geometrical orientation and thus promote cell adhesion [78, 127]. Similarly, incorporation of collagen-derived biopolymer gelatin (Gt) into PCL scaffolds can also increase the hydrophilicity of PCL scaffolds and further promote cell proliferation and infiltration into the scaffolds [128]. For nanofibers lacking in hydrophilic chemistries, surface modification could be used to optimize the nanofibers [129]. For example, Chuan et al. found that surface-aminated electrospun polyethersulfone nanofibers could enhance adhesion and expansion of human umbilical cord blood hematopoietic stem/progenitor cells [130]. In addition to the modulation of surface hydrophilicity, incorporation of biocompatible nanofibers into scaffolds can also provide recognition sites for cells and thus enhance biological affinity of the resulting composite scaffolds. For instance, it has been reported that chitosan can promote adhesion and functional expression of osteoblasts because of its similarity to glycosaminoglycan in structure [107, 131]. Besides, Grafahrend et al. [132] designed a type of functionalized PLGA nanofibers with both GRGDS adhesive peptide and star-shaped poly(ethylene oxide), the latter of which is a polymer capable of suppress non-specific protein adsorption. The resulting nanofibers promoted specific bioactivation and enabled adhesion of cells through exclusive recognition of the immobilized binding motifs. Similarly, potential cytotoxic effects associated with nanofibers such as carbon nanotube may be mitigated by chemically functionalizing the surface [133]. Meanwhile, the functionalized carbon nanotubes generally demonstrate better dispersibility in water compared to their intact ones, which further extends the potential of nanotubes for biomedical applications [94].

5.7 Summary

Addition of fibers or tubes in scaffolds reinforces the mechanical properties and alters the pore structures of scaffolds. The characteristics of fibers and tubes including their size, alignment and surface morphology impart new physiochemical properties to the reinforced scaffolds. The altered pore structures and physiochemical properties should affect the status of the adsorbed proteins on the scaffolds. Though extensive studies have been conducted about the amount or selective adsorption of proteins on the reinforced scaffolds, few studies paid attention to the conformation of the adsorbed proteins, which plays more crucial roles in protein functions. Instead, the protein conformation has been more often investigated by using some model structures, such as pure nanotubes or nanofibers or controlled pores. Understanding the collective contributions from wettability, pores, and fiber/tube size and alignment helps to purposefully control protein adsorption so that appropriate cell and tissue responses can be promised. To this end, both experimental and simulation methods should be required for this complex system. In this chapter, the summarized fundamentals on protein adsorption provide essential knowledge.

The mechanical properties of scaffolds are usually recognized to maintain the mechanical integrity of scaffolds. However, their significant roles in transmitting externally loadings and providing mechanical stimulation to cells/tissues are generally ignored in 3D scaffolds, though they have been widely studied and acknowledged in 2D systems and some 3D hydrogels. It should be reminded that one of the most important purposes for addition of fibers and tubes into scaffolds is to enhance the mechanical properties of the scaffolds. Therefore, more attentions should be paid to their contributions to cell/tissue responses. There is a large bank of literatures mentioning the enhanced mechanical properties and improved cell/tissue responses. Few of them, however, claimed their direct correlation. This chapter tried to find a certain correlation by providing several potential mechanisms for the mechanical properties to regulate cell/tissue responses. These mechanisms may be conductive to understand and utilize the mechanical properties of the fibers/tubes reinforced scaffolds. More importantly, this chapter tried to provoke concerns from biomaterials scientists for the regulation roles of mechanical properties in 3D scaffold systems to cell/tissue responses and for purposefully tailoring the mechanical properties of scaffolds to induce sound biocompatibility for specific applications.

The addition of fibers or tubes changes both the architecture and mechanical properties of scaffolds. Moreover, the latter is dependent on the former. In view of the fact that both the architecture and the mechanical properties influence cell/tissue responses, it is hard to discriminate their contributions. Fortunately, the mechanical properties and architecture seem to share common mechanics, namely focal adhesions and cytoskeletons, to transfer their effects on cell/tissue. This can be employed to harmonize them so as to induce sound biocompatibility for specific applications.

References

1. Williams DF. Definitions in biomaterials: proceedings of a consensus conference of the European Society for Biomaterials, Chester, England, March 3–5, 1986. Amsterdan: European Society for Biomaterials, Elsevier; 1987.
2. Salmerón-Sánchez M, Rico P, Moratal D, Lee TT, Schwarzbauer JE, García AJ. Role of material-driven fibronectin fibrillogenesis in cell differentiation. Biomaterials. 2011;32:2099–105.
3. Barrias C, Martins MPG, Barbosa M, Granja P. The correlation between the adsorption of adhesive proteins and cell behaviour on hydroxyl-methyl mixed self-assembled monolayers. Biomaterials. 2009;30:307–16.
4. Allen LT, Tosetto M, Miller IS, Connor DP O, Penney SC, Lynch I, Keenan AK, Pennington SR, Dawson KA, Gallagher WM. Surface-induced changes in protein adsorption and implications for cellular phenotypic responses to surface interaction. Biomaterials. 2006;27:3096–108.

5. González-García C, Sousa SR, Moratal D, Rico P, Salmerón-Sánchez M. Effect of nanoscale topography on fibronectin adsorption, focal adhesion size and matrix organisation. Colloids Surf B: Biointerfaces. 2010;77:181–90.
6. Anderson JM. Biological responses to materials. Annu Rev Mater Sci. 2003;31:81–110.
7. Keselowsky BG, Collard DM, García AJ. Surface chemistry modulates fibronectin conformation and directs integrin binding and specificity to control cell adhesion. J Biomed Mater Res A. 2003;66:247–59.
8. Llopishernández V, Rico P, Ballesterbeltrán J, Moratal D, Salmerónsánchez M. Role of surface chemistry in protein remodeling at the cell-material interface. PLoS One. 2011;6:e19610.
9. Geiger B, Bershadsky A, Pankov R, Yamada KM. Transmembrane crosstalk between the extracellular matrix-cytoskeleton crosstalk. Nat Rev Mol Cell Biol. 2001;2:793–805.
10. Pankov R, Cukierman E, Katz BZ, Matsumoto K, Lin DC, Lin S, Hahn C, Yamada KM. Integrin dynamics and matrix assembly: Tensin-dependent translocation of $\alpha 5\beta 1$ Integrins promotes early Fibronectin fibrillogenesis. J Cell Biol. 2000;148:1075–90.
11. Balaban NQ, Schwarz US, Riveline D, Goichberg P, Tzur G, Sabanay I, Mahalu D, Safran S, Bershadsky A, Addadi L. Force and focal adhesion assembly: a close relationship studied using elastic micropatterned substrates. Nat Cell Biol. 2001;3:466.
12. Goffin JM, Pittet P, Csucs G, Lussi JW, Meister JJ, Hinz B. Focal adhesion size controls tension-dependent recruitment of alpha-smooth muscle actin to stress fibers. J Cell Biol. 2006;172:259–68.
13. Choquet D, Dan PF, Sheetz MP. Extracellular matrix rigidity causes strengthening of integrin-cytoskeleton linkages. Cell. 1997;88:39–48.
14. Pelham R, Wang Y. Cell locomotion and focal adhesions are regulated by substrate flexibility. Proc Natl Acad Sci U S A. 1997;94:13661–5.
15. Katz BZ, Zamir E, Bershadsky A, Kam Z, Yamada KM, Geiger B. Physical state of the extracellular matrix regulates the structure and molecular composition of cell-matrix adhesions. Mol Biol Cell. 2000;11:1047–60.
16. Altankov G, Groth T. Reorganization of substratum-bound fibronectin on hydrophilic and hydrophobic materials is related to biocompatibility. J Mater Sci Mater Med. 1994;5:732–7.
17. Lin M, Wang H, Ruan C, Xing J, Wang J, Li Y, Wang Y, Luo Y. Adsorption force of fibronectin on various surface chemistries and its vital role in osteoblast adhesion. Biomacromolecules. 2015;16:973–84.
18. Asuri P, Bale SS, Karajanagi SS, Kane RS. The protein-nanomaterial interface. Curr Opin Biotechnol. 2007;17:562–8.
19. Roach P, Farrar D, Perry CC. Surface tailoring for controlled protein adsorption: effect of topography at the nanometer scale and chemistry. J Am Chem Soc. 2006;128:3939–45.
20. Karajanagi SS, Vertegel AA, Kane RS, Dordick JS. Structure and function of enzymes adsorbed onto single-walled carbon nanotubes. Langmuir. 2005;20:11594–9.
21. Vlasova I, Mikhalchik E, Barinov N, Kostevich V, Smolina N, Klinov D, Sokolov A. Adsorbed plasma proteins modulate the effects of single-walled carbon nanotubes on neutrophils in blood. Nanomed: Nanotechnol, Biol Medicine. 2016;12:1615–25.
22. Müller C, Lüders A, Hothhannig W, Hannig M, Ziegler C. Initial bioadhesion on dental materials as a function of contact time, pH, surface wettability, and isoelectric point. Langmuir. 2010;26:4136–41.
23. Vansteenkiste S, Corneillie S, Schacht E, Chen X, Davies M, Moens M, Van Vaeck L. Direct measurement of protein adhesion at biomaterial surfaces by scanning force microscopy. Langmuir. 2000;16:3330–6.
24. Sagvolden G, Giaever I, Feder J. Characteristic protein adhesion forces on glass and polystyrene substrates by atomic force microscopy. Langmuir. 1998;14:5984–7.
25. Michael KE. Adsorption-induced conformational changes in fibronectin due to interactions with well-defined surface chemistries. Nat Biotechnol. 2003;19:1335–6.
26. Sankar D, Shalumon KT, Chennazhi KP, Menon D, Jayakumar R. Surface plasma treatment of poly(caprolactone) micro, nano, and multiscale fibrous scaffolds for enhanced osteoconductivity. Tissue Eng A. 2014;20:1689–702.

27. Li L, Qian Y, Jiang C, Lv Y, Liu W, Zhong L, Cai K, Li S, Yang L. The use of hyaluronan to regulate protein adsorption and cell infiltration in nanofibrous scaffolds. Biomaterials. 2012;33:3428–45.
28. Shalumon KT, Anulekha KH, Chennazhi KP, Tamura H, Nair SV, Jayakumar R. Fabrication of chitosan/poly(caprolactone) nanofibrous scaffold for bone and skin tissue engineering. Int J Biol Macromol. 2011;48:571–6.
29. Jose MV, Thomas V, Xu Y, Bellis S, Nyairo E, Dean D. Aligned bioactive multi-component nanofibrous nanocomposite scaffolds for bone tissue engineering. Macromol Biosci. 2010;10:433–44.
30. Khang D, Kim SSP, Palmore G, Durbin S, Webster T. Enhanced fibronectin adsorption on carbon nanotube/poly(carbonate) urethane: independent role of surface nano-roughness and associated surface energy. Biomaterials. 2007;28:4756–68.
31. Binulal NS, Deepthy M, Selvamurugan N, Shalumon KT, Suja S, Mony U, Jayakumar R, Nair SV. Role of nanofibrous poly(caprolactone) scaffolds in human mesenchymal stem cell attachment and spreading for in vitro bone tissue engineering--response to osteogenic regulators. Tissue Eng A. 2010;16:393–404.
32. Vertegel AA, Siegel RW, Dordick JS. Silica nanoparticle size influences the structure and enzymatic activity of adsorbed lysozyme. Langmuir. 2004;20:6800–7.
33. Lord MS, Foss M, Besenbacher F. Influence of nanoscale surface topography on protein adsorption and cellular response. Nano Today. 2010;5:66–78.
34. Li X, Liu H, Niu X, Yu B, Fan Y, Feng Q, Cui FZ, Watari F. The use of carbon nanotubes to induce osteogenic differentiation of human adipose-derived MSCs in vitro and ectopic bone formation in vivo. Biomaterials. 2012;33:4818–27.
35. Akasaka T, Yokoyama A, Matsuoka M, Hashimoto T, Watari F. Thin films of single-walled carbon nanotubes promote human osteoblastic cells (Saos-2) proliferation in low serum concentrations. Mater Sci Eng C. 2010;30:391–9.
36. Firkowska I, Godehardt E, Giersig M. Interaction between human osteoblast cells and inorganic two-dimensional scaffolds based on multiwalled carbon nanotubes: a quantitative AFM study. Adv Funct Mater. 2008;18:3765–71.
37. Newman P, Minett A, Ellis-Behnke R, Zreiqat H. Carbon nanotubes: their potential and pitfalls for bone tissue regeneration and engineering. Nanomed Nanotechnol Biol Med. 2013;9:1139–58.
38. Goldberg-Oppenheimer P, Regev O. Exploring a nanotube dispersion mechanism with gold-labeled proteins via cryo-TEM imaging. Small. 2007;3:1894–9.
39. Zhong J, Song L, Meng J, Gao B, Chu W, Xu H, Luo Y, Guo J, Marcelli A, Xie S. Bio-nano interaction of proteins adsorbed on single-walled carbon nanotubes. Carbon. 2009;47:967–73.
40. Shen JW, Wu T, Wang Q, Kang Y. Induced stepwise conformational change of human serum albumin on carbon nanotube surfaces. Biomaterials. 2008;29:3847–55.
41. Yang DZ, Chen AZ, Wang SB, Li Y, Tang XL, Wu YJ. Preparation of poly(L-lactic acid) nanofiber scaffolds with a rough surface by phase inversion using supercritical carbon dioxide. Biomed Mater. 2015;10:35015–25.
42. Panahi-Joo Y, Karkhaneh A, Nourinia A, Abd-Emami B, Negahdari B, Renaud P, Bonakdar S. Design and fabrication of a nanofibrous polycaprolactone tubular nerve guide for peripheral nerve tissue engineering using a two-pole electrospinning system. Biomed Mater. 2016;11:25017.
43. Mukhatyar VJ, Salmerón-Sánchez M, Rudra S, Mukhopadaya S, Barker TH, García AJ, Bellamkonda RV. Role of fibronectin in topographical guidance of neurite extension on electrospun fibers. Biomaterials. 2011;32:3958–68.
44. Elter P, Lange R, Beck U. Electrostatic and dispersion interactions during protein adsorption on topographic nanostructures. Langmuir The ACS J Surf Colloids. 2011;27:8767–75.
45. Wang W, Liao S, Liu M, Zhao Q, Zhu Y. Polymer composites reinforced by nanotubes as scaffolds for tissue engineering. Int J Polym Sci. 2014;2014:1–14.

46. Perez RA, Mestres G. Role of pore size and morphology in musculo-skeletal tissue regeneration. Mater Sci Eng C Mater Biol Appl. 2015;61:922–39.
47. Ma H, Xue L. Carbon nanotubes reinforced poly(L-lactide) scaffolds fabricated by thermally induced phase separation. Nanotechnology. 2015;26:25701.
48. Wei G, Ma PX. Structure and properties of nano-hydroxyapatite/polymer composite scaffolds for bone tissue engineering. Biomaterials. 2004;25:4749–57.
49. Leong MF, Chian KS, Mhaisalkar PS, Ong WF, Ratner BD. Effect of electrospun poly(D,L-lactide) fibrous scaffold with nanoporous surface on attachment of porcine esophageal epithelial cells and protein adsorption. J Biomed Mater Res A. 2009;89:1040–8.
50. White AA, Best SM, Kinloch IA. Hydroxyapatite-carbon nanotube composites for biomedical applications: a review. Int J Appl Ceram Technol. 2006;4:1–13.
51. Bhattarai SR, Aryal S, Remant BKC, Bhattarai N, Hwang PH, Yi HK, Kim HY. Carbon nanotube-hydroxyapatite nanocomposite for DNA complexation. Mater Sci Eng C. 2008;28:64–9.
52. Ding Y, Liu J, Jin X, Lu H, Shen G, Yu R. Poly-L-lysine/hydroxyapatite/carbon nanotube hybrid nanocomposite applied for piezoelectric immunoassay of carbohydrate antigen 19-9. Analyst. 2008;133:184–90.
53. Hahn BD, Lee JM, Park DS, Choi JJ, Ryu J, Yoon WH, Lee BK, Shin DS, Kim HE. Mechanical and in vitro biological performances of hydroxyapatite–carbon nanotube composite coatings deposited on Ti by aerosol deposition. Acta Biomater. 2009;5:3205–14.
54. Singh MK, Shokuhfar T, Garmestani H, Ahzi S. Hydroxyapatite modified with carbon-nanotube-reinforced poly(methyl methacrylate): a nanocomposite material for biomedical applications. Adv Funct Mater. 2008;18:694–700.
55. Wahab R, Ansari SG, Kim YS, Mohanty TR, Hwang IH, Shin HS. Immobilization of DNA on nano-hydroxyapatite and their interaction with carbon nanotubes. Synth Met. 2009;159:238–45.
56. Zhao HY, Xu XX, Zhang JX, Zheng W, Zheng YF. Carbon nanotube-hydroxyapatite-hemoglobin nanocomposites with high bioelectrocatalytic activity. Bioelectrochemistry. 2010;78:124–9.
57. Sittichockechaiwut A, Scutt AM, Ryan AJ, Bonewald LF, Reilly GC. Use of rapidly mineralising osteoblasts and short periods of mechanical loading to accelerate matrix maturation in 3D scaffolds. Bone. 2009;44:822–9.
58. Rath B, Jin N, Knobloch TJ, Lannutti JJ, Agarwal S. Compressive forces induce osteogenic gene expression in calvarial osteoblasts. J Biomech. 2008;41:1095–103.
59. Gromova NB, Shelyubskii VI. Fluid flow increases mineralized matrix deposition in 3D perfusion culture of marrow stromal osteoblasts in a dose-dependent manner. Proc Natl Acad Sci U S A. 2002;99:12600–5.
60. Butler PJ, Norwich G, Weinbaum S, Chien S. Shear stress induces a time- and position-dependent increase in endothelial cell membrane fluidity. Am J Phys Cell Phys. 2001;280:C962–9.
61. Hahn C, Schwartz MA. Mechanotransduction in vascular physiology and atherogenesis. Nat Rev Mol Cell Biol. 2009;10:53–62.
62. White CR, Frangos JA. The shear stress of it all: the cell membrane and mechanochemical transduction. Philos Trans R Soc B Biol Sci. 2007;362:1459–67.
63. Hayakawa K, Tatsumi H, Sokabe M. Actin stress fibers transmit and focus force to activate mechanosensitive channels. J Cell Sci. 2008;121:496–503.
64. Ingber DE. Mechanosensation through integrins: cells act locally but think globally. Proc Natl Acad Sci. 2003;100:1472–4.
65. Myers KA, Rattner JB, Shrive NG, Hart DA. Osteoblast-like cells and fluid flow: cytoskeleton-dependent shear sensitivity. Biochem Biophys Res Commun. 2008;364:214–9.
66. Orr AW, Helmke BP, Blackman BR, Schwartz MA. Mechanisms of mechanotransduction. Dev Cell. 2006;10:11–20.

67. Wang N, Naruse K, Stamenovi D, Fredberg JJ, Mijailovich SM, Tolić-Nørrelykke IM, Polte T, Mannix R, Ingber DE. Mechanical behavior in living cells consistent with the tensegrity model. Proc Natl Acad Sci U S A. 2001;98:7765–70.
68. Xing J, Li Y, Lin M, Wang J, Wu J, Ma Y, Wang Y, Li Y, Luo Y. Surface chemistry modulates osteoblasts sensitivity to low fluid shear stress. J Biomed Mater Res A. 2014;102:4151–60.
69. Li Y, Luo Y, Xie Z, Xing J, Lin M, Yang L, Wang Y, Huang K. The optimal combination of substrate chemistry with physiological fluid shear stress. Colloids Surf B: Biointerfaces. 2013;112:51–60.
70. Li Y, Luo Y, Huang K, Xing J, Xie Z, Lin M, Yang L, Wang Y. The responses of osteoblasts to fluid shear stress depend on substrate chemistries. Arch Biochem Biophys. 2013;539:38–50.
71. Keselowsky BG, Collard DM, Garcia AJ. Surface chemistry modulates focal adhesion composition and signaling through changes in integrin binding. Biomaterials. 2004;25:5947–54.
72. Bottino MC, Thomas V, Janowski GM. A novel spatially designed and functionally graded electrospun membrane for periodontal regeneration. Acta Biomater. 2011;7:216–24.
73. Kumar G, Tison CK, Chatterjee K, Pine PS, Mcdaniel JH, Salit ML, Young MF, Simon CG. The determination of stem cell fate by 3D scaffold structures through the control of cell shape. Biomaterials. 2011;32:9188–96.
74. Pelipenko J, Kocbek P, Govedarica B, Rošic R, Baumgartner S, Kristl J. The topography of electrospun nanofibers and its impact on the growth and mobility of keratinocytes. Eur J Pharm Biopharm. 2012;84:401–11.
75. Yang JJ, Yu CH, Chang BS, Yeom JS, Lee JH, Lee CK. Subsidence and nonunion after anterior cervical interbody fusion using a stand-alone polyetheretherketone (PEEK) cage. Clin Orthop Surg. 2011;3:16–23.
76. Discher D, Janmey P, Wang Y. Tissue cells feel and respond to the stiffness of their substrate. Science. 2005;310:1139–43.
77. Engler AJ, Sen S, Sweeney HL, Discher DE. Matrix elasticity directs stem cell lineage specification. Cell. 2006;126:677–89.
78. Bacakova L, Filova E, Parizek M, Ruml T, Svorcik V. Modulation of cell adhesion, proliferation and differentiation on materials designed for body implants. Biotechnol Adv. 2011;29:739–67.
79. Chatterjee K, Lin-Gibson S, Wallace WE, Parekh SH, Lee YJ, Cicerone MT, Young MF, Simon CJ. The effect of 3D hydrogel scaffold modulus on osteoblast differentiation and mineralization revealed by combinatorial screening. Biomaterials. 2010;31:5051–62.
80. Bose S, Roy M, Bandyopadhyay A. Recent advances in bone tissue engineering scaffolds. Trends Biotechnol. 2012;30:546–54.
81. Murphy CM, O'Brien FJ, Little DG, Schindeler A. Cell-scaffold interactions in the bone tissue engineering triad. Eur Cell Mater. 2013;26:120–32.
82. Wake MC, Jr PC, Mikos AG. Pore morphology effects on the fibrovascular tissue growth in porous polymer substrates. Cell Transplant. 1994;3:339–43.
83. Jin WL, Ahn G, Kim JY, Cho DW. Evaluating cell proliferation based on internal pore size and 3D scaffold architecture fabricated using solid freeform fabrication technology. J Mater Sci Mater Med. 2010;21:3195–205.
84. Murphy CM, Haugh MG, O'Brien FJ. The effect of mean pore size on cell attachment, proliferation and migration in collagen-glycosaminoglycan scaffolds for bone tissue engineering. Biomaterials. 2010;31:461–6.
85. Pelipenko J, Kocbek P, Kristl J. Critical attributes of nanofibers: Preparation, drug loading, and tissue regeneration. Int J Pharm. 2015;484:57–74.
86. Chen L, Hu J, Shen X, Tong H. Synthesis and characterization of chitosan-multiwalled carbon nanotubes/hydroxyapatite nanocomposites for bone tissue engineering. J Mater Sci Mater Med. 2013;24:1843–51.
87. Edwards SL, Werkmeister JA, Ramshaw JA. Carbon nanotubes in scaffolds for tissue engineering. Expert Rev Med Devices. 2009;6:499–505.

88. Ogihara N, Usui Y, Aoki K, Shimizu M, Narita N, Hara K, Nakamura K, Ishigaki N, Takanashi S, Okamoto M, Kato H, Haniu H, Ogiwara N, Nakayama N, Taruta S, Saito N. Biocompatibility and bone tissue compatibility of alumina ceramics reinforced with carbon nanotubes. Nanomedicine (London). 2012;7:981–93.
89. Jell G, Verdejo R, Safinia L, Shaffer MSP, Stevens MM, Bismarck A. Carbon nanotube-enhanced polyurethane scaffolds fabricated by thermally induced phase separation. J Mater Chem. 2008;18:1865–72.
90. Lin C, Wang Y, Lai Y, Yang W, Jiao F, Zhang H, Ye S, Zhang Q. Incorporation of carboxylation multiwalled carbon nanotubes into biodegradable poly(lactic-co-glycolic acid) for bone tissue engineering. Colloids Surf B: Biointerfaces. 2011;83:367–75.
91. Vozzi G, Corallo C, Daraio C. Pressure-activated microsyringe composite scaffold of poly(L -lactic acid) and carbon nanotubes for bone tissue engineering. J Appl Polym Sci. 2013;129:528–36.
92. Sitharaman B, Shi X, Walboomers XF, Liao H, Cuijpers V, Wilson LJ, Mikos AG, Jansen JA. In vivo biocompatibility of ultra-short single-walled carbon nanotube/biodegradable polymer nanocomposites for bone tissue engineering. Bone. 2008;43:362–70.
93. Hirata E, Uo M, Takita H, Akasaka T, Watari F, Yokoyama A. Multiwalled carbon nanotube-coating of 3D collagen scaffolds for bone tissue engineering. Carbon. 2011;49:3284–91.
94. Venkatesan J, Ryu B, Sudha PN, Kim SK. Preparation and characterization of chitosan-carbon nanotube scaffolds for bone tissue engineering. Int J Biol Macromol. 2012;50:393–402.
95. Lahiri D, Rouzaud F, Richard T, Keshri AK, Bakshi SR, Kos L, Agarwal A. Boron nitride nanotube reinforced polylactide–polycaprolactone copolymer composite: mechanical properties and cytocompatibility with osteoblasts and macrophages in vitro. Acta Biomater. 2010;6:3524–33.
96. Lahiri D, Singh V, Benaduce AP, Seal S, Kos L, Agarwal A. Boron nitride nanotube reinforced hydroxyapatite composite: mechanical and tribological performance and in-vitro biocompatibility to osteoblasts. J Mech Behav Biomed Mater. 2011;4:44–56.
97. Li X, Feng Q. Porous poly-L-lactic acid scaffold reinforced by chitin fibers. Polym Bull. 2005;54:47–55.
98. Li X, Liu X, Dong W, Feng Q, Cui F, Uo M, Akasaka T, Watari F. In vitro evaluation of porous poly(L-lactic acid) scaffold reinforced by chitin fibers. J Biomed Mater Res B Appl Biomater. 2009;90:503–9.
99. Lafdi K, Fox W, Matzek M, Yildiz E. Effect of carbon nanofiber heat treatment on physical properties of polymeric nanocomposites: part I. J Nanomater. 2007;2007:6.
100. Rajzer I, Kwiatkowski R, Piekarczyk W, Biniaś W, Janicki J. Carbon nanofibers produced from modified electrospun PAN/hydroxyapatite precursors as scaffolds for bone tissue engineering. Mater Sci Eng C. 2012;32:2562–9.
101. Ma T, Li Y, Yang ST, Kniss DA. Effects of pore size in 3-D fibrous matrix on human trophoblast tissue development. Biotechnol Bioeng. 2001;70:606–18.
102. Karageorgiou V, Kaplan D. Porosity of 3D biomaterial scaffolds and osteogenesis. Biomaterials. 2005;26:5474–91.
103. Hollister SJ. Porous scaffold design for tissue engineering. Nat Mater. 2005;4:518–24.
104. Loh Q, Choong C. Three-dimensional scaffolds for tissue engineering applications: role of porosity and pore size. Tissue Eng Part B Rev. 2013;19:485–502.
105. Shokrgozar MA, Mottaghitalab F, Mottaghitalab V, Farokhi M. Fabrication of porous chitosan/poly(vinyl alcohol) reinforced single-walled carbon nanotube nanocomposites for neural tissue engineering. J Biomed Nanotechnol. 2011;7:276–84.
106. Kai D, Prabhakaran MP, Stahl B, Eblenkamp M, Wintermantel E, Ramakrishna S. Mechanical properties and in vitro behavior of nanofiber-hydrogel composites for tissue engineering applications. Nanotechnology. 2012;23:213–20.

107. Liu X, Zhang G, Hou C, Wang H, Yang Y, Guan G, Dong W, Gao H, Feng Q. Vascularized bone tissue formation induced by fiber-reinforced scaffolds cultured with osteoblasts and endothelial cells. Biomed Res Int. 2013;2013:854917.
108. Tiainen H, Lyngstadaas SP, Ellingsen JE, Haugen HJ. Ultra-porous titanium oxide scaffold with high compressive strength. J Mater Sci Mater Med. 2010;21:2783–92.
109. Venkatesan J, Qian ZJ, Ryu BM, Kumar NA, Kim SK. Preparation and characterization of carbon nanotube-grafted-chitosan-natural hydroxyapatite composite for bone tissue engineering. Carbohydr Polym. 2011;83:569–77.
110. Woo KM, Chen VJ, Ma PX. Nano-fibrous scaffolding architecture selectively enhances protein adsorption contributing to cell attachment. J Biomed Mater Res A. 2003;67:531–7.
111. Webster TJ, Ergun C, Doremus RH, Siegel RW, Bizios R. Enhanced functions of osteoblasts on nanophase ceramics. Biomaterials. 2000;21:1803–10.
112. Oh SH, Finõnes RR, Daraio C, Chen LH, Jin S. Growth of nano-scale hydroxyapatite using chemically treated titanium oxide nanotubes. Biomaterials. 2005;26:4938–43.
113. Oh S, Brammer KS, Li YS, Teng D, Engler AJ, Chien S, Jin S. Stem cell fate dictated solely by altered nanotube dimension. Proc Natl Acad Sci U S A. 2009;106:2130–5.
114. Chang HL, Shin HJ, Cho IH, Kang YM, Kim IA, Park KD, Shin JW. Nanofiber alignment and direction of mechanical strain affect the ECM production of human ACL fibroblast. Biomaterials. 2005;26:1261–70.
115. Liao CZ, Wong HM, Yeung KW, Tjong SC. The development, fabrication, and material characterization of polypropylene composites reinforced with carbon nanofiber and hydroxyapatite nanorod hybrid fillers. Int J Nanomedicine. 2014;9:1299–310.
116. Kolambkar YM, Bajin M, Wojtowicz A, Hutmacher DW, García AJ, Guldberg RE. Nanofiber orientation and surface functionalization modulate human mesenchymal stem cell behavior in vitro. Tissue Eng A. 2013;20:398–409.
117. Beachley V, Wen X. Fabrication of nanofiber reinforced protein structures for tissue engineering. Mater Sci Eng C. 2009;29:2448–53.
118. Albuschies J, Vogel V. The role of filopodia in the recognition of nanotopographies. Sci Report. 2013;3:1581–92.
119. Liu Y, Ji Y, Kaustabh G, Clark RAF, Huang L, Rafailovich MH. Effects of fiber orientation and diameter on the behavior of human dermal fibroblasts on electrospun PMMA scaffolds. J Biomed Mater Res A. 2009;90a:1092–106.
120. Whited BM, Rylander MN. The influence of electrospun scaffold topography on endothelial cell morphology, alignment, and adhesion in response to fluid flow. Biotechnol Bioeng. 2014;111:184–95.
121. Ghosh K, Ingber DE. Micromechanical control of cell and tissue development: implications for tissue engineering ☆. Adv Drug Deliv Rev. 2007;59:1306–18.
122. Provenzano P, Keely P. Mechanical signaling through the cytoskeleton regulates cell proliferation by coordinated focal adhesion and rho GTPase signaling. Ajn The Am J Nurs. 1984;84:1285–90.
123. Giancotti FG. Complexity and specificity of integrin signalling. Nat Cell Biol. 2000;2:13–4.
124. Ju GW, Miyazu M, Peng X, Shu NL, Sokabe M, Naruse K. Stretch-induced cell proliferation is mediated by FAK-MAPK pathway. Life Sci. 2005;76:2817–25.
125. Demali KA, Sun X, Bui GA. Force transmission at cell-cell and cell-matrix adhesions. Biochemistry. 2014;53:7706–17.
126. Allen LT, Blute I, Kelly ZD, Rochev Y, Keenan AK, Dawson KA, Gallagher WM. Interaction of soft condensed materials with living cells: phenotype/transcriptome correlations for the hydrophobic effect. Proc Natl Acad Sci. 2003;100:6331–6.
127. Yuan X, Liu D, Easteal A, Bhattacharyya D, Li J. Preparation of poly(lactic acid) scaffolds reinforced with bacterial cellulose nanofibers. In: Conference for International Committee on Composite Materials, Edinburgh, UK; 2009.

128. Zhang Y, Ouyang H, Lim CT, Ramakrishna S, Huang ZM. Electrospinning of gelatin fibers and gelatin/PCL composite fibrous scaffolds. J Biomed Mater Res B Appl Biomater. 2005;72:156–65.
129. Zhang Y, Su B, Venugopal J, Ramakrishna S, Lim C. Biomimetic and bioactive nanofibrous scaffolds from electrospun composite nanofibers. Int J Nanomedicine. 2007;2:623–38.
130. Chua KN, Chai C, Lee PC, Tang YN, Ramakrishna S, Leong KW, Mao HQ. Surface-aminated electrospun nanofibers enhance adhesion and expansion of human umbilical cord blood hematopoietic stem/progenitor cells. Biomaterials. 2006;27:6043–51.
131. Li X, Cui R, Liu W, Sun L, Yu B, Fan Y, Feng Q, Cui F, Watari F. The use of nanoscaled fibers or tubes to improve biocompatibility and bioactivity of biomedical materials. J Nanomater. 2013;8:4361–9.
132. Grafahrend D, Heffels KH, Beer MV, Gasteier P, Möller M, Boehm G, Dalton PD, Groll J. Degradable polyester scaffolds with controlled surface chemistry combining minimal protein adsorption with specific bioactivation. Nat Mater. 2011;10:67–73.
133. Harrison BS, Atala A. Carbon nanotube applications for tissue engineering. Biomaterials. 2007;28:344–53.

Chapter 6
The Potential Tissues and Their Properties

Yan Huang, Jing Ji, Xili Ding, and Xiaoming Li

Abstract Tissue engineering represents one of the most investigated biomedical fields, as the use of a combination of cells, engineering and materials methods, and suitable biochemical and physicochemical factors to improve or replace biological tissues. The development of biological substitutes that restore, maintain, or improve tissue function could significantly improve the life quality of numerous patients affected by diseases and yet untreatable health conditions. In recent years, numerous studies have reported the production of specialized materials able to pamper tissue engineering and to allow the development of novel therapeutic approaches for regenerative medicine. In order to precisely control of the scaffolds architecture like size, shape, inter connectivity, branching, geometry, and orientation to mimic the desired tissues, the need for knowledge of tissue properties remains essential so that each may be used with maximum understanding the properties of tissue and to greatest advantage. This chapter presents an overview of the biochemical and physical properties of main hard and soft tissues, including bones, teeth, cartilages, blood vessels, liver, nerves, tendons, and ligaments. For each case, the properties which are used to describe the tissue are reviewed, with emphasis being placed on their classification, structure, functions, and related diseases. Subsequently, tissue engineering concerning each tissue are also included. Intended as a broad-ranging reference, this chapter gives the bioengineer, physicist, or physiologist access to a literature which may not be known in detail. It will also be of value for those concerned with the study of tissue engineering.

Keywords Tissue engineering • Hard tissue • Soft tissue • Functions • Diseases

Y. Huang (✉) • J. Ji • X. Ding • X. Li
Key Laboratory for Biomechanics and Mechanobiology of Ministry of Education, School of Biological Science and Medical Engineering, Beihang University, Beijing 100191, China
e-mail: huangyan@buaa.edu.cn

X. Li (ed.), *Tissue Repair*, DOI 10.1007/978-981-10-3554-8_6

6.1 Introduction

Although significant advances have been made in medical techniques that reconstruct damaged organs or tissues as a result of trauma or cancer, tissues or organs transplantation is a generally accepted therapy to treat these patients. However, autologous transplantation is limited due to infection and donor site morbidity or pain to patients. Alternative tissue sources that have origins from other humans or animals remain problematic mainly because of immunological rejection by the patients upon implantation [1].

Tissue engineering has shown great potential in treating the damaged or lost tissues or organs [2]. Tissue engineering is an interdisciplinary and multi-disciplinary field that involve the expansion of cells, the design of three-dimensional scaffolds, and culturing of cells in scaffolds to repair or replace portions of or whole tissues. In addition, some vital factors should be taken into account including stem cells that can be differentiated into specific cell types, scaffolds that can support cell growth, and growth factors that can control cellular activities to gain damaged tissues or organs regeneration [3–5].

There are several requirements in the design of scaffolds for tissue engineering. Many of these requirements are complex and not yet fully understood. Ideally, a scaffold should meet several design criteria: (1) biocompatible and bioresorbable with controllable resorption and degradation rate to match tissue reconstruction; (2) the surface should support cell attachment, proliferation, and differentiation; (3) mechanical properties to match those of the desired tissues; (4) should with enough porosity for cell growth and flow transport of nutrients and metabolic waste; (5) be reproducible processed into variety of shapes and sizes by solid free form fabrication [6, 7]. Many porous scaffolds have been fabricated and used for tissue engineering of bone [8], cartilage [9–11], liver [12] nerve [13], and ligament [14], etc.

Many biodegradable biomaterials have been used and fabricated into various shapes for tissue engineering. These scaffolds showed promising results, guiding tissue development. However, tissue engineering is complex and specific to the structure and function of the tissue of interest. In order to precisely control of the scaffolds architecture like size, shape, inter connectivity, branching, geometry, and orientation to mimic the desired tissues, the need for knowledge of tissue properties remains essential so that each may be used with maximum understanding the properties of tissue and to greatest advantage. Thus, the aim of this chapter is to present an overview of the biochemical and physical properties of main hard and soft tissues, including bones, teeth, cartilages, blood vessels, liver, nerves, tendons, and ligaments. For each case, the properties which are used to describe the tissue are reviewed, with emphasis being placed on their classification, structure, functions, and related diseases. Subsequently, tissue engineering concerning each tissue are also included.

6.2 The Potential Tissues and Their Properties

6.2.1 Bones

6.2.1.1 Functions of Bones

Bone is a calcified, living and connective tissue forming the majority of the skeleton. Moreover, bone is the ultimate biomaterial because it is light, strong, which can adapt to its functional demands and repair itself. The functions of the bones in the human body are abundant and crucial which include the following aspects:

- Bones form the framework of the body. Like a house, we need our support "beams" to hold us upright and literally provide our body with a solid foundation.
- Bones are vital for protecting the most important and fragile organs in the body. For example the chest protects the heart, while the skull keeps the brain safe from harm.
- Bones work with joints, ligaments, tendons, and muscle to produce movement of limbs.
- Bone marrow, which is found in the inner part of the bones, is crucial for producing both red and white blood cells. Red blood cells can deliver oxygen and nutrients throughout the body and remove waste and white blood cells are involve in the body's immune system.
- Bones reserve a lot of calcium and phosphorus which are the most important minerals the body needs to function. They are stored or released over the body of calcium and phosphorus metabolism.
- Bones produce a hormone-signaling protein called osteocalcin to regulate blood sugar levels and manage weight. Additionally, the link between weak bone health and cardiovascular health is becoming increasingly evident.

Overall, bones play a much larger role in our bodies than simply keeping us stand on our own 2 ft. From blood cell production to mineral storage and so much more besides, bones are miraculous in the human body.

6.2.1.2 Classifications of Bones

The human adult skeleton is composed of 206 bones. There are two common classifications. Firstly, according to the site of its existence, bones can be divided into the axial and appendicular skeleton. The axial skeleton forms the central core or "axis" of the body's bony framework, which consists of the skull and trunk bones. The skull bones include 29 bones which can be divided into the cranial bones, facial bones, and auditory ossicles, while the trunk bones include 51 bones which are made up of vertebrae, ribs, and sternum. The appendicular skeleton is attached or "appended" to the axial skeleton, which includes the upper and lower limbs. The upper limbs include 64 bones, while the lower limbs include 62 bones. Secondly, based on the different shapes, bones can be divided into long bones, short bones, flat bones and irregular bones. A long bone is an elongated tubular bone that has a shaft and two ends, which is longer than it is wide. Long bones have a thick outside layer of compact bone and a spongy inner cancellous bone containing bone marrow. In addition, both ends of a long bone contain an epiphyseal line. The epiphyseal line is a remnant of an area that contained hyaline cartilage that grew during childhood to lengthen the bone. Femur, humerus, and tibia are typical long bones. Some of the smallest including the metacarpals, metatarsals, and phalanges are classified as long bones. This is due to the shape of the bones, not their size. On the contrary, short bones are short and cuboidal, which are generally distributed in solid and flexible connecting parts. Examples of short bones are the carpals and tarsals – the wrist and foot bones. Flat bones consist of two compact bone plates separated by spongy bone and the main function of spongy bone is to provide protection to the vital organs and be a base for muscular attachment. The classic examples of flat bones are scapula, cranium, sternum, and ribs. Irregular bones such as vertebra and bones of the face are bones with various shapes. They primarily consist of cancellous bone, with a thin outer layer of compact bone. Some irregular bones which contain cavities are known as pneumatic bones, such as maxilla. Some irregular bones which embedded in some tendons or ligaments and are small and flat round, are called sesamoid bone, such as patella and sesamoid bone at the base of the first metatarsal of the foot.

6.2.1.3 Structure of Bones

Types of tissue found in bones include periosteum, bone tissue, endosteum, bone marrow, blood vessels, and nerves. If you were to cut a cross-section through a bone, you would first come across a thin layer of dense connective tissue known as periosteum. The periosteum provides a good blood supply to the bone and a point for muscular attachment. Two types of bone tissue are cortical and cancellous bone. Under the periosteum is a thin layer of cortical bone also called compact bone, which provides the bones strength and accounts for 80% of the mass of bone in the human body. Cortical bone is solid bone with only a few small canals, which is covered by a periosteum on its outer surface, and an endosteum on its inner surface. Cancellous bone also called spongy bone is like scaffolding or a honey-comb, which accounts for roughly 20% of the total mass of the skeleton. In a long bone, cancellous bone is normally found at the ends of the bone, while in a flat or irregular bone, it is a thin layer found just inside the cortical bone. Moreover, the centre of the bone shaft is hollow and known as the medullary cavity which contains both red and yellow bone marrow. The red bone marrow produce blood cells, while yellow bone marrow is mainly a fatty tissue. In newborns, all bones are filled exclusively with red marrow, but as the child ages it is mostly replaced by yellow marrow. In adults, red marrow only remains in the vertebrae, ribs, sternum and cancellous bones of humerus and femur. Additionally, bone blood vessels mainly provide nutrition to the bones and many nerve fibers are associated with blood vessels in the bones.

6.2.1.4 Composition of Bones

Bone consists of organic and inorganic components [15]. The organic matrix is composed primarily of the protein collagen which provides flexibility and accounts for about 10% of adult bone mass. The main inorganic component is composed of hydroxyapatite, which is an insoluble salt of calcium and phosphorus and accounts for about 65% of adult bone mass. Moreover, bone also contains small amounts of magnesium, sodium, and bicarbonate. In addition, in living adult bone in the body contains between 10 and 20% water.

6.2.1.5 Bone Cells

There are three types of cell present in bone: osteoblasts, osteocytes, and osteoclasts, which are respectively responsible for the production, maintenance, and resorption of bone. The osteoblasts are cuboidal and columnar in shape with a central nucleus found near the surface of bones. They make the proteins that will form the organic matrix of bone and secrete alkaline phosphatase to create sites for calcium and phosphate deposition. The osteoblasts come from bone marrow precursor cells. These precursors are capable of turning into either osteoblasts or fat cells, and various factors determine which kind of cells will be made. One of the factors is called core binding factor alpha l (cbfa1), which will cause the cell to differentiate into an osteoblast [16]. Osteocytes live inside the bone. They can secrete growth factors to stimulate the osteoblasts and sense any mechanical strain on the bone. Osteoclasts are large cells with many nuclei, which resorb the bone. They travel to specific sites on the surface of bone and secrete acids and enzymes which degrade the bone. After they finish resorbing bone, they undergo apoptosis.

6.2.1.6 Development of Bones

The formation of bone is called ossification. There are two types of ossification: intramembranous and endochondral (see Fig. 6.1) [17]. Intramembranous ossification include the flat bones of the skull and some of the irregular bones, which are formed from connective tissue. Mesenchyme condensations differentiate into osteoblasts. Osteoblasts migrate to the membranes and deposit bony matrix around themselves without the requirement for a cartilage intermediate. When the osteoblasts are surrounded by matrix they are called osteocytes. Endochondral ossification occurs in long bones and most of the rest of the bones in the body, which involves the replacement of hyaline cartilage with bony tissue. Mesenchymal stem cells differentiate into chondrocytes that secrete cartilage matrix. Cartilage model is firstly

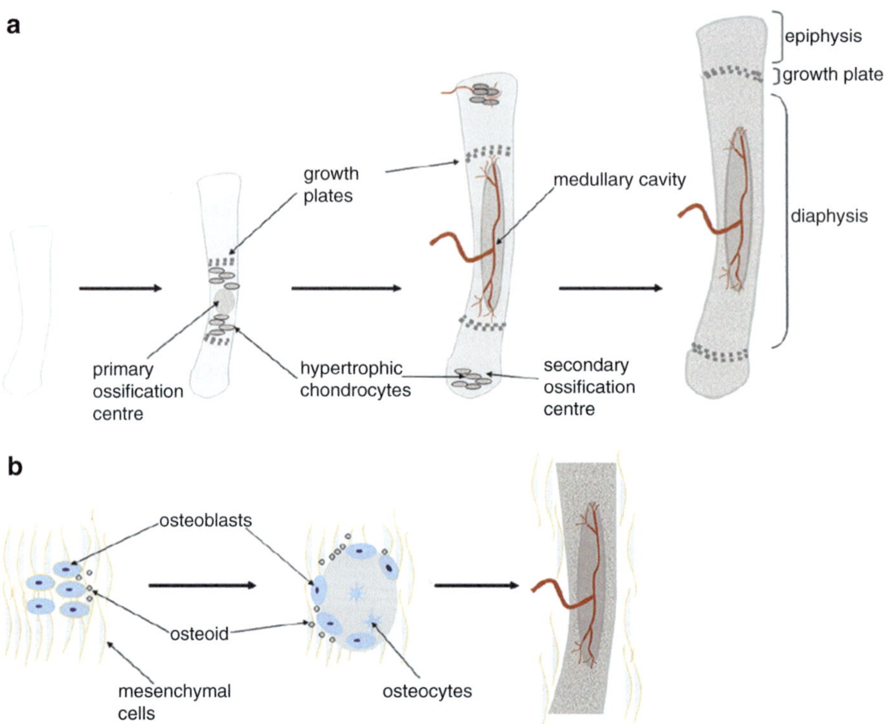

Fig. 6.1 Endochondral and intramembranous ossification. (**a**) Long bones form by endochondral ossification. (**b**) The skull vault forms by intramembranous ossification [17] (Copyright permission from Elsevier Science Ltd)

synthesized, the primary and secondary ossification center are then developed, articular cartilage and epiphyseal plate are finally formed [18]. Progression of endochondral ossification are regulated by multiple systemic hormones and various cytokines and growth factors [17].

6.2.1.7 Blood Vessel Formation and Function in Bone

The processes of bone development are coupled to the process of angiogenesis. For example, bone cells can secrete vascular endothelial growth factor (VEGF) that can trigger signaling responses in different cell populations expressing VEGF receptors, including the endothelial cells (ECs) that make up blood vessels as well as chondrocytes, osteoblasts, and osteoclasts. On the other hand, vascular cells can release factors that can act upon chondrocytes and osteoblasts. In addition, fibroblast growth factor (FGF) signaling pathway is also an important regulator of vessel growth and bone formation [19].

6.2.1.8 Diseases of Bones

A bone fracture is a medical condition in which there is a damage in the continuity of the bone. It can be the result of high stress, or a minimal trauma injury as a result of certain medical conditions that weaken the bones, such as osteoporosis and bone cancer. Bone fracture can breaking in the continuity of the periosteum, edema of nearby soft tissues and muscle spasms. The fracture healing is influenced by age, lifestyle, and comorbidities of bone healing [20].

Osteoporosis is a disease where decreased bone strength and increases the risk of a broken bone. It is the most common reason for bone fracture among the elderly and is becoming an increasingly important public health issue. Osteoporosis may be due to lower than normal peak bone mass and greater than normal bone loss. It can occur after menopause and diseases or treatments including alcoholism, anorexia, hyperthyroidism, surgical removal of the ovaries, and kidney diseases. Moreover, lack of enough exercise and smoking are also risk factors. Effective treatments to prevent fragility fractures are available. For example, osteoporosis imaging can identify individuals at risk for fractures who would require pharmacotherapy to reduce fracture risk and also monitor response to treatment [21]. Exercise can help prevent osteoporosis and are also benefit patients with osteoporosis. In addition, adequate calcium and vitamin are needed to prevent and treat osteoporosis [22].

A bone tumor is a neoplastic growth of tissue in bone which can be either benign or malignant. The most common symptom of bone tumors is pain. Some bone tumors may weaken the structure of the bone, causing fractures. Neoplastic transformation is a consequence of maladaptive alterations in the cellular processes normally involved in cell growth, proliferation, differentiation, and survival [23].

6.2.1.9 How to Measure Bone Mineral Density (BMD)?

If you take a sample of bone and wash out the liquid bone marrow, the remaining bone is about half mineral and half protein. The weight of mineral per volume of bone is the bone mineral density. The strength of the bone is largely determined by the bone mineral density. A BMD test can provide a snapshot of your bone health. It can diagnose osteoporosis, predict your risk of bone fractures and monitor of patients undergoing treatment. The most commonly and accurately way recognized BMD test is a dual-energy x-ray absorptiometry (DEXA) scan uses low-dose x-rays. There are two types of DEXA scans: central and peripheral DEXA. The density of the total hip is the best predictor for a hip fracture, while the lumbar spine is the best site for monitoring the effect of treatment [24]. For better characterizing fracture risk and treatment effects, there are some other techniques to analyze bone quality, including high-resolution peripheral quantitative computed tomography (CT), multidetector CT, magnetic resonance imaging, and quantitative ultrasonography [21].

6.2.1.10 Regeneration of Bones

Bone -since it is a live tissue – undertakes various insults and can modulate itself into a healthy bone. As a complex process of bone formation, the regeneration of bone involves continuous remodeling throughout adult life and can happen during normal fracture healing. The fracture healing process is modulated by the combination of a material for the new bone to grow into and cells that can allow osteoinduction. In addition, the combination of cells and signals and putting down the scaffolds for the new bone is also very important [25, 26]. Unfortunately, bone damage caused by reconstructive surgery, neoplasia, trauma, and osteoporosis remains a major clinical problem. Current bone regeneration strategies include auto- or xenogeneic bone grafting, gene therapy, the application of different bioactive factors, and the use of osteoprogenitor cells, osteoconductive scaffolds, and growth factors, alone or in various combinations [27]. Today, bone regeneration still has significant limitations.

Autologous bone grafting is considered as the 'gold standard' bone-grafting method, as it combines all properties required in a bonegraft material: osteoinduction [bone morphogenetic proteins (BMPs) and other growth factors], osteogenesis (osteoprogenitor cells) and osteoconduction (scaffold) [27]. But autologous bone graft has the severe shortcoming. In addition, allogeneic bone grafting has problems of immunogenicity, rejection reactions and so on. In recent years, tissue engineering and regenerative medicine has been providing exciting technologies for the development of bone-graft substitutes. Scaffolds, cells, and biological factors are the three key components of bone regeneration (see Fig. 6.2) [28].

Firstly, a scaffold that is similar in composition and structure to the natural bone forms the basic support frames of bone graft. A scaffold should have the following features in an ideal manner: (1) three-dimensional and highly porous; (2) biocompatible and bioresorbable with a controllable degradation and resorption rate; (3) suitable surface chemistry for advancing adhesion, growth and differentiation of cells; and (4) mechanical properties [27, 29]. Scaffolds are made of synthetic or natural biomaterials, including collagen, hydroxyapatite, tricalcium phosphate and calcium-phosphate cements, and glass ceramics [27]. Especially nano-biomaterials could elicit a higher degree of biological plasticity compared with conventional micro-materialsdue to their high surface-to-volume ratio. Moreover, nano-hydroxyapatite with chitin, chitosan, collagen, gelatin, fibrin, polylactic acid (PLA), poly (ε-caprolactone) (PCL), poly (lactic-co-glycolic acid) (PLGA), polyamide (PA), polyvinyl alcohol (PVA), polyurethane (PU), and polyhydroxybutyrate (PHB) proved to be promising biomaterials for bone tissue engineering [30]. In addition, metals including Zn, Ti, Zr, B, Sr, Mg, Ag, and Cu have the highest mechanical strength and play a major role in accelerating bone formation and promoting bone regeneration. They act as cofactors for enzymes, serving as a structural component of bone forming enzymes and proteins, increasing extra-cellular matrix synthesis, stimulating angiogenesis, promoting bone formation, and inhibiting bone resorption [31]. Silk fibroin produced mainly by silkworms and spiders has unique mechanical properties, tunable biodegradation rate and the ability to support the differentiation

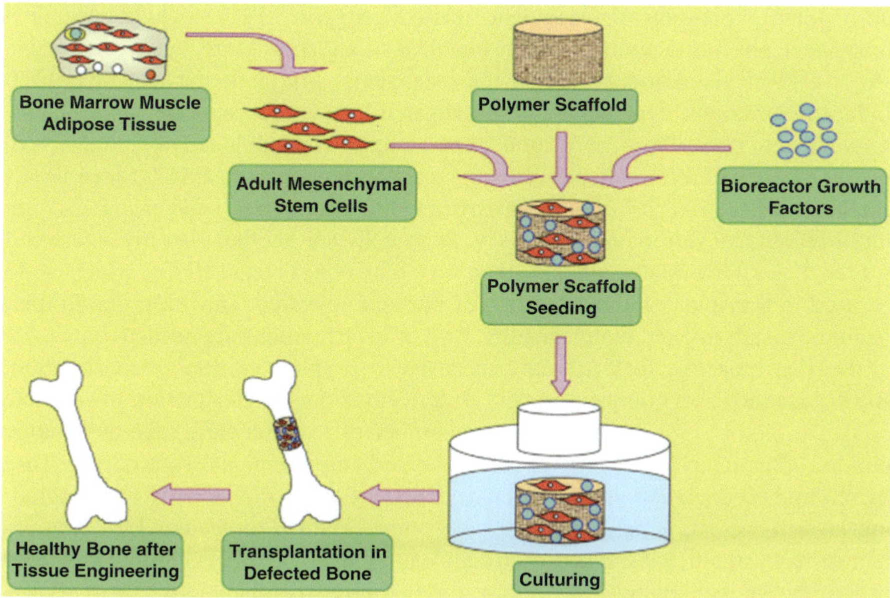

Fig. 6.2 Schematic presentation of bone regeneration by tissue engineering [28] (Copyright permission from Elsevier Science Ltd)

of mesenchymal stem cells (MSCs) along the osteogenic lineage. It can be processed into various scaffold forms, combined synergistically with other biomaterials to form composites and chemically modified which make it a favorable scaffold material for bone tissue engineering [32]. Graphene and its derivatives (graphene oxide and reduced graphene oxide) have high surface area and high mechanical strength. They has the ability to promote and enhance osteogenic differentiation making it an interesting material for bone regeneration research [33]. In addition, the combination of materials may be the most promising strategy for bone tissue regeneration [34].

Secondly, cells such as osteoblasts and stem cells constitute the biologically functional units of bone graft, and they have consistently been shown to promote bone formation inside scaffolds [35, 36]. Embryonic stem cells (ESCs) are an attractive stem cell source because they can self-renew over long periods of time and have a strong multilineage differentiation capability. But consideration of the ethics of using human embryos, there are a lot of controversy surrounded ESCs. In recent years, MSCs derived from cryopreserved umbilical cord blood, bone marrow, and adipose tissue have received extensive attention because they can be easily isolated, induce little immune response, have marked self-renewal properties and possess the biological capability to differentiate into osteoblasts in response to multiple environmental factors. For example, specific combinations of soluble factors including dexamethasone, ascorbic acid, and β-glycerophosphate disodium have been shown to induce osteoblastogenesis of MSCs. A variety of factors, like BMP and FGF can up-regulate expression of osteogenic related genes in MSCs. Besides chemical revulsants, physics factors such as mechanical strain, shear stress, and compressive stress also play important roles in the osteogenic differentiation of MSCs [37]. Induced pluripotent stem cells (also known as iPS cells or iPSCs) are a type of pluripotent stem cells. They can be generated directly from adult cells by the delivery of some genes to differentiated cells and are highly similar to ESCs. Since iPSCs can be generated directly from adult tissues without the embryo and have the strong multi-directional differentiation ability, these cells are another attractive stem cell source. Each individual could have their own pluripotent stem cell line which could be used to transplant without the risk of immune rejection. However, due to their genomic instability and tumorigenesis, further investigations are needed.

Thirdly, bioactive factors have been used to supplement bone reconstruction. Bone formation is a complex process that requires concerted function of multiple bioactive factors [38]. To date, a large number of technologies, such as osmotic pumps, bolus injection, release of surface-adsorbed protein, and controlled release by scaffold biodegradation, have been developed to deliver bioactive factors including growth factors, genetic material, and drugs or small molecules [39]. Growth factors have significant impacts on osteoblast behavior, and thus have been widely used in bone tissue engineering, alone and in various combinations. These factors include BMP-2, BMP-4, BMP-7, FGF-2, transforming growth factor (TGF)-β1, TGF-β2, TGF-β3, VEGF, insulin-like growth factor (IGF)-1, platelet derived growth factor (PDGF), and stromal cell-derived factor (SDF)-1 [39, 40]. However, growth factors have short half-lives, slow tissue penetration, and high toxicity, which suggest the efficiency of conventional routes is relatively low. Therefore, effective bone repair requires using multiple bioactive factors in different delivery systems [38].

Fourthly, bioreactor systems are key components for bone tissue engineering and medical regeneration, including perfusion, mechanical compression, and hydrostatic compression bioreactors. In particularly, perfusion bioreactors can provide dynamic environments with enhanced diffusion of nutrients and oxygen and removal of waste products [41, 42]. It is necessary to develop the automation of bioreactors to reach large-scale clinical application. Equally important is the generating of patient-specific bioreactors [43].

Moreover, vascularization of bone grafts exceeding a certain size is one of the main challenges of bone tissue engineering. Larger perfusable blood vessel grafts are needed for restoration of blood flow into the site of injury, while smaller microvascular beds are needed to provide thorough distribution of blood across the entire scaffold volume [36]. Scaffolds made of hydrogels and other materials have been performed for angiogenesis due to their capacity to be loaded with cells and signals at relatively high density and their ability to sustain cell viability for relatively long periods of time. Cells such as ECs, endothelial progenitor cells (EPCs), MSCs, human umbilical vein endothelial cells (HUVECs), and ESCs, have been used to assist vascularization inside scaffolds. Growth factors including VEGF and FGF-2 have been used to enhanced tissue and vascular growth in scaffolds [36]. Future bone grafts are hoped to have the surrounding vascular beds for complete restoration of bone function.

In addition, since results from *in vitro* studies can be difficult to extrapolate to the *in vivo* situation, animal models have been used extensively to investigate the biology of fracture healing and bone graft substitutes before clinical use in humans. There are several factors that should be considered when selecting an animal model, including availability of the animal, cost, ease of handling and care, size of the animal, the age of the animal, acceptability to society, resistance to surgery, infection and disease, biological properties analogous to humans, bone structure and composition, as well as bone modeling and remodeling characteristics. Small and large animals, including mice, rats, rabbits, dogs, pigs, goats and sheep have been used in animal experiments on bone tissue engineering and medical regeneration [44].

6.2.2 *Teeth*

6.2.2.1 Functions of Teeth

Teeth are part of the digestive system. Chewing is the main function of the human teeth. Food needs to be obtained, chewed and broken down before entering the digestive system so that our body can easily absorb nutrients from them. In addition, teeth can help us pronounce accurately and help us look better by giving us a good profile.

6.2.2.2 Classifications of Teeth

There are four different types of teeth, namely incisors, canines, premolars, and molars. Each type of tooth has a different job. The incisors are the eight front teeth in your mouth. They are the most visible of all teeth and have a pronounced aesthetic role, which cut food like a pair of scissors. The canines are the four teeth right next to the incisors in a full set of teeth. Their sharp point allows them to pierce through and tear food like a fork. They are also capable of grasping and retaining food. The premolars are located at the back of the canines. There are a total of eight premolars in the mouth. They help the canines to pierce and tear food and help the molars to crush and grind food during chewing. There are usually a total of 12 molars in the mouth, three behind each set of premolars. Molars are the most posterior and most complicated kind of teeth, which crush and grind food.

6.2.2.3 Anatomical Parts of Tooth

The tooth has three anatomical parts. The crown of a tooth is the visible, white part of the tooth. The root of a tooth is normally not visible, which extends below the gum line and anchors the tooth into the bone. Between the crown and root, it is the neck of a tooth.

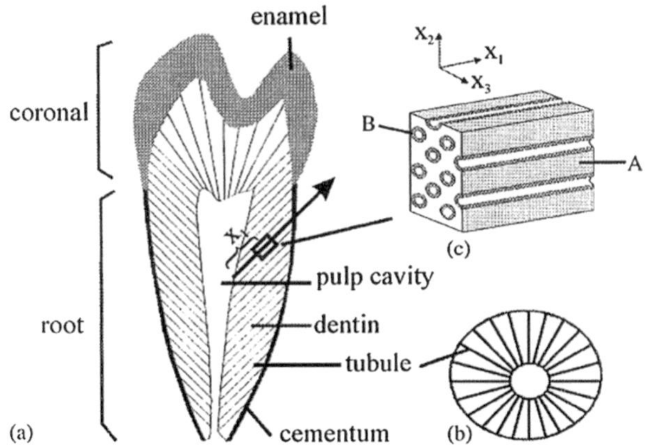

Fig. 6.3 Schematic figures of a premolar and the microstructure of its dentin: (**a**) the longitudinal section of the premolar; (**b**) the transverse section of the dentin; (**c**) a representative volume element (RVE). *A* – intertubular dentin, *B* – peritubular dentin [45] (Copyright permission from Elsevier Science Ltd)

6.2.2.4 Composition of Tooth

The tooth contains four kinds of tissue, namely enamel, dentin, cementum, and pulp (see Fig. 6.3) [45]. Each tissue does a different job. Enamel is the hard and visible substance that covers the tooth crown. Moreover, enamel is the hardest substance in the body and protects the tooth from decay. Their primary mineral is hydroxyapatite, which is a crystalline calcium phosphate. Dentin is underneath the enamel and not quite as hard as enamel. In dentin, fluid-filled microscopic tubules extend through the entire dentin from the pulp wall to the dentin–enamel junction or the cementum. The tubules in the root of tooth are approximately parallel, but the tubules in the crown of tooth are radial. The intertubular dentin occupies the volume outside of peritubular dentin and contains more collagen and less mineral than the peritubular dentin [45]. Cementum is a specialized bone like substance covering the root of a tooth. Cementum is softer than enamel and dentin and helps anchor the root into the bone. Pulp is found at the center of the tooth. It contains the blood vessels, nerves, and other soft tissues that deliver nutrients and signals to the teeth.

6.2.2.5 Structures Around Tooth

There are some structures around the tooth. Periodontal ligament is the supporting structure of a tooth. It is made up of thousands of fibres which help to attach the tooth to surrounding tissues and to allow sensations of touch and pressure. Oral Mucosa is the moist tissue that lines the mouth. Gingivae is the soft tissue that immediately surrounds the teeth and bone, which protects the bone and the roots of the teeth and provides an easily lubricated surface. Alveolar bone is the bone of the jaw which provides a socket to surround and supports the roots of the teeth. In orthodontics, a tooth can be moved through the alveolar bone when an appropriate orthodontic force is applied. The cellular responses in the area of bone under compressive force will result in bone resorption, whereas the cellular responses in the

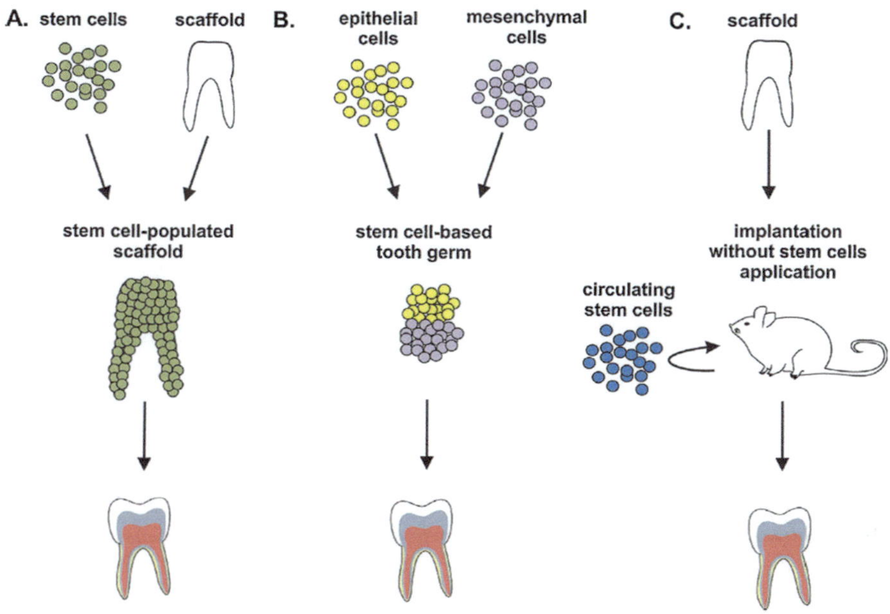

Fig. 6.4 Potential approaches for tooth regeneration. (**a**) Scaffold based approach. (**b**) Cell aggregation based approach. (**c**) Stem cell homing based approach [48] (Copyright permission from Elsevier Science Ltd)

area of bone receiving tension from periodontal ligaments will result in bone deposition [46]. Orthodontic tooth movement generates a change of the local vascularization and a biochemical adaptive response which leads to the release of arachidonic acid, growth factors, metabolites, cytokines, and various enzymes [47].

6.2.2.6 Dental Regeneration

Dental regeneration is to restore a tissue defect to its original form and function by biological substitutes. The possible experimental approaches to the creation of a dental prosthesis include scaffold based approach, cell aggregation based approach and stem cell homing based approach (see Fig. 6.4) [48]. Dental repair is by metal or artificial materials whereas regeneration is by biological restoration [49]. Ideally, biomaterial scaffolds in tooth regeneration should be biocompatible, be nontoxic, undergo biologically safe degradation, provide encapsulation of cells or surface adhesion for cells, and allow functionality of cells including ameloblasts, odontoblasts, cementoblasts, fibroblasts, vascular cells and neural endings [50]. Several classes of materials have been used for dental tissue engineering, such as PLGA, PLA and hydroxyapatite/tricalcium phosphate (HA/TCP) [48].

Damage of teeth, such as caries, dental wear lesions or trauma can result in rupture of the apical blood vessel network and dental pulp necrosis. Dental pulp therapy is currently limited to conventional restorations such as root canal therapy which allows for infection control, but does not facilitate the completion of root formation and protection against external root resorption. Dental pulp tissue engineering including potential cell sources and biocompatible materials have been considered an attractive strategy for dental pulp regeneration [51]. Responsive cells are generally stem cells. Dental pulp stem cells, which is therapeutically employed for direct and indirect pulp capping, are capable of differentiating into odontoblast-like cells or ECs. In addition, dental pulp stem cells have the lifelong regenerative cability and the potential to continue root development in necrotic immature teeth and transplanted/replanted teeth [52]. Studies in dental biomaterials research aim at the development of suitable biomimetic and degradable biomaterials with uniquely functional properties at a nanometer-scale. A variety of scaffolds, such as collagen, polyester, chitosan, or hydroxyapatite has been investigated. Self-assembling peptide hydrogels are an example of a smart material that can be modified to create customized matrices [53]. Pulp tissue is not particularly conducive to the rapid and effective vasculogenesis. VEGF is the prototypic pro-angiogenic factor. It could enhance the neovascularization of severed human dental pulps [54].

An ideal regenerative protocol would primarily include root decontamination with antimicrobial irrigant solutions, followed by insertion of a more cell-friendly bioactive scaffold containing antimicrobial substances to be released inside the canal. Once a bacteria-free environment conducive to tissue regeneration has been established, scaffolds containing growth factors and/or stem cells would be placed to induce development of a new pulp tissue containing odontoblasts that will form dentin-like tissue [55]. The new engineered pulp tissue could generate new dentin and establish tooth function, which is a critical step forward in the process of translating the concept of dental pulp tissue engineering to the clinic [51].

6.2.3 Cartilage

6.2.3.1 Overview of Cartilage

Cartilage is a strong, rubbery, and flexible connective tissue in the body. Articular cartilage covers the joint surfaces and is mainly responsible for the unique biomechanical properties of the joints. Joints are complex composites of different types of connective tissue including (subchondral) bone, cartilage surfaces, ligaments and the joint capsule (see Fig. 6.5) [56]. Because cartilage lacks blood vessels that provide the nutrient delivery to tissues, it does not possess the regenerative capacity of bone or the other connective tissue types. Cartilage is made up of specialized cells called chondrocytes. These chondrocytes produce large amounts of extracellular matrix (ECM) that is rich in collagen fibres, proteoglycan, and elastin fibers.

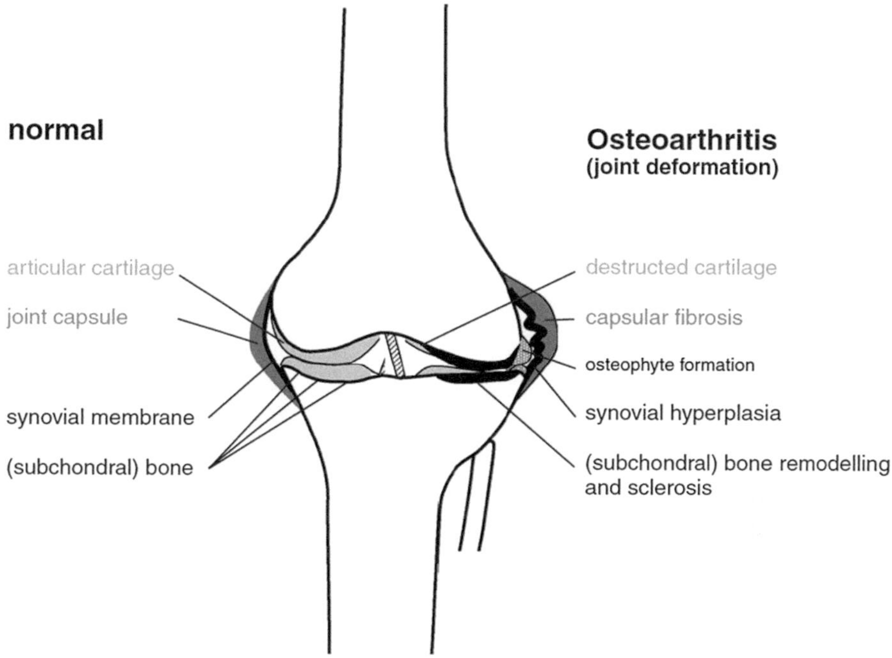

Fig. 6.5 Schematic representation of the main structures of a healthy (*left side*) and degenerated (*right side*) joint in osteoarthritis: in particular the articular cartilage is lost or severely thinned, the (subchondral) bone is sclerotic and the joint capsule thickened and the synovial membrane activated [56] (Copyright permission from Elsevier Science Ltd)

6.2.3.2 Development of Cartilage

Developmentally, cartilage can be followed histologically through three general phases [57, 58]. Prior to cell differentiation, it is a cellular condensation form which is known as cartilage condensation [58]. These cells proliferate and secrete the cartilage-specific matrix. The ECM calcifies as the cells enter the hypertrophic phase of differentiation. This calcified cartilage matrix is then replaced by bone [58].

6.2.3.3 Classifications of Cartilage

Cartilage is classified in three types, hyaline cartilage, fibrocartilage, and elastic cartilage, which differ in relative amounts of collagen and proteoglycan. The most abundant type is hyaline cartilage which is the typical cartilage that is generally referred to when one thinks of cartilage. It is widely distributed in the human fetus. Most of the fetal hyaline cartilage is replaced by bone, but some persists in the articular surface of bones, in the larynx, in the trachea, in the bronchi and on the ventral ends of ribs of the adult. Hyaline cartilage matrix is mostly made up of type

II collagen and chondroitin sulfate [59]. Fibrocartilage is a tough and inflexible form of cartilage, which makes up the intervertebral discs, the pubic symphysis, and appears in other high-stress areas. Fibrocartilage contains type I and II collagen [60]. Elastic can be found in the pinna of the ear, larynx, and epiglottis. Elastic cartilage is similar to hyamine cartilage in many respects, but has less matrix than hyaline cartilage. It shows expression of elastin in addition to type II collagen, which provides greater flexibility and thus have the ability to withstand repeated bending [61].

6.2.3.4 Functions of Cartilage

Human synovial joints are covered by articular cartilage to withstand physiological loading over a lifetime. It provides a smooth load-bearing surface to reduce friction and acts as shock-absorbing tissue. In the lower limbs, articular cartilage sustains physiological stresses estimated to be in the range of 1–6 MPa during level walking. However, these stresses increase up to and beyond 10 MPa with activities [62, 63]. Lubricin, a glycoprotein abundant in cartilage and synovial fluid, has multiple functions in articulating joints and tendons that include the protection of surfaces and the control of synovial cell growth [64]. The mechanical properties of articular cartilage include the response of cartilage in frictional, compressive, shear, and tensile loading. Cartilage is resilient and displays viscoelastic properties [65]. Besides, costal cartilage can attach ribs to the sternum.

6.2.3.5 Biomechanical Properties and Mechanobiology of the Articular Chondrocyte

The mechanical function of articular cartilage largely depends on the composition and structural integrity of the cartilage matrix produced by chondrocytes. The articular chondrocyte lives in a dynamic mechanical environment that is a complex combination of compression, hydrostatic pressure, osmotic stress, shear stress, and tensile stretch. The mechanical signals are transmitted through the complex network of the ECM, pericellular matrix, cytoskeleton, and nucleus of chondrocytes. Chondrocytes respond to these mechanical stimuli by altering gene expression and metabolism [66]. The precise mechanism that govern development of chondrocytes is still unknown, so future research is needed to decipher which signaling pathways are mainly affected by mechanical loads.

6.2.3.6 Diseases of Cartilage

Osteoarthritis (OA) is a type of joint disease that results from breakdown of joint cartilage and underlying bone. Knee osteoarthritis is the most common arthritic location and the impact of its symptomatic form is estimated at 240/100,000 people per year. It limits the movement of patients in 80% of cases and 25% find themselves unable to perform the tasks of daily life. The symptoms mainly include joint pain and stiffness which make patients difficult in walking and climbing stairs [67]. The interaction (crosstalk) between cartilage and subchondral bone is thought to be a central feature of the progression of OA [68]. There are multiple risk factors that are implicated in OA including previous joint injury, abnormal joint development, ageing, obesity, muscle weakness, and potentially sports participation [69]. Appropriate treatment includes exercise, biomechanical interventions, intra-articular corticosteroids, self-management, education, strength training, and weight management [70]. But effective early disease treatment for OA remain lacking.

Traumatic rupture or detachment is often found in athletes. The meniscus (a fibrocartilage structure) in the knee is frequently damaged. However, it can be partially repaired through knee cartilage replacement therapy.

6.2.3.7 Cartilage Tissue Engineering

Cartilage has limited repair capabilities. Because chondrocytes are bound in lacunae and cannot migrate to damaged areas. Moreover, cartilage is a hyaline tissue with no blood, lymphatic, or nerve supply. Therefore, cartilage damage is difficult to heal. Tissue engineering strategies involving the combination of cells, scaffolding biomaterials and bioactive agents to generate cartilage has emerged as a promising option for the repair and regeneration of damaged articular cartilage (see Fig. 6.6) [71].

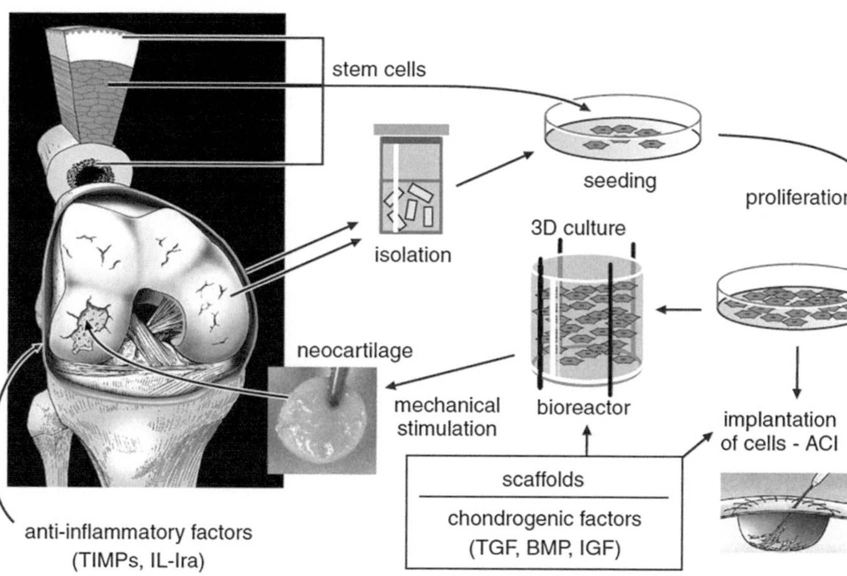

Fig. 6.6 Cartilage tissue re-engineering. Schematic presentation of a potential treatment for OA lesions [71] (Copyright permission from Elsevier Science Ltd)

The combination of cells, scaffolds and growth factors could be either directly injected in the lesion or further grown in bioreactors *in vitro* to form new tissue. Neocartilage would finally be implanted in the lesion [71].

The use of autologous chondrocyte implantation may represent a promising technology for cartilage repair because of avoiding immunogenic response or transferring diseases. MSCs have been regarded as an attractive candidate in the repair and regeneration of cartilage. MSCs can be derived from bone marrow, adipose tissue, trabecular bone, periosteum, synovial membrane, and skeletal muscle, as well as teeth and umbilical cord [72]. In addition, the differentiation of MSCs to obtain hyaline-like structures *in vitro*, requires seeding and culturing in 3D configurations, which can adopt many forms depending on the geometry and the physicochemical characteristics of the surrounding cells [73]. Besides, cell sheet technology is increasingly considered as a potential way to reconstruct cartilage tissues for its non-use of scaffolds and no destruction of matrix secreted by cultured cells [74].

Building a biomaterial is important in cartilage tissue engineering. This biomaterial needs to repair articular cartilage to reproduce the mechanical properties of the original cartilage and can be integrated into the joint. Fibrin/hyaluronic acid composite hydrogels have been regard as appropriate scaffolds for *in vivo* artificial cartilage implantation [75]. Polyvinylalcohol/hydroxyapatite cryogel is a potent source of artificial cartilage. Poly vinyl alcohol (PVA) hydrogels are promising implants for cartilage regeneration due to their similar properties for chondrocyte seeding [76].

Although signaling molecules and matrix modifying agents have been used as biochemical stimuli toward cartilage tissue engineering and have led to improvements in the functionality of engineered cartilage, mechanical stimulation is a key factor in articular cartilage generation and maintenance [77]. Bioreactor systems have been designed and built in order to delivery specific types of mechanical stimulation. Using different bioreactors, many researchers have studied the responses of the chondrocytes to well-defined mechanical forces, including compression, shear stress, and hydrostatic pressure. Appropriate mechanical stimulation can lead to increased chondrogenic properties in the bioreactors for articular cartilage tissue engineering [78].

Cartilage is a complex tissue type, therefore it is difficult to reproduce the finely balanced structural interactions of cartilage. OA is a complicated disease, with many factors giving rise to disturbed cartilage homeostasis and joint inflammation. A tissue-engineering approach combining cells, biomaterials, biofactors and disease-modifying OA drugs could be a relevant approach for improving OA treatment [71]. In addition, bioprinting technology may be the ultimate solution to engineer cartilage tissue.

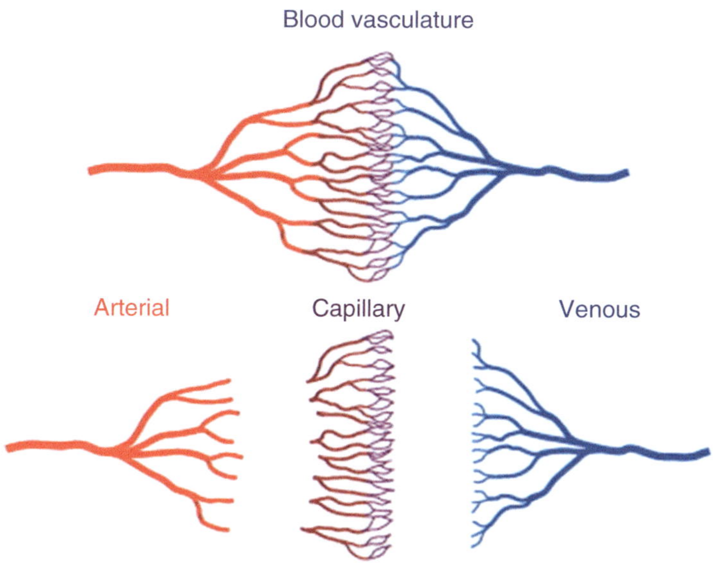

Fig. 6.7 Schematic diagram of the blood vasculature including arterial (arteries, larger arterioles), capillary (smaller arterioles, capillaries), and venous (venules, veins) compartments [79] (Copyright permission from Elsevier Science Ltd)

6.2.4 Blood Vessels

6.2.4.1 Types of Blood Vessels

The blood vessels are the part of the circulatory system. All blood is carried in these vessels. The blood vessels are responsible for passing nutrients, hormones, and other significant substances from and to body cells. There are five types of blood vessels: arterioles, arteries, capillaries, venules, and veins (see Fig. 6.7) [79].

Arteries deliver oxygen-rich blood from the heart to the tissues of the body. In addition, arteries hold a higher blood pressure than other parts of the circulatory system. Moreover, during the cardiac cycle the pressure in arteries varies. The pressure is lowest when heart relaxes and highest when the heart contracts and the variation in pressure generates a pulse. In order to keep blood pressure, the arteries can narrow passively when the heart is relaxing between beats. As the largest artery, aorta is the main high-pressure blood vessels connected to the heart's left ventricle. Moreover, the aorta branches into numerous smaller arteries that distribute throughout the body.

After travelling from the aorta, blood travels through peripheral arteries into smaller arteries named arterioles and finally to capillaries. The larger arteries have less smooth muscle than arterioles and are more elastic. When blood enters these muscular and smaller arteries do not expand as much. Then arterioles can supply the greatest resistance to blood flow to modulate the blood pressure. Due to the muscular walls, arteries and arterioles can regulate their diameter to decrease or increase blood flow to specific part of the body.

The capillary are the starting site for venous blood back to the heart and are the ultimate destination of arterial blood from the heart. The capillary walls have very thin flattened ECs that line the inner walls of the blood vessels. The capillaries can delivery waste products from tissues into the blood and provide oxygen and nutrients into tissues from the blood.

Veins are blood vessels that carry blood from various regions of the body to the heart. Compared to the arterial system, the venous system has a low pressure that depends on muscle contractions to return blood to the heart. Due to the lower pressure in veins, veins have much thinner walls than arteries. Moreover, veins can widen as the amount of fluid in them increases. Additionally, most veins are equipped with valves to prevent blood flowing in the reverse direction.

Venules are small branches that collect the blood from the various organs. Venules are groups of capillaries that unite to form small vessels that drain the capillary beds. Venules are very porous, but with scattered smooth muscle fibers in the tunica media.

6.2.4.2 Structure of Blood Vessels

The arteries and veins have the same structure and they are composed of three layers (see Fig. 6.8) [80]:

- *Tunica intima*: It consists of simple squamous endothelium, the subendothelial layer, and the internal elastic membrane. The tunica intima is in the innermost layer, as well as the thinnest layer.
- *Tunica media*: The tunica media is the middle layer, which is a thick layer of elastic fibers and smooth muscle. The tunica media is thicker in the arteries than it is in the veins. The second and third layer is separated by another thick elastic band called external elastic lamina.
- *Tunica adventitia*: The tunica adventitia is made of connective tissue that's mostly collagen fibers, which contains nerves that supply the vessel and nutrient capillaries in the larger blood vessels.

Capillaries are extremely small vessels located within the tissues of the body, which contain a layer of endothelium and occasional connective tissue.

6.2.4.3 Circulation

Blood is circulated through the body by the cardiovascular system, which contained the circulatory system and the heart. Oxygen-bearing blood is brought by the cardiovascular system to all the tissues of the body. The cardiovascular system pushes the blood out into two cycles when the heart contracts. The systemic loop begins when the oxygen-rich blood coming from the lungs flows into the left atrium. The

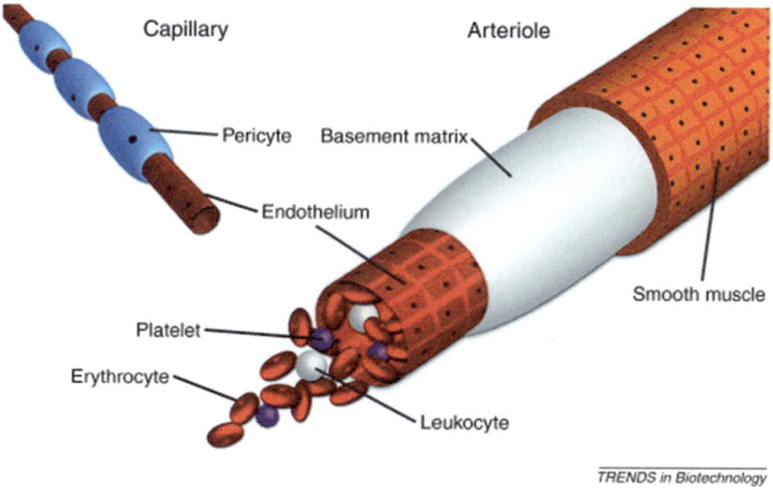

Fig. 6.8 Schematic diagram of blood vessel structures. An intact endothelium is contained in the lumen of the blood vessel. The blood contains cells such as platelets, leukocytes, and erythrocytes and plasma. The construction of the vessel lies on its location and size within the vasculature. Smaller vessels are surrounded by pericytes, which are regarded as the source of MSCs and preserve vessel maturity. On the other hand, larger vessels are surrounded by concentric layers of SMCs in order to maintain EC quiescence and vascular tone [80] (Copyright permission from ElsevierScience Ltd)

mitral valve will open when the chamber fills and the blood will flow into the left ventricle. Then the blood on the left flows into the aorta when the ventricles contract. Through the network of smaller arteries and capillaries, the blood brings oxygen to all organs. Through the network of veins, the used blood from the body come back to the heart [81].

6.2.4.4 Physiology

Arteries and veins do not actively take part in the delivery of blood since they have no peristalsis. However, blood vessels can partly adjust their diameter. Vasodilation is the widening of the blood vessels, while vasoconstriction is a narrowing of the vessels by contracting the vascular smooth muscle. Vasodilation and vasoconstrictionare not under conscious control, which are controlled by the autonomic nervous system. They are regulated by vasoconstrictors. These include a number of hormones, paracrine factors, and neurotransmitters from the nervous system. In addition, the most prominent vasodilator is NO. Vasodilation and vasoconstriction are also used antagonistically. Oxygen, which is the most vital nutrient carried by the blood, is bound to hemoglobin in red blood cells. In the arterial system, this is usually around 120 mmHg systolic and 80 mmHg diastolic. However, pressures in the venous system are constant and rarely exceed 10 mmHg.

6.2.4.5 Vascular Cells

ECs line the entire vascular system, from the heart to the smallest capillary. They form a one-cell-thick layer called the endothelium. Moreover, the ECs act as a semi-selective barrier between the surrounding tissue and vessel lumen to transport white blood cells out of and into the bloodstream. Molecules can pass through the endothelium by diffusion and enter the surrounding tissues. In addition, waste products diffuse back into the blood.

The endothelium also plays a vital role in modulating blood flow. ECs provide an anticoagulant barrier between the vessel wall and blood. Perturbations disrupt these activities and promote ECs to create an antifibrinolytic and prothrombotic microenvironment. Secretion and uptake of vasoactive substances by the endothelium may also partly regulated blood flow [82].

ECs also have several other specialized functions. Firstly, ECs can produce nitric oxide (NO) and NO can bring down inflammation and prevent platelets from sticking to vessel walls as well as dilate blood vessels to increase blood flow. Secondly, the endothelium allows white blood cells to the infected tissue and increases blood flow. Thirdly, ECs prevent blood clotting from happening inside the blood vessel. Fourthly, ECs release clotting proteins like factor VII to help in blood clotting.

Vascular smooth muscle cells (VSMCs) are the main constitutive cells of the vascular, which are subjected to biochemical components and mechanical stress. They participate in all the pathological changes and the physiological functions. VSMCs withstand the high pressure of the circulating blood compartment and generate the ECM during development. Moreover, VSMCs are also took part in arterial repair after injury in adult life and participate in the regulation of blood pressure due to their contractile tonus. VSMCs respond to hypertension-induced increases by an increase in cell protein synthesis and ECM secretion [83].

6.2.4.6 Blood Vessels Diseases

Blood vessel disease, which is also known as vascular or artery disease, is related to our circulatory system. Narrowing down or constriction of blood vessels can cause decreased transport of oxygen-rich blood to the body. It is caused by a build-up of fatty deposits -or plaque-that narrows the vessels, depleting the supply of oxygen-rich blood to affected parts of the body. This can cause tissue and cell death, which makes blood vessel disease.

Blood vessel disease is usually caused because of narrowing down of blood vessels by plaque with the fatty diets over a long period of time. Saturated fats collect on the walls of blood vessels, finally restricting the blood flow. Age and gender, as well as certain conditions such as high blood pressure, obesity, diabetes, and smoking can lead to blood vessel problems.

Blood vessel diseases are grouped into different types relying on the area of blockage. The four types of blood vessel diseases are the following:

Coronary Artery Disease: The main cause for this condition is the plaque aggregation on the walls of the coronary arteries. The fatty from the blood collect on the walls of the coronary arteries when a person has a high-fat diet. Over a period of time, the excess fat deposits will block the blood vessels, which can lead to a heart attack or stroke.

Aneurysm: The generation of aneurysm happens as a weakness in the wall of the blood vessel. This weak portion is prone to balloon out due to the force of blood in the blood vessel. The aneurysm gets larger gradually, which make it prone to rupture. Aneurysm can occur in the heart or the arteries of the brain, especially at the abdominal aorta and the basal portion of the brain. This disease can be life-threatening, which can lead to brain hemorrhage [84].

Peripheral Arterial Disease: Peripheral arterial disease can affect the blood supply to the legs and arms. Owning to the plaque aggregation in the arteries, peripheral arterial disease takes place when the arteries in the legs narrow down. Atherosclerosis is the thickening of arterial walls. The blood flow to the tissues is reduced and the arteries become narrow.

Hypertension: High blood pressure is caused by the increased pressure of the blood, which is one very common artery diseases. This disease can result in heart failure and stroke as well. Hypertension and coronary artery disease are the two most common forms of blood vessel problems.

6.2.4.7 Regeneration of Blood Vessels

Cardiovascular disease is the leading cause of mortality in today's world, necessitating the development of vascular substitutes with long-term patency [85]. Common disorders are connected with the stenosis or occlusion of blood vessels result in tissue damage and reduced blood flow because of inadequate nutrientsupply. Although autologous vessels such as internal thoracic artery and the saphenous vein represent the gold standard for small diameter grafts, patients do not have a vessel suitable for use because of vascular disease, amputation, or previous harvest [86]. The development for vascular materials has far been a half-century endeavor. Synthetic vascular grafts are also available as an alternative to autologous vessels. While large diameter (>6 mm) synthetic vascular grafts are successfully used to bypass arteries in high flow regions (e.g. the thoracic and abdominal aorta) [87], grafts with a small diameter (<6 mm) fail earlier because of thrombus formation and intimal hyperplasia [88].

Vascular graft failures are most commonly result from intimal hyperplasia, atherosclerosis, thrombosis, or infection. The absence of ECs lining the graft lumen can result in thrombosis, leading to the activation of clotting mechanisms and the adherence of blood proteins [89]. Intimal hyperplasia attributed to the migration of VSMCs from the vessel media to the intima. Intimal hyperplasia may happen in the native vessel or in the graft vessel around the anastomosis [86]. There are multiple causes, including (1) compliance mismatch between native vessel and the graft; (2) damage to, or a lack of ECs; (3) vessel diameter mismatch; (4) hemodynamic factors

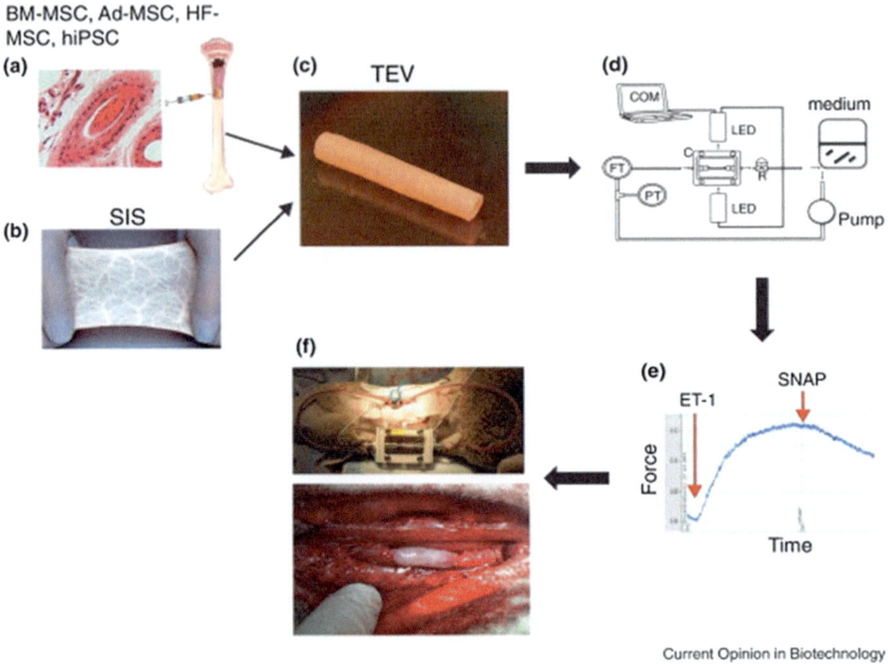

Fig. 6.9 Biomaterials, stem cells, and bioreactors for tissue engineering vascular grafts. Mesenchymal stem cells from various sources such as hair follicles (*HF*), adipose tissue (*AT*), bone marrow (*BM*), or iPSC have been developed to generate functional vascular tissue. Small intestinal submucosa (*SIS*) was combined with fibrin glue to embed HF-MSC or BM-MSC into the vascular wall. The tissue engineering vascular grafts were subjected to pulsatile pressure and shear stress in a dynamic bioreactor after seeding with ECs [87] (Copyright permission from Elsevier Science Ltd)

causing blood flow disturbances; (5) trauma during surgery; and (6) suture line stress concentrations [90–95]. The main cause of graft failure after 1 year is atherosclerosis [96]. Atheroma formation is connected with the same factors as in the native arteries and happens through a similar process [86]. The vessel neointima was invaded by monocytes and result in the generation of atherosclerotic plaque [97, 98].Graft infection is very common in synthetic grafts, which can release toxins and give rise to chronic inflammation, resulting in sepsis and an astomotic rupture or failure [99–101].

To address this, a tissue-engineered vascular graft (TEVG) presents an attractive potential solution for the future of vascular surgery (see Fig. 6.9) [87]. A tissue-engineered vessel with the ability to grow, remodel, and repair *in vivo*, has clear advantages and would be of great benefit [86]. Thus, the development of tissue-engineered blood vessels has gained much attention in the field of vascular tissue engineering. The aim of vascular tissue engineering is to develop biological vascular tissues with functional properties of native vessels, combining various cells with biodegradable scaffolds. Three tissue engineering process are developed to regenerate blood vessels: *in vivo*, *in vitro*, and in situ. For *in vitro* vascular tissue engineering, cells, scaffolds and bioreactors are used to construct functional vascular tissues outside the body. *In vivo* vascular tissue engineering is a process that the tissue environment of the body is used as a bioreactor to generate an autologous vascular tissue. For in situ vascular tissue engineering, aacellular vascular graft is fabricated and a blood vessel is generated with the host cells to avoid the in vitro culture time period in situ. In situ vascular tissue engineering is a promising process to generate blood vessels since it can reduce the in vitro cultivation time and makes the grafts available off-the-shelf [86].

As an integrated part of the vascular network, a TEVG must satisfy a number of design criteria to be fit for purpose [102]. Fundamentally, it must possess appropriate mechanical properties, including the ability to withstand long-term hemodynamic stress, physiological compliance, and no susceptibility to permanent creep [103–106]. In addition, the graft should also be biocompatible, ie, nonimmunogenic, nonthrombogenic, and resistant to infection. In addition, the graft should possess appropriate water permeability [106]. Additionally, a TEVG should be suitable for implantation; with kink resistance and the ability to be sutured, manipulated, and handled. Ultimately, the graft should be able to remodel, grow, and self-repair *in vivo* [86].Although different methods try to meet these criteria, three factors are necessary: (1) a biocompatible elastic component to provide recoil and prevent aneurysm formation; (2) a nonactivated, confluent endothelium to prevent thrombosis; (3) a biocompatible component with high tensile strength to provide mechanical support [106, 107].

6.2.5 *Liver*

6.2.5.1 Anatomy of Liver

The liver is both the largest and heaviest internal organ in the human body, occupying 2.5% of total body weight and providing a host of functions necessary for maintaining normal physiological homeostasis. It weighs about 1.4 kg in an adult female and about 1.8 kg in the male [108].

The liver is located in the right upper quadrant of the abdomen, against the inferior surface of the diaphragm. Most part of the liver is behind the rib cage but the lower edge may be palpated along the right costal margin during inspiration. A fibrous membrane known as Glisson's capsule covers the surface of the liver. The capsule extends to invest all but the smallest the vessels within the liver [109]. The liver is the largest organ in the body. Its domed upper surface relates entirely to the diaphragm while its postero-inferior, or visceral, surface rests against the abdominal oesophagus, stomach, upper duodenum, hepatic flexure of the colon, right kidney and suprarenal gland, as well as carrying the gall bladder (see Fig. 6.10) [110]. To the right, these layers fuse to form the short right triangular ligament. The falciform ligament runs to the liver from the umbilicus carrying the ligamentum teres,

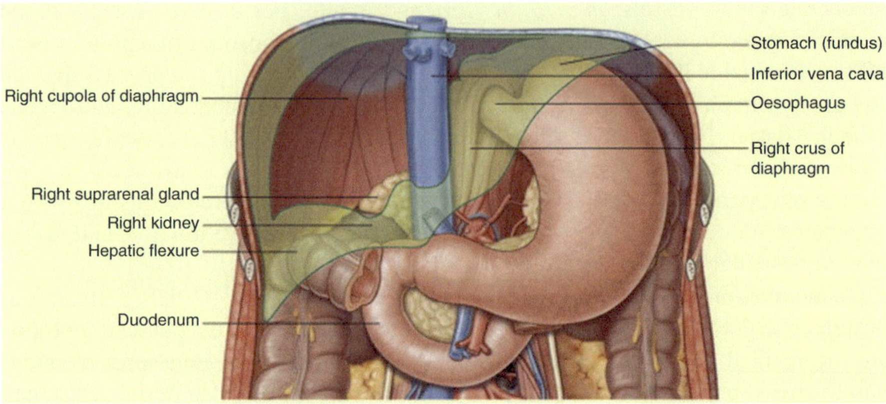

Fig. 6.10 The 'bed' of the liver. The outline of the liver is *shaded green*. The central bare area is *unshaded* [110] (Copyright permission from Elsevier Science Ltd)

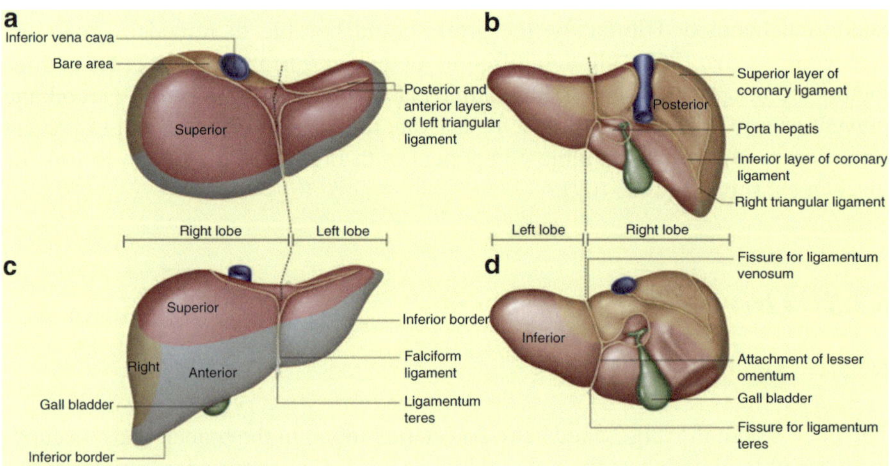

Fig. 6.11 The surface features, ligaments and peritoneal attachments of the liver [110] (Copyright permission from Elsevier Science Ltd)

in its free border. The ligamentum teres passes into its fissure on the inferior surface of the liver, while the falciform ligament passes over the dome of the liver and then divides; its right limb joins the upper layer of the coronary ligament, while its left limb stretches out as the long narrow left triangular ligament, which joins the lesser omentum as this arises from the fissure for the ligamentum venosum. (see Fig. 6.11) [110].

The connective tissues divide the liver lobes into hexagon-shaped lobules with a portal triad at each corner and a central vein in the center of each lobule. Lobules are the functional units of the liver. Each lobule is made up of millions of hepatocytes. Branches of the hepatic artery, portal vein, and bile duct are arranged at the periphery of each lobule to form the portal triad [111]. Hepatic cords composed with many layers of cuboidal hepatocytes radiate out from the central vein of each lobule. Sinusoids are the spaces between the plates of hepatocytes, which are enlarged capillaries through which blood from the portal triads drains to the central vein.

The liver is primarily composed of five types of specialized cells classifiable into parenchymal and nonparenchymal types. The main cell type of the liver that carries out most hepatic functions is the parenchymal cell, or hepatocyte, which makes up ~80% of hepatic volume or mass. The other 20% comprises extracellular spaces and the non-parenchymal cells including ECs, Kupffer cells, lymphocytes and Stellate cells [112]. The liver parenchyma is arranged in hexagonal lobules as presented before. Each sinusoidal capillary which drains the portal triad toward the central vein is lined with specialized fenestrated ECs, through which the portal blood can communicate with the adjacent hepatocytes [113]. Hepatocytes dominate the liver's mass and are responsible for most functions of the liver including synthesis and storage functions and filtration of the portal venous blood [114]. Kupffer cells are specialized liver macrophages with immunological and phagocytic functions [115]. Hepatic stellate cells, which are liver fibroblasts, are typically quiescent but are integral to the pathological liver remodeling process associated with hepatic fibrosis [116]. Bile duct epithelial cells transport bile and maintain its pH [117].

Figure 6.12 is a schematic illustration of the liver microstructure and the fabrication procedure of the hepatic cord-like micro-organoid by using the microfluidic system and the hydrogel microfiber [118].

6.2.5.2 Blood Supply of Liver

The main blood supply of the liver comes from the portal vein, which gives it a special place in the control of metabolism. About 25–30% of total cardiac output at rest enters the sinusoids via terminal portal and arterial vessels. Of this, 70% comes from the portal vein, the remaining 30% is oxygenated blood, comes from the hepatic artery. The blood mixes, passes through the sinusoids and drains into the central vein. About 1.5 L of blood exit the liver every minute [119].

The viscera and liver are supplied by the coeliac, superior mesenteric, and inferior mesenteric arteries. The hepatic artery is a branch of the coeliac trunk. The portal vein is formed from the superior mesenteric and splenic veins in front of the inferior vena cava, behind the neck of the pancreas. Thus, it receives, almost all the products of digestion, absorption, and the hormones secreted by the pancreas. It is about 7–8 cm long and enters the liver via the porta hepatis. It differs from other veins since it has no valves and, in conditions such as cirrhosis, the flow can reverse.

The blood in hepatic portal veins is nutrient-rich but oxygen-poor, so it supplies the hepatocytes with nutrients. The oxygen supply is come from the hepatic artery, although the blood mixes while passing through the sinusoids. Blood in the sinusoids picks up plasma proteins, glycogens, and other processed or waste molecules produced by the hepatocytes, and enters central veins which eventually connect to inferior vena cava [120].

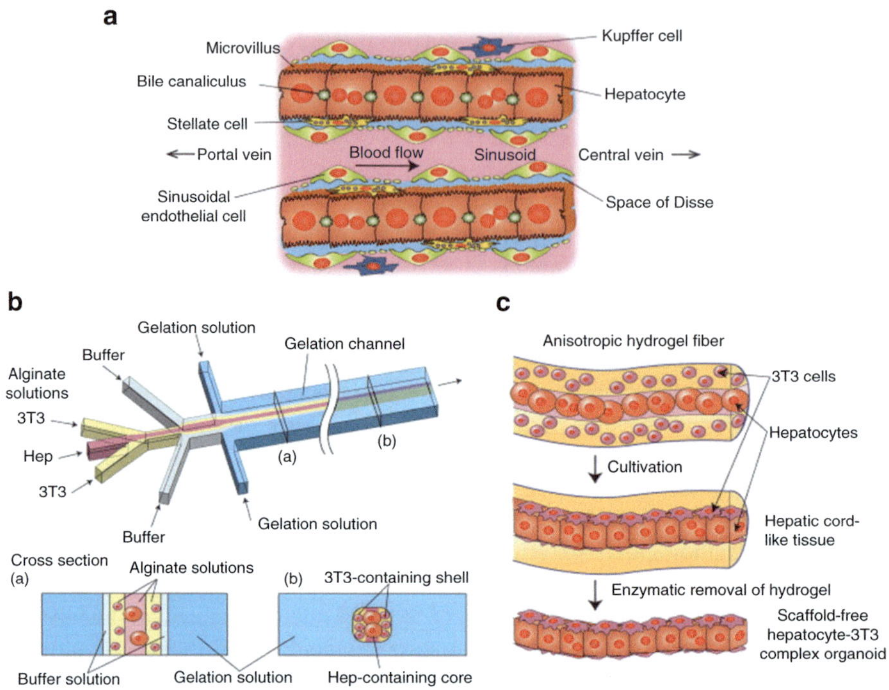

Fig. 6.12 Schematic illustration of the liver microstructure and the fabrication procedure of the hepatic cord-like micro-organoid by using the microfluidic system and the hydrogel microfiber. (**a**) Structure of the hepatic cord in the liver lobule. (**b**) Microfluidic system for fabricating sandwich-type alginate hydrogel microfibers that incorporate hepatocytes and 3T3 cells. (**c**) The formation process of hepatocyte-3T3 complex micro-organoids in the microfiber [118] (Copyright permission from Elsevier Science Ltd)

6.2.5.3 Bile Production

The liver is central to a multitude of physiologic functions, including bile production, sugar storage in the form of glycogen, nutrient inter-conversion among proteins, lipids and carbohydrates, detoxification and drug metabolism, phagocytosis, synthesis of many blood proteins, usually in combination with other systems and organs.

Hepatocytes excrete bile into bile canaliculi that join the bile duct in the portal triad, and finally coalesce to form the right and left bile ducts. They join outside the liver to form the common hepatic duct. The common hepatic duct unites with the cystic duct which drains the gallbladder to form the common bile duct. The latter one enters the duodenum at the duodenal papilla along with the pancreatic duct. Bile has several functions, mainly involved in fat absorption. When food enters the mouth, the sphincter around the duodenal papilla relaxes and when gastric contents enter the duodenum, the hormone from the small intestine causes the gallbladder to contract [119].

The liver produces and secretes about 600–1000 ml of bile per day, which circulates three times. Gallbladder, a small pouch that sits just under the liver, can store about 40–70 ml of bile that is five to ten times more concentrated than that of freshly secreted bile. There are no digestion enzymes in bile; the main components are bile salts, which neutralize and dilute stomach acid and emulsify fat. Cholesterol and alkaline phosphatase are excreted in the bile, many of which are subsequently reabsorbed into the enterohepatic circulation.

6.2.5.4 Nutrient Management

Glucose is a vital energy source for cells and levels in the blood stream, which must remain constant. The liver is central to blood glucose management because the liver is the only organ that can store and release glucose into the blood for use by other organs. Glycogen is a complex molecule composed of thousands of glucose units. Hepatocytes store glucose as intracellular glycogen granules. When energy intake exceeds energy output, the liver stores the surplus glucose as glycogen. When energy output exceeds energy intake, the liver releases glucose into the blood by performing glycogenolysis, the breakdown of glycogen into glucose. These processes response to the pancreatic hormones insulin and glucagon.

The liver is also responsible for gluconeogenesis, which is the synthesis of glucose from certain amino acids, lactate or glycerol. Adipose and liver cells produce glycerol by breakdown of fat, which the liver uses for gluconeogenesis.

The liver plays several roles in lipid metabolism: it performs cholesterol synthesis, lipogenesis, the production of triglycerides, and a bulk of the body's lipoproteins are synthesized in the liver, including: chylomicrons remnants, very low density lipoproteins (VLDL), low density lipoproteins (LDL), high density lipoproteins (HDL) and fatty acids [121].

The liver also uses dietary amino acids and those released during normal tissue catabolism to synthesize its own proteins and enzymes as well as plasma proteins. Plasma proteins produced by hepatocytes include: albumin, fibrinogen, prothrombin, a-fetoprotein, a2-macroglobin, hemopexin, transferrin, complement components C3, C6 and C1, a1-antitrypsin, caeruloplasmin [122]. Up to 50% of the livers' energy requirements can be supplied by amino acid oxidation. Oxidative deamination breaks amino acids into keto acid and an ammonia molecule. The keto acid is used in the Kreb's cycle to produce adenosine triphosphate (ATP). The liver combines ammonia with CO_2 to form urea and H_2O.

The liver also plays an important role in vitamin and mineral storage. About 80% of the body's vitamin A stores are concentrated in fat droplets within the stellate cells of the liver. In pathological conditions like hepatic fibrosis or liver cirrhosis the stellate cells lose vitamin A, transform into fibroblasts or myofibroblasts and begin producing large amounts of collagen and adhesive glycoproteins [123]. Normal vitamin A reserves are enough to prevent a deficiency for about 10 months. The liver also contains about a year supply of B12. Vitamin D stores equal about 3–4 months. Small amounts of Vitamins E and K and Vitamin C are stored in the liver to facilitate liver functions.

6.2.5.5 Phagocytosis and Detoxification

Kupffer cells are reticuloendothelial cells resident in the liver sinusoids that scavenge damaged red blood cells (RBCs) and bacteria as they pass through. Therefore, they are also known as liver macrophages. While hundreds of millions of RBCs are removed by the reticuloendothelial system every minute, Kupffer cellslyse RBCs into heme and globin. Globin is further catabolized into polypeptide components for reuse. Heme is broken into biliverdin and iron. Biliverdin is converted to bilirubin and finally excreted in the bile. Iron is transported to the liver and spleen for storage and to the bone for hematopoiesis.

Hepatocytes help the body eliminate xenobiotics (toxins) by transforming them into more excretable polar compounds through a two-step process. In PHASE I (Biotransformation), cytochrome p450 liver enzymes and NADPH-cytochrome c reductase catalyse hydrolysis, reduction and hydroxylation of the xenobiotic. These chemical reactions add or expose functional groups on the substance. This prepares it for conjugation with a polar, water soluble substance. PHASE II (Conjugation) liver enzymes such as glucuronosyltransferase, sulfotransferase, glutathione-S-transferase and acetyltransferase add glucuronide, sulfate, glutathione or acetyl adducts to the functional group making the conjugate more water soluble and more easily excreted in the urine and feces. In general, Phase I and Phase II reactions serve to minimize the accumulation of foreign substances by increasing their water solubility and excretion. However, these reactions may also transform some xenobiotics from a non-toxic substance to a toxic metabolite. Production of toxic metabolites within the liver can cause hepatic injury.

6.2.5.6 Biomaterials for Liver Tissue Engineering

End-stage liver diseases such as acute and chronic liver failure, cirrhosis, and liver cancer are serious health threats with high morbidity and mortality. There is currently no way to compensate for the absence of liver function in the long term, although liver dialysis techniques can be used in the short term. Artificial livers are yet to be developed to promote long term replacement in the absence of the liver. As of now, liver transplantation is the only option for complete liver failure, but limited by a severe shortage of donor organs, high cost of treatment, and lifetime immunosuppression [124]. Inquiry into materials for liver transplantation research has been necessary. Researchers intend to reconstruct the liver by tissue engineering, specifically, reseeding hepatocytes or stem cells in scaffolds to reconstruct liver like tissue to compensate liver function [125]. One of the main limiting factors hindering liver tissue engineering application is the lack of an ideal scaffold, which not only has all the necessary microstructural and ECM such as collagen, mucopolysaccharide and growth factor for cell adhesion, functioning, proliferation, migration, differentiation, but also provides microvascular networks for oxygen and nutrient transport, as well as metabolite excretion [126, 127]. So far, the types of biomaterials used in liver tissue engineering mainly include synthetic/natural polymeric scaffolds, decellularized scaffolds, and hydrogels [128].

Three-dimensional synthetic polymer scaffolds could facilitate liver tissue regeneration. Several types of biocompatible and biodegradable polymers, including PLGA, PCL, poly (glycolic acid) (PGA), and poly (Llactic acid) (PLLA) have been used to fabricate 3D scaffolds [129]. These synthetic polymers have advantages in pro-accessibility, good mechanical properties and manipulating degradation rate. Importantly, they have been approved by the Food and Drug Administration for specific applications in the human body. However, synthetic polymer scaffolds manufactured with current technology lack cell recognition signals and hinder successful cell seeding because of their hydrophobic trait [130]. Therefore, to encourage cell ingrowths, the effective hybridization with bioactive molecules is needed [131]. Moreover, synthetic polymers have the problem of potential toxicity from acidic degradation products [132].

Three-dimensional scaffolds fabricated with naturally derived polymers are suitable for cell activity and interaction because natural polymers contain numerous integrin-binding sites and growth factors [130]. Natural polymers used for liver tissue engineering include hyaluronate, collagen, alginate, and chitosan. However, scaffolds fabricated purely from these molecules exhibit poor mechanical strength and are not easy to handle. Meanwhile, controlling the rate and mode of degradation is difficult when compared with synthetic polymer scaffolds [133].

Injectable hydrogels may be useful for efficient and minimally invasive cell transplantation, especially for defects that are irregular in shape because of their in situ polymerization. The advantages of hydrogels include their viscoelastic characteristics, biocompatibility, easiness of fabrication in different shapes, and quickly mass transfer between nutrients and cells. Various types of hydrogel systems have been prepared from synthetic or natural polymers and applied as scaffold platforms for hepatocyte culture and transplantation. Due to the variability, the manufacturing techniques of hydrogels need to be further studied and developed [134, 135].

Hepatocytes and hepatic linage cells require 3D scaffolds with perfect ECM structures in order to maintain their hepatic stability and liver-specific functions. Thus, decellularized organ scaffold became one of the most popular biomaterials in liver tissue engineering because of its advantages of favorable biocompatibility, native ECM and vascular system which provide incomparable convenience for tissue culture and *in vivo* implantation. Researchers from different groups demonstrated the feasibility of recellularization of hepatocytes or stem cells on decellularized liver scaffold and compensatory liver function of organoids in rodent, goat and swine, that creates a decellularization upsurge all over the world [136–140]. There has also been report to use three-dimensional decellularized splenic scaffold as potential liver scaffold, considering the more extensive organ source [141].

6.2.6 Nerves

6.2.6.1 Composition of Nerves

The human nervous system consists of two main parts, the central nervous system (CNS) and the peripheral nervous system (PNS). The CNS is made up of the brain, spinal cord and nerves. It acts as the central control region of the human nervous system, integrating information and issuing commands. The PNS includes the somatic motor nervous system and the sensory nervous system, which connect the CNS to every other part of the body. The PNS consists mainly of nerves that transmit signals from the brain which are called motor or efferent nerves, while those nerves that transmit information from the body to the CNS are called sensory or afferent. Most nerves serve both functions and are called mixed nerves.

6.2.6.2 Main Categories of Cells in the Nervous System

The nervous system contains two main categories of cells: neurons and glial cells. Neurons generate and propagate electrical and chemical signals, whereas glia function mainly to modulate neuron function and signaling [142].

Neurons are the basic information-processing units of the nervous system (see Fig. 6.13) [143]. The main portion of the neuron is the cell body which contains a nucleus and cytoplasm. Over most of the cell there are numerous projections that branch out into still finer extensions. These branching threads are called dendrites. At one end of an elongated structure is a branching mass that is the axon. Along the axon are a number of protuberances labeled as myelin sheaths. The axon and pre-

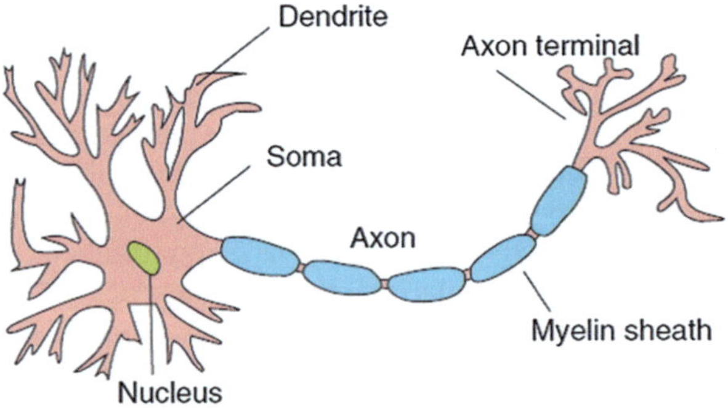

Fig. 6.13 Features of a generalized neuron. A neuron comprises the soma (cell body), dendrites, and axon. The soma gives rise to numerous dendrites which receive synaptic signals from other neurons. The soma has only one long thin axon which is covered by the myelin sheath and transmits signals to other neurons [143] (Copyright permission from Elsevier Science Ltd)

synaptic nerve terminal contain a large population of mRNAs and an active protein synthetic system which play a critical role in the development, function, and long-term viability of the neuron [144]. Mutations in RNA-binding proteins or perturbations of the mRNA metabolism can cause neurodegenerative disorders [145].

Glial cells are non-neuronal cells. There are many types of glia that perform various functions [142]. For example, glia form myelin, participate in signal transmission, maintain homeostasis, and provide support and nutrition in the nervous system [146]. In the human brain, the total number of glia is roughly estimated to equal the number of neurons [147]. Glial cells have various functions since they have unique protein compositions that can change depending on activity [142]. The main functions of glial cells are to destroy pathogens and remove dead neurons; to insulate neurons electrically; to supply nutrients to neurons; to support neurons; and to provide guidance cues directing the axons of neurons to their targets [146].

6.2.6.3 Function of Nerves

Nerve has the function of sensory stimulation and conduction excitatory. Information flow commences at the somatodendritic domain, which receives synaptic input from neighboring or distant neurons. It is subsequently transmitted to the axonal domain, which specializes in sending signals to the next neuron in the neural circuitry or to a target tissue [148].

6.2.6.4 Cytoskeleton of Nerves

The axon initial segment resides in between axonal and somatodendritic domains serves as the site of action potential firing and helps to maintain neuron polarity. The cytoskeleton is fundamental to the functions of axon initial segment [148]. Neurons use their microtubule arrays as architectural elements required for the elaboration of elongated axons and dendrites [149]. Microtubules and their associated proteins play important roles during the development of neuronal morphology, the plasticity of neurons, and neurodegenerative processes [150]. Microtubules also act as directional railways for organelle transport [149].

6.2.6.5 Neurodegenerative Diseases

Neurodegeneration is the umbrella term for the progressive loss of structure or function of neurons, including death of neurons. Neurodegenerative diseases result in progressive degeneration and/or death of neuron cells, such as Parkinson's disease, Alzheimer's disease, Huntington's disease and Amyotrophic lateral sclerosis.

Alzheimer's disease is characterised by loss of neurons and synapses in the cerebral cortex and certain subcortical regions. This loss results in gross atrophy of the affected regions, including degeneration in the temporal lobe and parietal lobe, and parts of the frontal cortex and cingulate gyrus [151].

Parkinson's disease is the second most common neurodegenerative disorder characterized by prominent loss of nigro-striatal dopaminergic neurons and synaptic dysfunction [152, 153]. It manifests as bradykinesia, rigidity, resting tremor and posture instability. Parkinson's disease is associated with alterations to many cell types, most notably microglia [154]. Many studies implicate enhanced inflammatory response in the development and progression of Parkinson's disease pathology. Understanding the inflammatory contributions to Parkinson's disease may facilitate the development of other strategies such as stem cell therapy [155].

6.2.6.6 Neural Tissue Engineering

Injuries of the nervous system occur commonly among people of many different ages and backgrounds. Recently, liver growth factor (LGF) has been used as a neural tissue regenerator. LGF administration stimulates neurogenesis and neuron survival and promotes migration of newly generated neurons. LGF treatment induces the outgrowth of striatal dopaminergic terminals in 6-hidroxydopamine-lesioned rats and protects dopaminergic neurons in hemiparkinsonian animals [156]. However, there are still no effective strategies to improve neural regeneration. Currently, Neural tissue engineering provides a promising avenue for regeneration of nerves [157]. Neural tissue engineering combining customized biomaterial scaffolds with cells offers an integrative approach to tackle the trauma, stroke, or degenerative diseases [158].

Since nanomaterials can be tailored at the molecular level and scaffold morphology may more closely resemble features of ECM components in terms of porosity, framing, and biofunctionalities, nanostructured scaffolds recently showed great promise in tissue engineering [159]. Both carbon nanotube and graphene hold great potential in the biomedical field because of their extraordinary properties [160]. Many novel biomaterials are being developed, such as multi-faceted nanoscaffolds built by the approaches and applications of self-assembling peptides and peptide amphiphiles for direct application to neural injury [161]. Besides, among all the biomaterials, hyaluronic acid is a promising candidate for central neural tissue engineering because of its unique physico-chemical and biological properties [162].

Most popular cell candidates used for neural tissue engineering include glial cells and stem/stem-like cells. Glial cells can secrete neuroprotective agents that stimulate natural neural repair. Stem/stem-like cells have the additional potential to replace lost tissue via differentiating into neuronal cells [158, 163]. Most studies have examined the effects of soluble pharmacological factors on the cellular neurogenesis. On the other hand, it is now recognized that mechanical cues such as stretch and shear stress also have a strong potential to induce cellular neurogenesis [164]. In addition, because patient-specific iPSC can be differentiated into disease-relevant cell types, including neurons, cellular reprogramming of somatic cells to human iPSC represents an efficient tool for *in vitro* modeling of human brain diseases and provides an innovative opportunity in the identification of new therapeutic drugs [153].

6.2.7 *Connective Tissue: Tendon and Ligament*

6.2.7.1 Function of Tendon and Ligament

Connective tissue is widely dispersed with diverse forms, including ligaments, tendons, fascia, and dermis. Ligaments and tendons are dense connective tissues. A ligament most commonly refers to the fibrous connective tissue that connects two bones to form a joint which is called articular ligaments. It allow a body part to flex or extend, help control their range of motion, and stabilize them so that the bones move in proper alignment. If a ligament primarily strengthens or supports other ligaments, it is called an accessory ligament. There are additional structures that are called ligaments even though they do not connect bones. For example, peritoneal ligaments surround a number of blood vessels, including the hepatic portal vein to the liver, and support significant parts of the reproductive system in women. A tendon is a tough band of fibrous connective tissue that is capable of withstanding tension. Tendons are similar to ligaments. Ligaments join one bone to another bone, while tendons connect muscle to bone.

6.2.7.2 Composition of Tendon and Ligament

Most tendons and ligaments are made of protein (collagen and elastin) and polysaccharide (proteoglycans). The composition allows them to stretch significantly when they move, such as when the elbow is bent or straightened. The main cell type found in tendon and ligament tissue is the fibroblast. Fibroblasts have a spindle-shaped morphology, are sparsely distributed throughout the tissue, and align in columns along the direction of the collagen fibers (see Fig. 6.14) [165].

6.2.7.3 Diseases of Tendon and Ligament

Achilles tendon rupture is a frequent injury with an increasing incidence. Surgical treatment of tendon ruptures is currently the most common therapeutic modality. Until now, there is no consensus regarding optimal treatment of tendon ruptures [166]. Tendon repair is a slow process in order to reestablish the tendon fiber continuity and the normal tissue function. The low cell density (5%), scarce oxygen and nutrient supply, and abnormal collagen deposition lead to scar tissue formation and adhesion [167, 168].

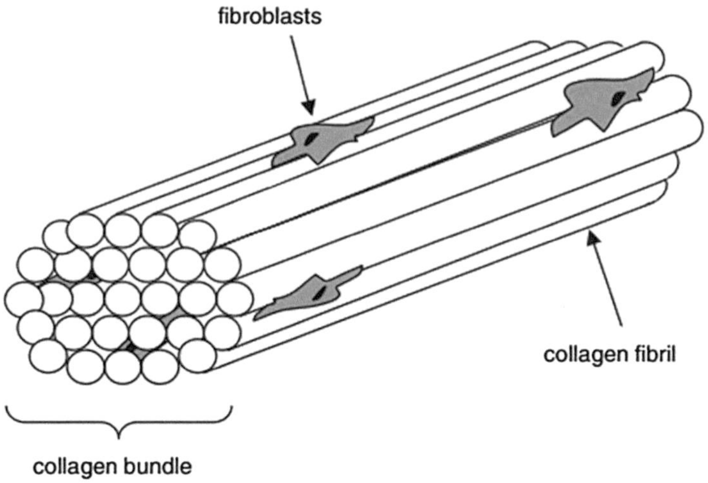

Fig. 6.14 Relationship between a collagen bundle and fibroblasts. The collagen bundle is comprised of individual collagen fibrils that tightly surround the cells [165] (Copyright permission from Elsevier Science Ltd)

6.2.7.4 Tendon and Ligament Tissue Engineering

To reproduce the biomechanical and biochemical properties similar to the native tissue structure, scaffolds for tendon and ligament treatment should have mechanical properties, biocompatibility, absence of inflammatory or immune response, and a close interaction with tendon cells [169, 170]. Nanoscaffold manufactured using electrospinning technology may provide great improvement in future practice [169]. Among biological scaffolds, the use of decellularized tendon-derived matrix increasingly represents an interesting approach to obtain medical devices for tendon substitution [168].

Stem cell therapy has been shown to a promising cell resource of tendon and ligament tissue engineering. From commonly investigated bone marrow-derived MSCs to emerging anterior cruciate ligament (ACL)-derived CD34+ stem cells, multiple stem cell types have been proven to be effective in accelerating tendon-bone healing [171]. Mechanical loading is recognized to play an important role in regulating the behaviors of cells in bone and surrounding tissues *in vivo* [172]. Tenocytes respond to mechanical deformation via force-sensitive membrane receptor activity, changes in cytoskeletal contractility, and transcriptional regulation [173]. Bioreactors is to provide a both an expandable source of readily available pluripotent cells and to facilitate their controlled differentiation into a clinically applicable ligament or tendon like neotissue [174]. Growth factors and small molecules are also be used in tendon and ligament tissue engineering [175].

6.3 Conclusions and Future Directions

An overview of the biochemical and physical properties of hard and soft tissues was presented, including bones, teeth, cartilages, blood vessels, liver, nerves, tendons, and ligaments. Moreover, tissue engineering concerning each tissue are also included. Engineered tissues consisting of biomaterials and cells can mimic the physiological function of the native tissues and be cultured on demand. Even though the recent advances toward the development of biomimetic materials for tissue regeneration applications, several challenges still remain including the precise design of the scaffold with the desired characteristics such as porosity, pore size, shape, the degradation rate, and mechanical properties.

Despite problems remain, the use of stem cells in tissue regeneration shows promise for tissue engineering and regenerative medicine. It is vital to determine whether stem cells are existed in all tissues, promote the differentiation efficiency and understand the signals required to direct the committed differentiation of stem cells. Although many bioreactors have been designed to optimize the conditions applied to the engineered tissues, flexible bioreactors are required to switch to specialized bioreactors to meet the transition from laboratory- to industrial-scale. The resulting devices will provide an economically viable approach to the automated manufacture of functional grafts.

Many technical challenges are still needed to overcome before we create "off-the-shelf" tissues, which can translate scientific discoveries into treatments for millions of patients. The optimization of scaffolds and an adequate source of healthy expandable cells as well as the creation of bioreactors are needed to produce large-scale engineered tissues. In addition, the preservation of the product is also challenged to have a long shelf-life and the successful use of diverse approaches to prevent tissue rejection is also needed to address. Along with the advancement of biomaterials, cells and bioreactors, the resulting higher quality tissues will lead to more complex strategies for potential therapeutic applications.

References

1. Shin H, Jo S, Mikos AG. Biomimetic materials for tissue engineering. Biomaterials. 2003;24(24):4353–64.
2. Lee KY, Mooney DJ. Hydrogels for tissue engineering. Chem Rev. 2001;101(7):1869–80.
3. Putnam AJ, Mooney DJ. Tissue engineering using synthetic extracellular matrices. Nat Med. 1996;2(7):824.
4. Heath CA. Cells for tissue engineering. Trends Biotechnol. 2000;18(1):17–9.
5. Chen GP, Takashi U, Tetsuya T. Scaffold design for tissue engineering. Macromol Biosci. 2002;2(2):67–77.
6. Hutmacher DW, GohJ CH, Teoh SH. An introduction to biodegradable materials for tissue engineering applications. Ann Acad Med Singap. 2001;30(2):183–91.
7. Yang S, Leong KF, Du Z, Chua CK. The design of scaffolds for use in tissue engineering. Part I Traditional factors. Tissue Eng. 2001;7(6):679–89.

8. Ishaug-Riley SL, Crane GM, Gurlek A, Miller MJ, Yasko AW, Yaszemski MJ, Mikos AG. Ectopic bone formation by marrow stromal osteoblast transplantation using poly (DL-lactic-co-glycolic acid) foams implanted into the rat mesentery. J Biomed Mater Res. 1997;36(1):1–8.
9. Sims DC, Butler PEM, Cao YL, Randolph MA, Black A, Yaremchuk MJ. Tissue engineered neocartilage using plasma derived polymer substrates and chondrocytes. Plast Reconstr Surg. 1998;101(6):1580–5.
10. Cao Y, Vacanti JP, Paige KT, Upton J, Vacanti CA. Transplantation of chondrocytes utilizing a polymer-cell construct to produce tissue-engineered cartilage in the shape of a human ear. Plast Reconstr Surg. 1997;100(2):297–302.
11. Freed LE, Grande DA, Lingbin Z, Emmanual J, Marquis JC, Langer R. Joint resurfacing using allograft chondrocytes and synthetic biodegradable polymer scaffolds. J Biomed Mater Res. 1994;28(8):891–9.
12. Kneser U, Kaufmann PM, Fiegel HC, Pollok JM, Kluth D, Herbst H, Rogiers X. Long-term differentiated function of heterotopically transplanted hepatocytes on three-dimensional polymer matrices. J Biomed Mater Res. 1999;47(4):494–503.
13. Hadlock T, Sundback C, Hunter D, Vacanti JP. A polymer foam conduit seeded with Schwann cells promotes guided peripheral nerve regeneration. Tissue Eng. 2000;6(2):119–27.
14. Lin VS, Lee MC, O'Neal S, McKean J, Sung KP. Ligament tissue engineering using synthetic biodegradable fiber scaffolds. Tissue Eng. 1999;5(5):443–51.
15. Boskey AL. Bone composition: relationship to bone fragility and antiosteoporotic drug effects. Bonekey Rep. 2013;2:447.
16. Dalle Carbonare L, Innamorati G, Valenti MT. Transcription factor Runx2 and its application to bone tissue engineering. Stem Cell Rev. 2012;8(3):891–7.
17. Wojcicka A, Bassett JH, Williams GR. Mechanisms of action of thyroid hormones in the skeleton. Biochim Biophys Acta. 2013;1830(7):3979–86.
18. Berendsen AD, Olsen BR. Bone development. Bone. 2015;80:14–8.
19. Sivaraj KK, Adams RH. Blood vessel formation and function in bone. Development. 2016;143(15):2706–15.
20. Foulke BA, Kendal AR, Murray DW, Pandit H. Fracture healing in the elderly: a review. Maturitas. 2016;92:49–55.
21. Link TM. Osteoporosis imaging: state of the art and advanced imaging. Radiology. 2012;263(1):3–17.
22. Gupta A, March L. Treating osteoporosis. Aust Prescr. 2016;39(2):40–6.
23. Monument MJ, Jones KB. Molecular characterization of bone tumors and implications for treatment and prognosis. J Natl Compr Cancer Netw. 2014;12(2):214–20.
24. Sheu A, Diamond T. Bone mineral density: testing for osteoporosis. Aust Prescr. 2016;39(2):35–9.
25. Panetta NJ, Gupta DM, Longaker MT. Bone regeneration and repair. Curr Stem Cell Res Ther. 2010;5(2):122–8.
26. Schroeder JE, Mosheiff R. Tissue engineering approaches for bone repair: concepts and evidence. Injury. 2011;42(6):609–13.
27. Dimitriou R, Jones E, McGonagle D, Giannoudis PV. Bone regeneration: current concepts and future directions. BMC Med. 2011;9:66.
28. Singh RS, Kaur N, Rana V, Kennedy JF. Recent insights on applications of pullulan in tissue engineering. Carbohydr Polym. 2016;153:455–62.
29. Hutmacher DW. Scaffolds in tissue engineering bone and cartilage. Biomaterials. 2000;21(24):2529–43.
30. Venkatesan J, Kim SK. Nano-hydroxyapatite composite biomaterials for bone tissue engineering–a review. J Biomed Nanotechnol. 2014;10(10):3124–40.
31. Dhivya S, Ajita J, Selvamurugan N. Metallic nanomaterials for bone tissue engineering. J Biomed Nanotechnol. 2015;11(10):1675–700.

32. Melke J, Midha S, Ghosh S, Ito K, Hofmann S. Silk fibroin as biomaterial for bone tissue engineering. Acta Biomater. 2016;31:1–16.
33. Dubey N, Bentini R, Islam I, Cao T, Castro Neto AH, Rosa V. Graphene: a versatile carbon-based material for bone tissue engineering. Stem Cells Int. 2015;2015:804213.
34. de Misquita MR, Bentini R, Goncalves F. The performance of bone tissue engineering scaffolds in in vivo animal models: a systematic review. J Biomater Appl. 2016;31(5):625–36.
35. Romagnoli C, Brandi ML. Adipose mesenchymal stem cells in the field of bone tissue engineering. World J Stem Cells. 2014;6(2):144–52.
36. Mercado-Pagán ÁE, Stahl AM, Shanjani Y, Yang Y. Vascularization in bone tissue engineering constructs. Ann Biomed Eng. 2015;43(3):718–29.
37. Huang Y, Niu X, Song W, Guan C, Feng Q, Fan Y. Combined effects of mechanical strain and hydroxyapatite/collagen composite on osteogenic differentiation of rat bone marrow derived mesenchymal stem cells. J Nanomater. 2013;2013(2013):343909.
38. Farokhi M, Mottaghitalab F, Shokrgozar MA, Ou KL, Mao C, Hosseinkhani H. Importance of dual delivery systems for bone tissue engineering. J Control Release. 2016;225:152–69.
39. Samorezov JE, Alsberg E. Spatial regulation of controlled bioactive factor delivery for bone tissue engineering. Adv Drug Deliv Rev. 2015;84:45–67.
40. Yun YR, Jang JH, Jeon E, Kang W, Lee S, Won JE, Kim HW, Wall I. Administration of growth factors for bone regeneration. Regen Med. 2012;7(3):369–85.
41. Gardel LS, Serra LA, Reis RL, Gomes ME. Use of perfusion bioreactors and large animal models for long bone tissue engineering. Tissue Eng Part B Rev. 2014;20(2):126–46.
42. Ceccarelli G, Bloise N, Vercellino M, Battaglia R, Morgante L, De Angelis MG, Imbriani M, Visai L. In vitro osteogenesis of human stem cells by using a three-dimensional perfusion bioreactor culture system: a review. Recent Pat Drug Deliv Formul. 2013;7(1):29–38.
43. Costa PF, Martins A, Neves NM, Gomes ME, Reis RL. Automating the processing steps for obtaining bone tissue-engineered substitutes: from imaging tools to bioreactors. Tissue Eng Part B Rev. 2014;20(6):567–77.
44. Bigham-Sadegh A, Oryan A. Selection of animal models for pre-clinical strategies in evaluating the fracture healing, bone graft substitutes and bone tissue regeneration and engineering. Connect Tissue Res. 2015;56(3):175–94.
45. Huo B. An inhomogeneous and anisotropic constitutive model of human dentin. J Biomech. 2005;38(3):587–94.
46. Murshid SA. The role of osteocytes during experimental orthodontic tooth movement: a review. Arch Oral Biol. 2016;73:25–33.
47. Isola G, Matarese G, Cordasco G, Perillo L, Ramaglia L. Mechanobiology of the tooth movement during the orthodontic treatment: a literature review. Minerva Stomatol. 2016;65(5):299–327.
48. Steindorff MM, Lehl H, Winkel A, Stiesch M. Innovative approaches to regenerate teeth by tissue engineering. Arch Oral Biol. 2014;59(2):158–66.
49. Yildirim S, Fu SY, Kim K, Zhou H, Lee CH, Li A, Kim SG, Wang S, Mao JJ. Tooth regeneration: a revolution in stomatology and evolution in regenerative medicine. Int J Oral Sci. 2011;3(3):107–16.
50. Yuan Z, Nie H, Wang S, Lee CH, Li A, Fu SY, Zhou H, Chen L, Mao JJ. Biomaterial selection for tooth regeneration. Tissue Eng Part B Rev. 2011;17(5):373–88.
51. Rosa V, Zhang Z, Grande RH, Nör JE. Dental pulp tissue engineering in full-length human root canals. J Dent Res. 2013;92(11):970–5.
52. Surendran S, Sivamurthy G. Current applications and future prospects of stem cells in dentistry. Dent Update. 2015;42(6):556–8. 560–1
53. Galler KM, D'Souza RN, Hartgerink JD, Schmalz G. Scaffolds for dental pulp tissue engineering. Adv Dent Res. 2011;23(3):333–9.
54. Piva E, Silva AF, Nör JE. Functionalized scaffolds to control dental pulp stem cell fate. J Endod. 2014;40(4 Suppl):S33–40.

55. Albuquerque MT, Valera MC, Nakashima M, Nör JE, Bottino MC. Tissue-engineering-based strategies for regenerative endodontics. J Dent Res. 2014;93(12):1222–31.
56. Aigner T, Sachse A, Gebhard PM, Roach HI. Osteoarthritis: pathobiology-targets and ways for therapeutic intervention. Adv Drug Deliv Rev. 2006;58(2):128–49.
57. Cole AG. A review of diversity in the evolution and development of cartilage: the search for the origin of the chondrocyte. Eur Cell Mater. 2011;21:122–9.
58. Hall BK, Miyake T. All for one and one for all: condensations and the initiation of skeletal development. BioEssays. 2000;22:138–47.
59. Lee P, Tran K, Chang W, Shelke NB, Kumbar SG, Yu X. Influence of chondroitin sulfate and hyaluronic acid presence in nanofibers and its alignment on the bone marrow stromal cells: cartilage regeneration. J Biomed Nanotechnol. 2014;10(8):1469–79.
60. Rothrauff BB, Tuan RS. Cellular therapy in bone-tendon interface regeneration. Organ. 2014;10(1):13–28.
61. Nam S, Cho W, Cho H, Lee J, Lee E, Son Y. Xiphoid process-derived chondrocytes: a novel cell source for elastic cartilage regeneration. Stem Cells Transl Med. 2014;3(11):1381–91.
62. Temple DK, Cederlund AA, Lawless BM, Aspden RM, Espino DM. Viscoelastic properties of human and bovine articular cartilage: a comparison of frequency-dependent trends. BMC Musculoskelet Disord. 2016;17(1):419.
63. Yao JQ, Seedhom BB. Mechanical conditioning of articular cartilage to prevalent stresses. Br J Rheumatol. 1993;32:956–65.
64. Rhee DK, Marcelino J, Baker M, Gong Y, Smits P, Lefebvre V, Jay GD, Stewart M, Wang H, Warman ML, Carpten JD. The secreted glycoprotein lubricin protects cartilage surfaces and inhibits synovial cell overgrowth. J Clin Invest. 2005;115(3):622–31.
65. Hayes WC, Mockros LF. Viscoelastic properties of human articular cartilage. J Appl Physiol. 1971;31(4):562–8.
66. Chen C, Tambe DT, Deng L, Yang L. Biomechanical properties and mechanobiology of the articular chondrocyte. Am J Phys Cell Physiol. 2013;305(12):C1202–8.
67. Mahir L, Belhaj K, Zahi S, Azanmasso H, Lmidmani F, El Fatimi A. Impact of knee osteoarthritis on the quality of life. Ann Phys Rehabil Med. 2016;59S:e159.
68. Findlay DM, Kuliwaba JS. Bone-cartilage crosstalk: a conversation for understanding osteoarthritis. Bone Res. 2016;4:16028.
69. Tran G, Smith TO, Grice A, Kingsbury SR, McCrory P, Conaghan PG. Does sports participation (including level of performance and previous injury) increase risk of osteoarthritis? A systematic review and meta-analysis. Br J Sports Med. 2016;pii:bjsports-2016-096142.
70. McAlindon TE, Bannuru RR, Sullivan MC, Arden NK, Berenbaum F, Bierma-Zeinstra SM, Hawker GA, Henrotin Y, Hunter DJ, Kawaguchi H, Kwoh K, Lohmander S, Rannou F, Roos EM, Underwood M. OARSI guidelines for the non-surgical management of knee osteoarthritis. Osteoarthr Cartil. 2014;22(3):363–88.
71. Nesic D, Whiteside R, Brittberg M, Wendt D, Martin I, Mainil-Varlet P. Cartilage tissue engineering for degenerative joint disease. Adv Drug Deliv Rev. 2006;58(2):300–22.
72. Baghaban Eslaminejad M, Malakooty PE. Mesenchymal stem cells as a potent cell source for articular cartilage regeneration. World J Stem Cells. 2014;6(3):344–54.
73. Panadero JA, Lanceros-Mendez S, Ribelles JL. Differentiation of mesenchymal stem cells for cartilage tissue engineering: individual and synergetic effects of three-dimensional environment and mechanical loading. Acta Biomater. 2016;33:1–12.
74. Ge Y, Gong YY, Xu Z, Lu Y, Fu W. The application of sheet technology in cartilage tissue engineering. Tissue Eng Part B Rev. 2016;22(2):114–24.
75. Gonzalez JS, Alvarez VA. Mechanical properties of polyvinylalcohol/hydroxyapatite cryogel as potential artificial cartilage. J Mech Behav Biomed Mater. 2014;34:47–56.
76. Rampichová M, Filová E, Varga F, Lytvynets A, Prosecká E, Koláčná L, Motlík J, Nečas A, Vajner L, Uhlík J, Amler E. Fibrin/hyaluronic acid composite hydrogels as appropriate scaffolds for in vivo artificial cartilage implantation. ASAIO J. 2010;56(6):563–8.

77. Kwon H, Paschos NK, Hu JC, Athanasiou K. Articular cartilage tissue engineering: the role of signaling molecules. Cell Mol Life Sci. 2016;73(6):1173–94.
78. Khozoee B, Mafi P, Mafi R. Khan WS. Mechanical stimulation protocols of human derived cells in articular cartilage tissue engineering – a systematic review. Curr Stem Cell Res Ther. 2017;12(3):260–70.
79. Fantini S. Dynamic model for the tissue concentration and oxygen saturation of hemoglobin in relation to blood volume, flow velocity, and oxygen consumption: implications for functional neuroimaging and coherent hemodynamics spectroscopy (CHS). NeuroImage. 2014;85(Pt 1):202–21.
80. Khan OF, Sefton MV. Endothelialized biomaterials for tissue engineering applications in vivo. Trends Biotechnol. 2011;29(8):379–87.
81. Hendry C, Farley A, McLafferty E. Blood vessels, circulation and blood pressure. Nurs Stand. 2012;27(11):35–40.
82. Cines DB, Pollak ES, Buck CA, Loscalzo J, Zimmerman GA, McEver RP, Pober JS, Wick TM, Konkle BA, Schwartz BS, Barnathan ES, McCrae KR, Hug BA, Schmidt AM, Stern DM. Endothelial cells in physiology and in the pathophysiology of vascular disorders. Blood. 1998;91(10):3527–61.
83. Lacolley P, Regnault V, Nicoletti A, Li Z, Michel J-B. The vascular smooth muscle cell in arterial pathology: a cell that can take on multiple roles. Cardiovasc Res. 2012;95(2):194–204.
84. Ladich E, Yahagi K, Romero ME, Virmani R. Vascular diseases: aortitis, aortic aneurysms, and vascular calcification. Cardiovasc Pathol. 2016;25(5):432–41.
85. Murray CJ, Lopez AD. Mortality by cause for eight regions of the world: global burden of disease study. Lancet. 1997;349(9061):1269–76.
86. Pashneh-Tala S, MacNeil S, Claeyssens F. The tissue-engineered vascular graft-past, present, and future. Tissue Eng Part B-Rev. 2016;22(1):68–100.
87. Swartz DD, Andreadis ST. Animal models for vascular tissue-engineering. Curr Opin Chem Biol. 2013;24(5):916–25.
88. Gupta B, Kasyanov V. Biomechanics of human common carotid artery and design of novel hybrid textile compliant vascular grafts. J Biomed Mater Res. 1997;34(3):341–9.
89. Eslami MH, Gangadharan SP, Belkin M, Donaldson MC, Whittemore AD, Conte MS. Monocyte adhesion to human vein grafts: a marker for occult intraoperative injury? J Vasc Surg. 2001;34(5):923–8.
90. Sarkar S, Salacinski HJ, Hamilton G, Seifalian AM. The mechanical properties of infrainguinal vascular bypass grafts: their role in influencing patency. Eur J Vasc Endovasc. 2006;31(6):627–36.
91. Tiwari A, Cheng KS, Salacinski H, Hamilton G, Seifalian AM. Improving the patency of vascular bypass grafts: the role of suture materials and surgical techniques on reducing anastomotic compliance mismatch. Eur J Vasc Endovasc. 2003;25(4):287–95.
92. Greenwald SE, Berry CL. Improving vascular grafts: the importance of mechanical and haemodynamic properties. J Pathol. 2000;190(3):292–9.
93. Haruguchi H, Teraoka S. Intimal hyperplasia and hemodynamic factors in arterial bypass and arteriovenous grafts: a review. J Artif Organs. 2003;6(4):227–35.
94. Ballyk PD, Walsh C, Butany J, Ojha M. Compliance mismatch may promote graft–artery intimal hyperplasia by altering suture-line stresses. J Biomech. 1997;31(3):229–37.
95. Davies MG, Hagen P-O. Pathophysiology of vein graft failure: a review. Eur J Vasc Endovasc. 1995;9(1):7–18.
96. Fitz Gibbon GM, Leach AJ, Kafka HP, Keon WJ. Coronary bypass graft fate: long-term angiographic study. J Am Coll Cardiol. 1991;17(5):1075–80.
97. Van der Wal A, Becker A, Elbers J, Das P. An immunocytochemical analysis of rapidly progressive atherosclerosis in human vein grafts. Eu J Cardio-Thorac. 1991;6(9):469–73.
98. Cox JL, Chiasson DA, Gotlieb AI. Stranger in a strange land: the pathogenesis of saphenous vein graft stenosis with emphasis on structural and functional differences between veins and arteries. Prog Cardiovasc Dis. 1991;34(1):45–68.

99. Padberg FT, Calligaro KD, Sidawy AN. Complications of arteriovenous hemodialysis access: recognition and management. J Vasc Surg. 2008;48(5):S55–80.
100. Zetrenne E, McIntosh BC, McRae MH, Gusberg R, Evans G, Narayan D. Prosthetic vascular graft infection: a multi-center review of surgical management. Yale J Biol Med. 2007;80(3):113–21.
101. Chiesa R, Astore D, Frigerio S, Garriboli L, Piccolo G, Castellano R, Scalamogna M, Odero A, Pirrelli S, Biasi G, Mingazzini P, Biglioli P, Polvani G, Guarino A, Agrifoglio G, Tori A, Spina G. Vascular prosthetic graft infection: epidemiology, bacteriology, pathogenesis and treatment. Acta Chir Belg. 2002;102(4):238–47.
102. Chlupac J, Filova E, Bacakova L. Blood vessel replacement: 50 years of development and tissue engineering paradigms in vascular surgery. Physiol Res. 2009;58:S119.
103. Owida AA, Do H, Morsi YS. Numerical analysis of coronary artery bypass grafts: an over view. Comput Methods Prog Biomed. 2012;108(2):689–705.
104. Scharn D, Daamen W, van Kuppevelt T, van der Vliet J. Biological mechanisms influencing prosthetic bypass graft patency: possible targets for modern graft design. Eur J Vasc Endovasc. 2012;43(1):66–72.
105. Weinberg CB, Bell E. A blood vessel model constructed from collagen and cultured vascular cells. Science. 1986;231(4736):397–400.
106. Isenberg BC, Williams C, Tranquillo RT. Small-diameter artificial arteries engineered in vitro. Circ Res. 2006;98(1):25–35.
107. Mitchell SL, Niklason LE. Requirements for growing tissue-engineered vascular grafts. Cardiovasc Pathol. 2003;12(2):59–64.
108. Kelly DA. Diseases of the liver and biliary system in children. 3rd ed. Chichester: Wiley; 2008.
109. Tortora G, Derrickson B. Principles of anatomy and physiology. 12th ed. Wiley: New York; 2008.
110. Ellis H. Anatomy of the liver. Surgery (Oxford). 2011;29(12):589–92.
111. Si-Tayeb K, Lemaigre FP, Duncan SA. Organogenesis and development of the liver. Dev Cell. 2010;18(2):175–89.
112. Morin O, Goulet F, Normand C. Liver sinusoidal endothelial cells: isolation, purification, characterization and interaction with hepatocytes. Revisiones Sobre Biología Celular Rbc. 1988;15:1–85.
113. Malarkey DE, Johnson K, Ryan L, Boorman G, Maronpot RR. New insights into functional aspects of liver morphology. Toxicol Pathol. 2005;33(1):27–34.
114. Alberts B, Johnson A, Lewis J, Raff M, Roberts K, Walter P. Specialized tissues, stem cells, and tissue renewal. Molecular biology of the cell. 5th ed. New York: Garland Science, Taylor & Francis Group; 2008.
115. Naito M, Hasegawa G, Takahashi K. Development, differentiation, and maturation of Kupffer cells. Microsc Res Techn. 1997;39(39):350–64.
116. Geerts A. History, heterogeneity. Developmental biology, and functions of quiescent hepatic stellate cells. Semin Liver Dis. 2001;21(3):311–35.
117. Tietz PS, Larusso NF. Cholangiocyte biology. Curr Opin Gastroenterol. 2006;22(3):279–87.
118. Yamada M, Utoh R, Ohashi K, Tatsumi K, Yamato M, Okano T, Seki M. Controlled formation of heterotypic hepatic micro-organoids in anisotropic hydrogel microfibers for long-term preservation of liver-specific functions. Biomaterials. 2012;33(33):8304–15.
119. Campbell I. Liver: functional anatomy and blood supply. Anaesthesia Intens Care Med. 2006;7(2):49–51.
120. Seely RR, Stephens TD, Tate P. Anat Physiol. 7th ed. New York: McGraw-Hill; 2006.
121. Kuntz E, Kuntz H-D. Hepatology: textbook and atlas. 3rd ed. New York: Springer; 2008.
122. Sherlock S, Dooley J. Diseases of the liver and biliary system. 11th ed. Chichester: Blackwell Science; 2002.

123. Senoo H, Yoshikawa K, Mezaki Y, Morii M, Fujiwara M. Regulation of vitamin a homeostasis by the stellate cell (Vitamin A-storing cell) system. New York: Nova Science Publishers Inc; 2011.
124. Kim WR, Stock PG, Smith JM, Heimbach JK, Skeans MA, Edwards EB, Harper AM, Snyder JJ, Israni AK, Kasiske BL. OPTN/SRTR 2011 annual data report: liver. Am J Transplant. 2013;13(Suppl s1):73–102.
125. Badylak SF, Weiss DJ, Caplan A, Macchiarini P. Engineered whole organs and complex tissues. Lancet. 2012;379(9819):943–52.
126. Gilbert TW, Sellaro TL, Badylak SF. Decellularization of tissues and organs. Biomaterials. 2006;27(19):3675–83.
127. Badylak SF. The extracellular matrix as a biologic scaffold material. Biomaterials. 2007;28(25):3587–93.
128. Lee JS, Cho SW. Liver tissue engineering: recent advances in the development of a bio-artificial liver. Biotechnol Bioprocess Eng. 2012;17(3):427–38.
129. Lee NK, Oh HJ, Hong CM, et al. Comparison of the synthetic biodegradable polymers, polylactide (PLA), and polylactic- co-glycolic acid (PLGA) as scaffolds for artificial cartilage. Biotechnol Bioprocess Eng. 2009;14(2):180–6.
130. Tibbitt MW, Anseth KS. Hydrogels as extracellular matrix mimics for 3D cell culture. Biotechnol Bioeng. 2009;103(4):655–63.
131. Venugopal J, Zhang YZ, Ramakrishna S. Fabrication of modified and functionalized polycaprolactone nanofibre scaffolds for vascular tissue engineering. Nanotechnology. 2005;16(16):2138–42.
132. Kim BS, Mooney DJ. Development of biocompatible synthetic extracellular matrices for tissue engineering. Trends Biotechnol. 1998;16(5):224–30.
133. Li YS, Harn HJ, Hsieh DK, Wen TC, Subeq YM, Sun LY, Lin SZ, Chiou TW. Cells and materials for liver tissue engineering. Cell Transplant. 2013;22(4):685–700.
134. Wang S, Nagrath D, Chen PC, Berthiaume F, Yarmush ML. Three-dimensional primary hepatocyte culture in synthetic self-assembling peptide hydrogel. Tissue Eng A. 2008;14(2):227–36.
135. Chen AA, Thomas DK, Ong LL, Schwartz RE, Golub TR, Bhatia SN. Humanized mice with ectopic artificial liver tissues. Proc Natl Acad Sci. 2011;108(29):11842–7.
136. Ji R, Zhang N, You N, Li Q, Liu W, Jiang N, Jiang N, Liu J, Zhang H, Wang D, Tao K, Dou K. The differentiation of MSCs into functional hepatocyte-like cells in a liver biomatrix scaffold and their transplantation into liver-fibrotic mice. Biomaterials. 2012;33(35):8995–9008.
137. Yagi H, Fukumitsu K, Fukuda K, Kitago M, Shinoda M, Obara H, Itano O, Kawachi S, Tanabe M, Coudriet GM, Piganelli JD, Gilbert TW, Soto-Gutierrez A, Kitagawa Y. Human-scale whole-organ bioengineering for liver transplantation: a regenerative medicine approach. Cell Transplant. 2013;22(2):231–42.
138. Nakamura S, Ijima H. Solubilized matrix derived from decellularized liver as a growth factor-immobilizable scaffold for hepatocyte culture. J Biosci Bioeng. 2013;116(6):746–53.
139. Uygun BE, Soto-Gutierrez A, Yagi H, Izamis ML, Guzzardi MA, Shulman C, Milwid J, Kobayashi N, Tilles A, Berthiaume F, Hertl M, Nahmias Y, Yarmush ML, Uygun K. Organ reengineering through development of a transplantable recellularized liver graft using decellularized liver matrix. Nat Med. 2010;16(7):814–20.
140. Soto-Gutierrez A, Zhang L, Medberry C, Fukumitsu K, Faulk D, Jiang H, Reing J, Gramignoli R, Komori J, Ross M, Nagaya M, Lagasse E, Stolz D, Strom SC, Fox IJ, Badylak SF. A whole-organ regenerative medicine approach for liver replacement. Tissue Eng Part C Methods. 2011;17(6):677–86.
141. Xiang JX, Zheng XL, Gao R, Wu WQ, Zhu XL, Li JH, Lv Y. Liver regeneration using decellularized splenic scaffold: a novel approach in tissue engineering. Hepatobiliary Pancr Dis Int. 2015;14(5):502–8.
142. Rasband MN. Glial contributions to neural function and disease. Mol Cell Proteomics. 2016;5(2):355–61.

143. Moore FB, Baleja JD. Molecular remodeling mechanisms of the neural somatodendritic compartment. Biochim Biophys Acta. 2012;1823(10):1720–30.
144. Kaplan BB, Kar AN, Gioio AE, Aschrafi A. MicroRNAs in the axon and presynaptic nerve terminal. Front Cell Neurosci. 2013;7:126.
145. Lenzken SC, Achsel T, Carrì MT, Barabino SM. Neuronal RNA-binding proteins in health and disease. Wiley Interdiscip Rev RNA. 2014;5(4):565–76.
146. Allen NJ, Barres BA. Neuroscience: glia – more than just brain glue. Nature. 2009;457(7230):675–7.
147. Azevedo FA, Carvalho LR, Grinberg LT, Farfel JM, Ferretti RE, Leite RE, Jacob Filho W, Lent R, Herculano-Houzel S. Equal numbers of neuronal and nonneuronal cells make the human brain an isometrically scaled-up primate brain. J Comp Neurol. 2009;513(5):532–41.
148. Jones SL, Svitkina TM. Axon initial segment cytoskeleton: architecture, development, and role in neuron polarity. Neural Plast. 2016;2016:6808293.
149. Baas PW, Rao AN, Matamoros AJ, Leo L. Stability properties of neuronal microtubules. Cytoskeleton (Hoboken). 2016;73(9):442–60.
150. Penazzi L, Bakota L, Brandt R. Microtubule dynamics in neuronal development, plasticity, and neurodegeneration. Int Rev Cell Mol Biol. 2016;321:89–169.
151. Wenk GL. Neuropathologic changes in Alzheimer's disease. J Clin Psychiatry. 2003;64(Suppl 9):7–10.
152. Bellucci A, Mercuri NB, Venneri A, Faustini G, Longhena F, Pizzi M, Missale C, Spano P. Review: Parkinson's disease: from synaptic loss to connectome dysfunction. Neuropathol Appl Neurobiol. 2016;42(1):77–94.
153. Torrent R, De Angelis RF, Dell'Era P, Memo M, Raya A, Consiglio A. Using iPS cells toward the understanding of Parkinson's disease. J Clin Med. 2015;4(4):548–66.
154. Vivekanantham S, Shah S, Dewji R, Dewji A, Khatri C, Ologunde R. Neuroinflammation in Parkinson's disease: role in neurodegeneration and tissue repair. Int J Neurosci. 2015;125(10):717–25.
155. Stojkovska I, Wagner BM, Morrison BE. Parkinson's disease and enhanced inflammatory response. Exp Biol Med (Maywood). 2015;240(11):1387–95.
156. Gonzalo-Gobernado R, Calatrava-Ferreras L, Perucho J, Reimers D, Casarejos MJ, Herranz AS, Jimenez-Escrig A, Diaz-Gil JJ, Bazan E. Liver growth factor as a tissue regenerating factor in neurodegenerative diseases. Recent Pat CNS Drug Discov. 2014;9(3):173–80.
157. Zhu W, O'Brien C, O'Brien JR, Zhang LG. 3D nano/microfabrication techniques and nanobiomaterials for neural tissue regeneration. Nanomedicine (London). 2014;9(6):859–75.
158. Wong FS, Chan BP, Lo AC. Carriers in cell-based therapies for neurological disorders. Int J Mol Sci. 2014;15(6):10669–723.
159. Saracino GA, Cigognini D, Silva D, Caprini A, Gelain F. Nanomaterials design and tests for neural tissue engineering. Chem Soc Rev. 2013;42(1):225–62.
160. Kumar S, Chatterjee K. Comprehensive review on the use of graphene-based substrates for regenerative medicine and biomedical devices. ACS Appl Mater Interf. 2016;8(40):26431–57.
161. Koss KM, Unsworth LD. Neural tissue engineering: bioresponsive nanoscaffolds using engineered self-assembling peptides. Acta Biomater. 2016;44:2–15.
162. Wang X, He J, Wang Y, Cui FZ. Hyaluronic acid-based scaffold for central neural tissue engineering. Interf Focus. 2012;2(3):278–91.
163. Moshayedi P, Carmichael ST. Hyaluronan, neural stem cells and tissue reconstruction after acute ischemic stroke. Biomatter. 2013;3(1):pii: e23863.
164. Stoll H, Kwon IK, Lim JY. Material and mechanical factors: new strategy in cellular neurogenesis. Neural Regen Res. 2014;9(20):1810–3.
165. Doroski DM, Brink KS, Temenoff JS. Techniques for biological characterization of tissue-engineered tendon and ligament. Biomaterials. 2007;28(2):187–202.
166. Holm C, Kjaer M, Eliasson P. Achilles tendon rupture–treatment and complications: a systematic review. Scand J Med Sci Sports. 2015;25(1):e1–10.

167. James R, Kesturu G, Balian G, Chhabra AB. Tendon: biology, biomechanics, repair, growth factors, and evolving treatment options. J Hand Surg [Am]. 2008;33(1):102–12.
168. Lovati AB, Bottagisio M, Moretti M. Decellularized and engineered tendons as biological substitutes: a critical review. Stem Cells Int. 2016;2016:7276150.
169. Chen J, Xu J, Wang A, Zheng M. Scaffolds for tendon and ligament repair: review of the efficacy of commercial products. Exp Rev Med Devices. 2009;6(1):61–73.
170. Font Tellado S, Balmayor ER, Van Griensven M. Strategies to engineer tendon/ligament-to-bone interface: biomaterials, cells and growth factors. Adv Drug Deliv Rev. 2015;94:126–40.
171. Hao ZC, Wang SZ, Zhang XJ, Lu J. Stem cell therapy: a promising biological strategy for tendon-bone healing after anterior cruciate ligament reconstruction. Cell Prolif. 2016;49(2):154–62.
172. Yu HS, Kim JJ, Kim HW, Lewis MP, Wall I. Impact of mechanical stretch on the cell behaviors of bone and surrounding tissues. J Tissue Eng. 2015;6:2041731415618342.
173. Youngstrom DW, Barrett JG. Engineering tendon: scaffolds, bioreactors, and models of regeneration. Stem Cells Int. 2016;2016:3919030.
174. Mace J, Wheelton A, Khan WS, Anand S. The role of bioreactors in ligament and tendon tissue engineering. Curr Stem Cell Res Ther. 2016;11(1):35–40.
175. Hiong Teh TK, Hong Goh JC, Toh SL. Controlled bioactive molecules delivery strategies for tendon and ligament tissue engineering using polymeric nanofibers. Curr Pharm Des. 2015;21(15):1991–2005.

Chapter 7
Scaffolds Reinforced by Fibers or Tubes for Hard Tissue Repair

Jianrong Wei, Tianxiao Zhao, Jie Liao, Yan Liu, Linhao Li, and Xiaoming Li

Abstract Formed by biological mineralization, hard tissues not only play a supporting role *in vivo*, but also contribute to the protection of internal organs. Currently, hard tissue defects have already become a type of common disease owing to fracture, trauma, pathological biomineralization or degeneration, etc. It has been well recognized that scaffolds reinforced by fibers or tubes can overcome the shortages of traditional tissue grafts to repair hard tissues. However, to achieve satisfactory scaffolds, their composition, physicochemical properties, biodegradability, biocompatibility and bioactivity should be comprehensively studied. In this chapter, we mainly introduce those properties and their influential factors of scaffolds reinforced by fibers or tubes, mainly focusing on special design and fabrication of novel satisfactory scaffolds reinforced by fibers or tubes for hard tissue repair, and how the addition of fibers or tubes affects the special functions *in vitro* and *in vivo*, which suggests that the reinforcement using fibers or tubes can enhance significantly hard tissue repair or regeneration efficacy, and should be one promising means to develop new high-performanced scaffolds for hard tissue repair.

Keywords Scaffold • Mechanical properties • Reinforce • Fiber • Tube • Hard tissue repair

7.1 Introduction

Hard tissues play indispensable roles in the support, mastication and protection of the internal organs in the human being. To date, hard tissue-related diseases and defects have been developed into a type of common cases in the clinic. Over the past few decades, the use of autogenous and allogenous tissue grafts has been the traditional method to repair hard tissues. However, the harvest of bone grafts inevitably

---

Jianrong Wei, Tianxiao Zhao and Jie Liao contributed equally to this work.

J. Wei • T. Zhao • J. Liao • Y. Liu • L. Li • X. Li (✉)
Key Laboratory for Biomechanics and Mechanobiology of Ministry of Education, School of Biological Science and Medical Engineering, Beihang University, Beijing 100191, China
e-mail: x.m.li@hotmail.com

X. Li (ed.), *Tissue Repair*, DOI 10.1007/978-981-10-3554-8_7

leads to the increase of additional trauma parts, and the available tissue source is not enough in many cases. Recently, using biomaterials to repair hard tissue have been widely studied due to their special properties which can overcome or avoid the deficiencies of bone grafts [1, 2].

Currently, traditional hard tissue repair materials are metals, polymers, ceramic materials and their composites [3, 4]. The metal replacement materials are one kind of the most frequently used substitutes in orthopedic and dental fields, such as artificial joint, artificial centrum, fracture fixation plate, bone nails, dental implants [5, 6]. Polymers, including polyester, polyethylene, polytetrafluoroethylene, polylactic acid, etc., have been used as another kind of artificial bone and joint materials [7, 8]. Moreover, biological ceramics play an important role in the repair and reconstruction of the hard tissues [9]. However, the current hard tissue repair materials cannot satisfy all the clinical requirements. Therefore, numerous novel biomaterials have been tried to be applied in hard tissue repair, one of which is the scaffolds reinforced by fibers or tubes [10]. After the study of past 20 years, this kind of reinforced scaffolds have very wide application range, not only concerning small hard tissue defect repair but also having the ability to repair big ones based on their excellent mechanical properties.

To achieve the satisfactory-performaced scaffolds, their physical and mechanical properties, such as porosity, pore interconnectivity, pore size, mechanical strength and the surface-area-to-volume ratio(SA:V) should be comprehensively studied [11]. It has been shown that those properties also have significant effects on the biocompatibility and biodegradability of the scaffold, thereby indirectly influencing the promotion of cellular functions in the scaffold within the surrounding tissues [12]. The availability of nutrients and oxygen required for cell viability and infiltration are strongly dependent on the scaffold's porosity, pore interconnectivity and pore size [13–15]. Sufficient mechanical strength is required for the scaffold to support itself against the surrounding tissues during cell proliferation. Also, it has been shown that the attachment and proliferation of cells can be readily improved by increasing the scaffold SA:V [16–19]. Recently, fibers or tubes based structures applied in hard tissue repair attract much attention by providing large functional surfaces for desired biological interactions and cell adhesion [20–22]. Moreover, the mechanical properties of tissue engineering scaffolds reinforced by fibers or tubes can be well controlled to meet the requirements of different hard tissue repair.

In this chapter, we mainly introduce the special properties and their influential factors of scaffolds reinforced by fibers or tubes for hard tissue repair, mainly focusing on their special design and fabrication and how the addition of fibers or tubes affects the special functions *in vitro* and *in vivo*. In addition, the recent studies to improve the mechanical properties of the scaffolds reinforced by fibers or tubes will be discussed. Moreover, the current applications of this kind of reinforced scaffolds in hard tissue repair are reviewed from two aspects: dental repair and bone repair. Each aspect will be detailedly introduced from *in vitro* and *in vivo* studies, which indicates that the scaffolds reinforced by fibers or tubes have great potential to be more widely used for hard tissue repair in near future.

7.2 The Mechanical Properties of the Scaffold Reinforced by Fibers or Tubes for Hard Tissue Repair

In order to provide a suitable biomimetic extracellular matrix for the growth of osteoblast-related cells, the mechanical properties of scaffolds should be close to the human hard tissues. Generally, the Young's modulus value of natural bone varies from 0.1 to 27 GPa, which exhibits hyperelastic performance and high mechanical properties. In particular, human cancellous bone with an elastic modulus in the range of 0.1–0.5 GPa, and compressive strength in the range of 4–12 MPa. However, the elastic modulus value of human compact bone is between 12 and 18 GPa, and its compressive strength is between 130 and 180 MPa. Once the mechanical properties of the scaffold materials cannot meet the requirements, it could lead to bone resorption and even implant failure [23]. Therefore, the desired scaffold should have appropriate strength, stiffness and mechanical behaviors. In addition, the physical properties of scaffolds can affect their mechanical properties, such as the porosity, pore interconnectivity, pore size. Generally, if biological scaffolds possess porous structure, they can offer large, interconnected pore spaces for cell adhesion and the nutrients interchange to supply the cells or metabolic waste to be removed, respectively. However, this structure tends to reduce the mechanical properties of the scaffold. Because of the good mechanical properties of the fiber and tubular materials, they are usually used in the preparation of scaffold materials in the field of hard tissue to overcome the shortcoming. As long as the design and preparation conditions are there, fiber or tube reinforced scaffold materials can meet the requirements of mechanical properties. Meanwhile, the scaffolds may show different structural properties, functional characteristics through appropriate preparation methods, which may be more suitable for hard tissue repair [24, 25]. To our knowledge, many kinds of fibers and tubes have already served as reinforcing component in scaffolds based on their excellent mechanical properties. Moreover, they can provide large specific surface areas to facilitate the improvement in the biological functions of the scaffolds [26].

Carbon fibers have been utilized as biological scaffold mechanical support materials due to their nontoxic property, good biocompatibility, high strength and modulus [27]. Moreover, they can guide the proliferation of fibrous tissue by making the tissue cells follow the orientation of their filaments; therefore, they can promote the tissue regeneration process [28]. Shi et al. prepared a novel scaffold for hard tissues by introducing activated carbon fibers (ACF) with high adsorption capacity into poly(lactic-co-glycolic) acid (PLGA) [29]. The ACF and PLGA all were the materials with excellent bioactivity and biocompatibility. They obtained the ACF by high-temperature processing of carbon fibers to get richly distributed pore structures. In addition, the ACF/PLGA composite scaffolds were prepared by switched capacitor/piecewise-linea (SC/PL) method with sodium chloride particles as a leachable component. SC/PL method can change the porosity and pore size of the scaffolds by respectively controlling the initial salt weight fraction and sieving salt particle size [30]. Their results showed that the scaffolds had good mechanical properties.

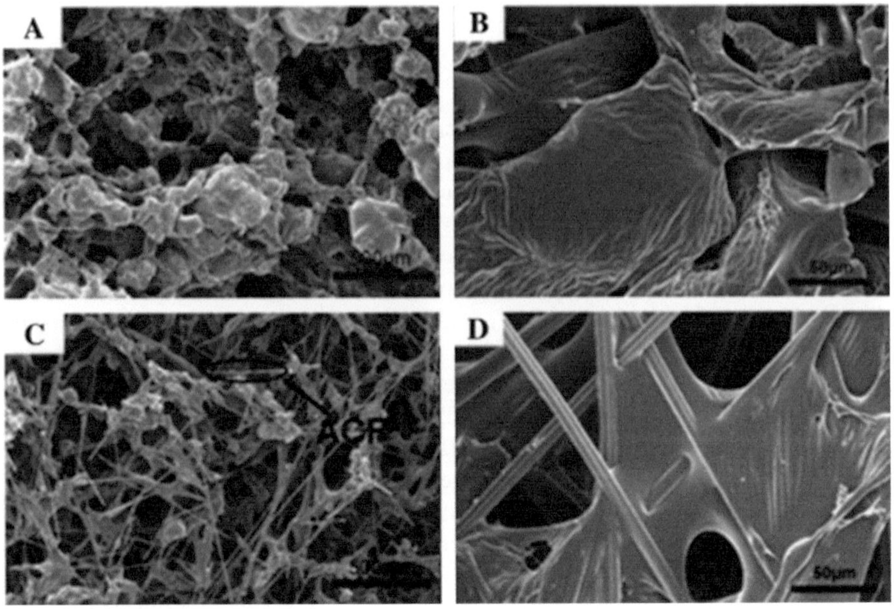

Fig. 7.1 Scanning electron microscope images of pure PLGA (**a** and **b**) and ACF/PLGA scaffolds(**c** and **d**), showing that both ACF/PLGA scaffolds and pure PLGA were porous, and the ACF distributed randomly inside scaffolds (Adapted with permission from Ref. [30]. Copyright 2012 Elsevier Ltd)

The scanning electron microscope images of the scaffold were Fig. 7.1. Meanwhile, the cell experiments showed that the ACF/PLGA scaffolds had higher cell adhesion and proliferation, which demonstrated that the ACF/PLGA scaffolds had great potential to be applied in hard tissue repair.

The low Yang's moduli of dental composites restrict their application in the dental research field. Sabbagh et al. investigated the dynamic elastic modulus of the scaffold containing biological molecule and nanotubes. They found that the 3% biofunctionalized nanotubes composite possessed the greatest Young's modulus (16.8 GPa) compare to control composite. The result likely contributed to the reinforcement of the composite by nanotube bundles and strong adhesion between n-TiO2 and the resin matrix [31]. Therefore, these reinforced composites, serving as restorative materials, could provide an excellent mechanical match with tooth structure. In addition, the fibers or tubes also can be used for coating materials to reinforce the mechanical properties of scaffolds [32]. Collagen is one of the main components of the extracellular matrix [33, 34]. To make use of it, the chitosan scaffolds can be dip coated in an ultrasonic collagen bath. Collagen in the bath can fibrillate on the fiber surfaces and cause a coating with nanofibrous extensions. The addition of nanofiber and collagen enlarged the inner surfaces, and collagen provided good conditions for cell adhesion and growth.

To date, many fabrication techniques are used in the manufacture of hard tissue scaffolds, which can fabricate into meet the needs of hard tissue scaffold materials. Among these preparation methods, meltspinning, which is based on the extrusion process of a polymer melt, is regarded as a suitable method for the production of fiber or tube reinforced scaffold used in hard tissue scaffolds [35]. This method does not require the use of volatile organic solvents, which could possibly affect cell viability or a removal process for the remnant solvent. Park et al. prepared melt-spun shaped fibers with enhanced surface effect. The report showed the woven scaffolds composed of the shaped fibers had good stiffness and increased significantly the proliferation of human osteosarcoma MG63 cells [36]. However, conventional fabrication techniques are not sufficiently suitable to control scaffold structure to modulate mechanical properties, which put forward the urgent need for the new preparation process. Within novel scaffold fabrication processes, rapid prototyping technique has already attracted great attention as powerful tools for the fabrication of hard tissue scaffolds through a stage material deposition, these scaffolds can be built layer by layer in two different circumstances—a molten phase (referred to as fused deposition modeling (FDM)) [37–39] or droplets with a binding agent (known as three-dimensional (3D) printing) [40–42]. During this fabrication, the layers are deposited as interpenetrating networks, leading to the 3D outcomes of 100% interconnected porous scaffolds every definition. Scaffolds can be built in a customized shape through computer-aided design techniques, achieving a defined structure and architecture. Among these devices, 3D fiber deposition (3DF) [44, 45] and 3D plotting [43] have been recently developed and used for tissue engineering. 3DF is a fused deposition technique, during the processes of which extrudated molten polymer is deposited from the XYZ motor drive syringe on a stationary stage under the applying pressure. 3D fiber deposition is characterized by 100% interconnected pores with various sizes and shapes, which evidently affects mechanical properties, performing great potential for tissue engineering applications because of the precision in making reproducible 3D scaffolds. Furthermore, the versatility and flexibility of scaffolds allow us to use rapid prototyping devices to generate improved scaffolds, as well as to study the effects of different structural phenomena on tissue reconstruction. In this respect, FDM techniques offer appealing solutions for scaffold fabrication, having been recently used for tissue engineering purposes.

Electrospinning technique has also been widely used in the fabrication of fiber scaffolds for hard tissue, which can enlarge the functional surface of the scaffolds and get nanofiber scaffolds that have better mechanical strength [46]. The nanofibrous networks interweave in the total inner structure of the scaffolds, and then constitute a uniform hierarchical structure so that the microfibers form an interconnected pore space for cell adhesion, migration, growth, and proliferation [47, 48]. Lee and his group designed the biointerface of biopolymer electrospun fibrous matrices with a biofunctional protein including fibronectin 9–10 domain (FNIII9–10) and osteocalcin (OCN), where FNIII9–10, the 9-to-10 sequence of FN domain III, was used with OCN protein by vector expressions. They found the biopolymer fiber scaffolds have a strong affinity, binding to the mineralized surface in large quantities while preserving stability over a long period, which was not readily achievable in the bare polymer fiber surface [49].

Horner et al. fabricated electrospun core-shell scaffolds by biocompatible polymer systems, the core materials of which are polyetherketoneketone (PEKK) and gelatin, maintaining the shell polymer with polycaprolactone (PCL). In their study, they found the mechanical properties of individual core-shell fibers were decided by core-shell composition and structure. As their study showed, the modulus of individual fiber is related to the increase core size, which rangs from 0.55 ± 0.10 GPa to 1.74 ± 0.22 GPa and 0.48 ± 0.12 GPa to 0.53 ± 0.12 GPa for the PEKK-PCL and gelatin-PCL fibers, respectively [50]. In addition, Liu et al. developed a novel approach to encapsulate and distribute the cells in the nano-fibrillated cellulose based matrices during the swelling process. The nanocellulose hydrogels made through this way with a resultant aerogels, with porosity up to 99.7% and specific surface area up to 308 m^2/g, and the various swelling degree, which up to 500 times, by tuning the material processing approach, the swelling media conditions, and nanocellulose charge density [51].

From a mechanical perspective, mechanical properties are intimately correlated with the porosity of porous structures. Specifically, a stiffer and less porous scaffold shows a better integration with the surrounding natural tissue, while a flexible and more porous one allows cells to attach and proliferate more efficiently. It is rapid prototyping that offers probabilities to compromise such varying requirements into one scaffold, because it adds the freedom of different structural parameters to the non-variable bulk mechanical properties of the material used.

7.3 Biomedical Applications: Fiber or Tube Reinforced Scaffolds for Hard Tissue

7.3.1 Dental Applications of the Scaffolds Reinforced by Fibers or Tubes

Tooth of human being are highly calcified and the hardest tissues in the body. However, the diseases of the tooth can be seen in all ages and their prevalence rate is high [52]. In the last few decades, dentistry has developed rapidly. Novel techniques have improved the conventional treatment methods, and new dental materials have been applied. Traditionally, the polymer matrix and dispersed reinforcing inorganic filler particles are mainly used as dental materials for directly aesthetic restoration [53].

However, the wear of dental materials needs to be improved, which can be achieved by reducing the size of filler particles and using fiber or tube fillers [54]. Therefore, fiber or tube reinforced composites have been extensively applied in dental repair in recent decades [55]. Fiber-reinforced composites are typically made of fine thin fibers reinforced polymer matrix. The polymeric matrix can hold the fibers together in the composite structure and influence the internal laminar shear, compressive strength, in-plate shear properties and the interaction of the polymeric matrix and fibers and defects [56, 57]. In oral regenerative medicine, bone reconstruction is an essential requirement for functional rehabilitation of the stomatognathic system. Indeed, recent studies demonstrated tubes and fibers to be feasible materials that can increase the formation of bone in tooth sockets of rats. Numerous manufacturing methods have been applied tofiber or tube reinforced composites, such as compressive molding [58], injection molding [59], hydrostatic extrusion and self-reinforced [60, 61].

7.3.1.1 *In Vitro* Studies

For tooth repair and regeneration, biomedical scaffolds must meet the following requirements *in vitro*: good mechanical properties, biocompatibility, biodegradability and non-toxic; they can promote cell adhesion, proliferation and growth. These scaffolds should allow various of cells including ameloblasts, odontoblasts, cementoblasts, to carry out its function and further promote regeneration for single or multiple tooth tissue. Therefore, the performance of the scaffolds should be as close as possible to the above requirements. In fact, fracture toughness, which represents material absorbed energy before the fraction for crack propagation, is one of the most significant mechanical properties for the dental materials. In order to acquire excellent fatigue resistance and fracture toughness, many novel fabrication techniques have been used in the proceeding of synthesis of fiber or tube reinforced scaffolds. With regard to fiber reinforced composites, they can be prepared from thermosets reinforced by electronic-grade glass (E-glass) fibers are promising to be used as metallic implants. Bioactive glass granules can be combined with a polymer matrix to improve the osteointegration of fiber-reinforced composites implants.

However, it is difficult to use the bioactive glass granules to create a viable surface layer. Kulkova et al. investigated the potential application of excimer laser ablation to achieve selectively remove the matrix to expose the surface of bioactive glass granules [62]. To study the differences of light in the thermoset matrix and bioactive glass during the penetration process, they studied ultra-violet-visible spectroscopy. In addition, they established the optimal excimer laser ablation parameters. The research verified that a calcium phosphate layer formed on the surface of the laser-ablated specimens in simulated body fluid. Moreover, they investigated the proliferation of human osteosarcoma cells on the surfaces of the laser-ablated specimens. Their results showed that excimer laser ablation had the potential to create a bioactive surface on fiber-reinforced composites implants.

Air jet spinning technique has been successfully developed to fabricate 3D porous and micro-nano plain organic or organic-inorganic hybrid/composite fibers [63]. Although the initial and operational costs of air jet spinning are very low, the production rate is relative high. Air jet spinning can cause a high-speed compressed gas onto the suitable solution to get a polymer jet. Abdalla et al. used hybrid nanofibers to improve the biocompatibility and biodegradation rate of AM50 magnesium

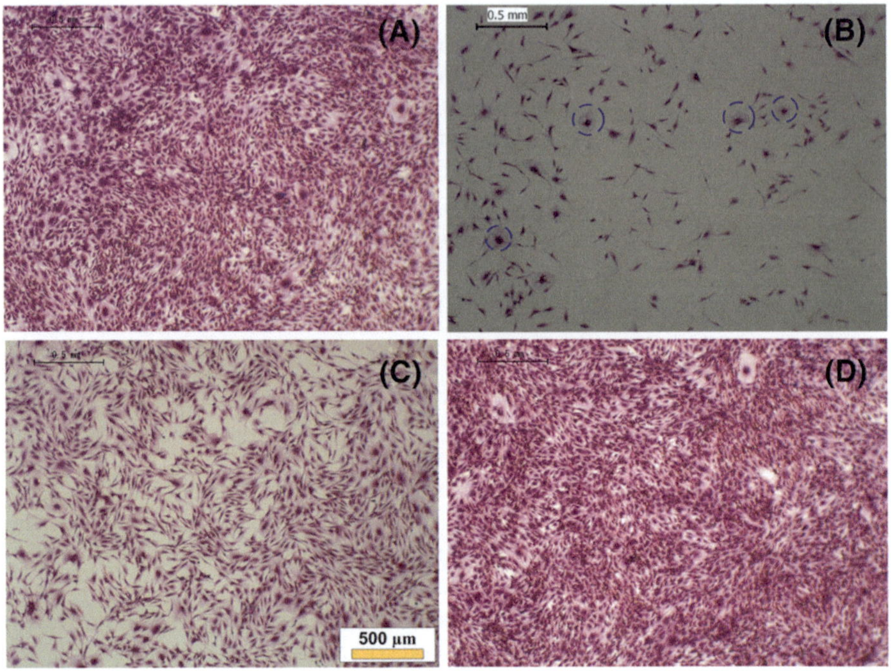

Fig. 7.2 An optical morphology observation ofMC3T3-E1 pre-osteoblast cell culture 2 days at (**a**) negative control; (**b**) magnesium alloys; (**c**) magnesium alloys treated with plain PLA using air jet spinning; (**d**) magnesium alloys treated withPLA and 3% nHA using air jet spinning, showing a higher cytocompatibilityof air jet spinning-coatings compared to neat Mg alloys (Adapted with permission from Ref. [64]. Copyright 2016 Elsevier Ltd)

alloy [64]. They prepared biodegradable hybrid membrane fiber layers by nano-hydroxyapatite particles and poly(lactide) nanofibers and coated layer-by-layer on AM50 coupons by air jet spinning approach. The corrosion performance of coated and uncoated coupon samples was investigated by means of electrochemical measurements. The results showed that compared to the neat AM50 coupon samples, the air jet spinning 3D membrane fiber layers effectively decreased the initial degradation rate and induced a higher biocorrosion resistance. Due to the presence of nano-hydroxyapatite particles in the air jet spinning layer, the adhesion strength improved highly. In addition, the air jet spinning-coatings significantly controlled the long biodegradation rate of AM50 alloy in Hank's balanced salt solution. Moreover, the cytocompatibility of air jet spinning-coatings was higher than that for neat Mg alloys (Fig. 7.2). Therefore, the nanostructured hydroxyapatite embedded hybrid poly(lactide) nanofibers coating could be a suitable coating material for Mg alloy as a potential material for biodegradable metallic orthopedic implants.

Another great advantage of fiber reinforced composites in implant prostheses is that they can solve several problems associated with metal alloy frameworks such as high cost, the complexity of fabrication and corrosion. Erkan et al. built a 3D finite element analysis model to evaluate the stress distribution in implant-abutment complex, bone and prosthetic structures [65]. Two distinctly different models, composed of fiber-reinforced composite and particulate composite (FRC-FPD model) or Cr-Co and porcelain(M-FPD model) were studied. The results showed that all the stress values in the M-FPD model were higher than in the FRC-FPD model, except for the stress values in the implant-abutment complex. They concluded that the implant supported FRC-FPD could maintain normal physiological loading of the surrounding bone and eliminate the excessive stresses in the bone-implant interface, thus reducing the risk of peri-implant bone loss. Similarly, Boudeau et al. injected model of polyetheretherketone reinforced with carbon fibers (CF/PEEK) to obtain composite stems used for hip replacement [66]. They combined process simulations and structural analysis to explore different compositions of CF/PEEK and injection conditions. The obtained implants were compared to the bone-implant system based on conventional metallic materials and the bone alone under walking load. Comparisons were done from four objective criteria: stress deficiency, stress shielding, global deformation and debonding. CF/PEEK injected implant might reduce three of the criteria. The carbon fibers presented in the interface bone-implant could exhibit health problem for the patient with bone fragments due to wear and friction between implant and bone.

To investigate the effects of short fibers such as silk microfibers (SMFs), Ceria nanofibers (CNFs) and alumina microfibers (AMFs) as reinforced materials in 2,2-bis[4-(20-hydroxy-30-ethacryloyloxypropoxy)phenyl]propane(Bis-GMA) and triethyleneglycol dimethacrylate (TEGDMA) resin on the development of composite dental filler. Rameshbabu et al. studied the morphology, phase, degree-of-conversion, viscosity, flowability and physical properties of AMFs, SMFs, CNFs and their representative fracture surfaces of the reinforced dental resins/composites [67]. Their results showed that addition of short fibers into base resin during formulation of composites reduced the degree of conversion and depth of cure. The resultant composites had significantly better mechanical properties compared to the base resin for both alumina and silk as additives. SMFs, CNFs and AMFs reinforced composite resins significantly improved the physical and mechanical properties. Therefore, they may be used as additives in composite dental filler for the plausible application.

Cellulosic fibers with low cost, low density, specific strength and high elasticity modulus have been extensively used for reinforcing the physical structure of some composites. Silva et al. applied cellulosic fibers to modify a conventional restorative glass ionomer cement (GIC), and used scanning electron microscopy, light microscopy, and energy dispersive X-ray spectroscopy to characterize the composite and its precursors [68]. The addition of cellulosic fibers reinforced the solubility in water, compressive strength, resistance to abrasion and bond strength of the fiber-reinforced glass ionomer cement. Similarly, Anderson et al. studied the physical, chemical and degradation properties of three different kinds of epoxide-based materials used for restorative dental applications by Raman, Fourier transform infrared, X-ray fluorescence spectroscopy, Brunauer-Emmett-TellerAnalysis, and degradation experiments [69]. Physicochemical characterization showed that the materials had similar chemical composition, the differences between them mainly depended on the different phase distribution, not chemical composition. Maryam et al. further used polyurethane combined with hydroxyapatite and fibers to prepare composites for dental applications by solution casting technique [70]. The concentrations of hydroxyapatite and polyurethane hard segment accompanied with the duration of immersion in artificial saliva were two main factors of the modification of solid-state properties of hydroxyapatite. The use of an optimum amount of hydroxyapatite and polyurethane hard segment in composites and fibers could respectively improve the properties of the dental filling material.

To study the combined effect of dental luting cement and posts on fracture resistance and failure mode of the endodontically-treated tooth, Mohamed et al. randomly restored 60 endodontically-treated upper central incisors with fiber, titanium, or stainless steel posts and luted by zinc phosphate or composite resin cement [71]. They also analyzed a 3D tooth model by the finite element method to contrast the stress distributions caused by different post-core systems. Their results indicated that zinc phosphate cement showed reversely higher fracture resistances, however, luting with composite resin led to failures that are more restorable. What is more, compared to the tooth restored by titanium posts or stainless steel, those treated by fiber posts showed satisfactory fracture resistances. In addition, to enhance the mechanical performance of fiber-reinforced composite for dental applications, structural optimization was used to obtain an alternative design for a 3D inlay-retained fiber-reinforced composite dental bridge [72]. Chen et al. tried to validate *in vitro* the improved adhesive fixed partial denture design obtained by the stress-induced material transformation technique by simulating occlusal loading [73]. The 3D inlay-retained fiber-reinforced composite dental bridge was made with glass fibers as the substructure and a veneering composite. The samples were loaded to 400 N on a universal test machine and recorded the force and displacement. At the same time, they used a two-channel acoustic emission system to monitor the development of cracks during loading. The results showed that the optimized fiber-reinforced composite bridge design had a higher fracture resistance. Similarly, Baaran et al. compared the load bearing capacity of fiber-reinforced or unreinforced composites for fixed dental prostheses by simulated manufacturing fabrication [74]. The results showed that fixed partial dental prostheses with fiber-reinforced composite resin blocks existed higher load-bearing capacities than that unreinforced composite resin blocks. The shock-absorbing mechanism in commercial dental implants may be the factor caused bone loss and possible implant failure. Therefore, Sheikhhassani et al. tried to generate a shock-absorbing dental implant that is similar to the periodontal ligament by using polyethylene fibers to reinforce polycarbonate-urethane [75]. To further study the implant type, tests based on mechanical testing and finite element analysis was proposed. The reinforced composite dental implant showed good shock-absorbing properties.

To enhance osseointegration and decrease healing time after implantation, Lee et al. studied the physiology interaction of drug (N-acetyl cysteine) loaded nanotube titanium (NLN-Ti) implants *in vitro*. MC-3T3 E1 osteoblast-like cells were cultured on the surfaces of drug loading Ti nanotubes. They demonstrated that the cellular viability was evidently enhanced. Additionally, osteogenesis molecule tumor necrosis factor-alpha and interleukin were observed in MC-3T3-E1 cells by means of reducing the inflammatory response [76]. The biological effects of the drug on Ti surfaces have been reported to improve osteoconduction and osseointegrate capacity [77, 78].

To clarify fracture toughness of tubular materials, Dafar et al. investigated two different kinds of composites consisted of titania (TiO_2) nanotubes. The research found that functionalized n-TiO_2 modified composites showed greater values of fracture toughness than control composite, and n-TiO_2-reinforced composites exhibited outstanding biocompatibility *in vitro* [79].

Chen and his group prepared a novel dental resin containing silanized halloysite nanotubes (HNTs). The HNTs were mixed with 2.2′-bis-[4-(methacryloxypropoxy) -phenyl]-propane/tri(ethyleneglycol) dimethacrylate (Bis-GMA/TEGDMA) resin by three different techniques to synthesize uniform scaffold containing different content of HNTs. They found that mechanical properties of the composites improved substantially, when the weight content of the HNTs is low (e.g., 1 and 2.5%). However, the mechanical properties never further enhance even if adding their amount (e.g., 5%). This result showed that the different content of HNTs into dental composites could lead to opposite enhancement effect [80].

As for tooth materials, it is well known that CNTs could serve as collagen fibers during the tooth formation. They always control crystal nucleation and inorganic component growth [81]. Many studies have shown that CNTs could boost the growth of osteoblasts and osteogenesis. Cheng et al. studied a composite scaffold consisted of CNT and PLGA. The research indicated that the addition of CNT into PLGA facilitated the proliferation and differentiation of osteoblast cell. In addition, CNT modified composite scaffolds exhibited excellent mechanical strength. The results could be ascribed to higher surface roughness and qualified mechanical properties of CNTs, which further promotes the adsorbent of growth factors on the surface of CNT. Moreover, the incorporation of CNT is conducive to control phenotypic expression and lineage specification of stem cells [82].

In recent studies, CNT-reinforced scaffolds have exhibited great potential in the repair of dental defects. Mohamed et al. investigated the biocompatibility of carbon nanotube reinforced tetragonal zirconia polycrystalline (TZP) composite *in vitro*. In their study, osteoblast-like cell MG63 was cultured on the scaffold. The result showed that CNTs enhanced differentiation and proliferation of MG63 cell, which could ascribe to their large specific surface area that provides more opportunities to absorb osteogenic-related proteins. In addition, the addition of CNTs enhanced fracture toughness for the CNTs reinforced materials compare to monolithic zirconia [83].

7.3.1.2 *In Vivo* Studies

Repairing temporary fixed dental prostheses can cause inconvenience and lost productivity to patients. To choose the suitable biomaterials for the special needs of different patients, clinicians must know about the strengths and weaknesses of numerous biomaterials [84]. Barry et al. used the technique of glass fiber reinforcement to improve the strength, longevity and curative effect of the definitive fixed prosthodontic treatment [85]. Fiber-reinforced interim restorations were applied for longer-term periods in clinical situations indicating multiple pontic spaces or in patients with heavy occlusal function. Kumbuloglu et al. evaluated the clinical performance of indirect, anterior, surface-retained, fiber-reinforced composite restorations (FRCR) [86]. There were 134 patients (51 males, 83 females, 6–8 years old) received the FRCR tests. The restorations were used unidirectional E-glass fibers combined with a laboratory resin composite to cement indirectly on a plaster model. All the patients were followed at 6 months and then annually up to 7.5 years after baseline recordings. The evaluation protocol involved technical (chipping, fracture or debonding of tooth/restoration) and biological failures (caries). The results were 13 failures, 1 catastrophic fracture, eight partial debonding and four delaminations of veneering composite. The restorations FRCR showed similar clinical survival rate after cemented with the resin cement. Experienced failures generally were owing to delamination of the veneering composite or debonding of the restoration. The FRCR could be considered cost-effective alternatives to conventional tooth- or implant-bonded fixed dental prosthesis.

To understand the osteogenesis effect of the tubular nanostructure, Lee et al. compared the osteogenesis bioactivity of nanotube Ti and pure Ti loaded with N-acetyl cysteine. After 4 weeks implantation, a large number of osteogenic growth factor, like BMP-2 and BMP-7, were shown on the surface of implants (Fig. 7.3). Moreover, the bone inflammatory marker (Receptor Activator for Nuclear Factor-κ B Ligand, RANKL) was clearly less in the regenerated bone around nanotube Ti filled composites compare to pure Ti filled implants [76].

Mendes et al. studied the influence of single-walled carbon nanotube (SWCNT) reinforced sodium hyaluronate (HY-SWCNT) scaffold on tooth sockets regeneration *in vivo*. The scaffolds, which increased nearly three folds, obviously enhanced the formation of bone trabecula compare with the control sockets. Furthermore, the expression level of collagen I was dramatically improved after a week of tooth extraction [87]. This result further demonstrated that SWCNT plays a positive effect on accelerating the formation of new bone in sockets. Likewise, Sá et al. came to similar conclusions on the osteogenesis effects of HY-SWCNT acted as tooth socket scaffolds. In this study, HY-SWCNT scaffold was implanted in rats with type I diabetes. The implantation sharply enhanced the formation of bone trabecula, whose final values approach to those obtained in non-diabetic rats [88]. In a word, SWCNT reinforced scaffolds have great advantages in dental regeneration, whether normal or unhealthy metabolic condition.

7.3.2 Scaffolds Reinforced by Fibers or Tubes for Bone Repair

Bone defect has been a common disease due to fracture, trauma, pathological biomineralization or degeneration, tumor and deformity, etc., especially under the current situation of growing population aging. For achieving the complete bone

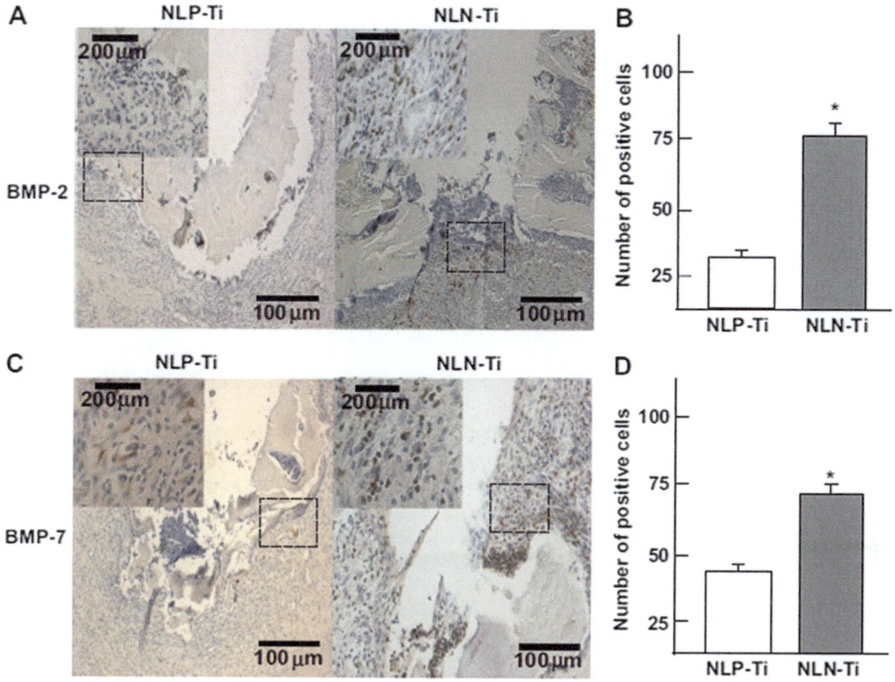

Fig. 7.3 Theexpressions of BMP-2 (**a**, **b**), and BMP-7(**c**, **d**) analyzed by immunohistochemistry, with date expressed by mean scanning electron microscope (*$p < 0.05$), showing that the expression of BMP-2 and BMP-7, as the markers for osteogenic differentiation, was more intense around N-acetyl cysteine-loaded nanotube titanium (NLN-Ti) implants than around N-acetyl cysteine-loaded pure Ti (NLP-Ti) implants (Adapted with permission from Ref. [76]. Copyright 2013 Elsevier Ltd)

healing, grafts (autografts or allografts) are traditionally used in surgery. It is reported that more than 500,000 bone-grafting procedures are performed in the United States every year. However, The harvest of bone grafts inevitably leads to the increase of additional trauma parts, and the available tissue source is not enough in many cases. Blazing new trails of therapies for enhancing angiogenesis and osteogenesis had been a critical challenge in the generation of a large bone for long time. Fortunately, it has been well recognized that tissue engineering scaffolds can overcome or avoid those deficiencies of bone grafts, providing a promising method for bone repair. In recent years, more and more attention has been paid to the scaffolds reinforced by fibers or tubes based on their enhanced bone repair capability.

7.3.2.1 *In Vitro* Studies

In the bone tissue engineering applications, biomimetic nanofibrous scaffolds derived from natural biopolymers require good mechanical and biological performances including biomineralization. Pangon et al. demonstrated the utility of chitin

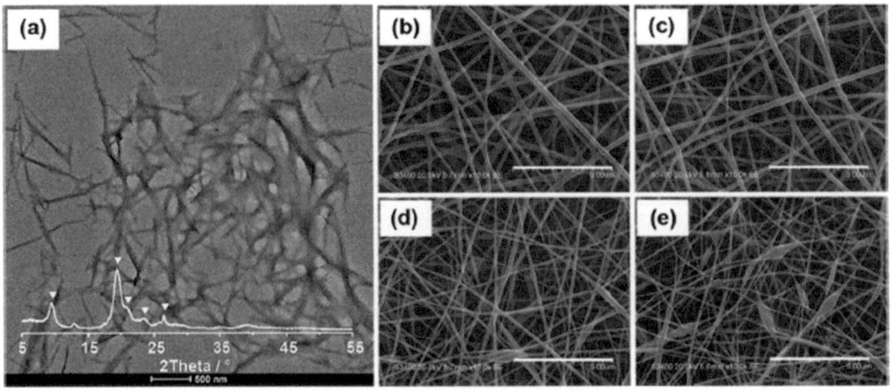

Fig. 7.4 Scanning electron microscope images of representative bio-nanocomposite fibers before (*top*) and after (*bottom*) immersion in deionized water for 4 h, showing electrospun CS/PVA nanofibrous membranes containing CTWK with different content (**a**) as-spun and (**b**)GA-crosslinked CS/PVA/CTWK-2 nanofibers for (**c**) 8 wt% (CS/PVA/CTWK-1), (**d**) 17 wt% (CS/PVA/CTWK-2), and (**e**) 26 wt% (CS/PVA/CTWK-3) (Adapted with permission from Ref. [89]. Copyright 2016 Elsevier Ltd)

whisker (CTWK) to promote mechanical properties of chitosan/poly(vinyl alcohol) (CS/PVA) nanofibers and to provide osteoblast cell growth with hydroxyapatite (HA) mineralization [89]. In their study, electrospun CS/PVA nanofibrous membranes containing CTWK can be easily gained (Fig. 7.4) by using diacid as a solvent. The dimension stability of nanofibrous CS/PVA/CTWK bio-nanocomposite can be further controlled by exposing to glutaraldehyde vapor.

The nanofibrous membranes obtain the mineralization of HA in concentrated simulated body fluid resulting in an improvement of Young's modulus and tensile strength. The CTWK combined with HA in bio-nanocomposite is a key to promoting osteoblast cell adhesion and proliferation. The present work, for the first time, demonstrates the use of CTWKs for bio-nanocomposite fibers of chitosan and its hydroxyapatite with the function of bio-mineralization in osteoblast cell culture. Figure 7.5 was the scanning electron microscope images of the MC3T3-E1 osteoblast cells adhered on different bionanocomposite fibers. These hydroxyapatite-hybridized CS/PVA/CTWK nanocompositefibers (CS/PVA/CTWK-HA) offer a great potential for bone tissue engineering applications. In their present study, chitosan/CTWK nanocomposites in the form of HA-hybridized nanofibers were developed for bone tissue engineering scaffolds. The well-defined CS/PVA/CTWK nanocomposite fibers were electrospun from CS/PVA blend containing CTWK for 8 wt%, 17 wt%, and up to 26 wt% in succinic acid solution. The fiber stability was improved remarkably, after cross-linked using glutaraldehyde vapor. When subjected to 10× simulated body fluid solution for 2 h, the cross-linked nanofibers were randomly mineralized with Ca-deficient HA. The thermal gravimetric analysis results indicated that such mineralization process resulted in HA deposition for 17.6–20.2 wt%, independent of CTWK content. The addition of CTWK could

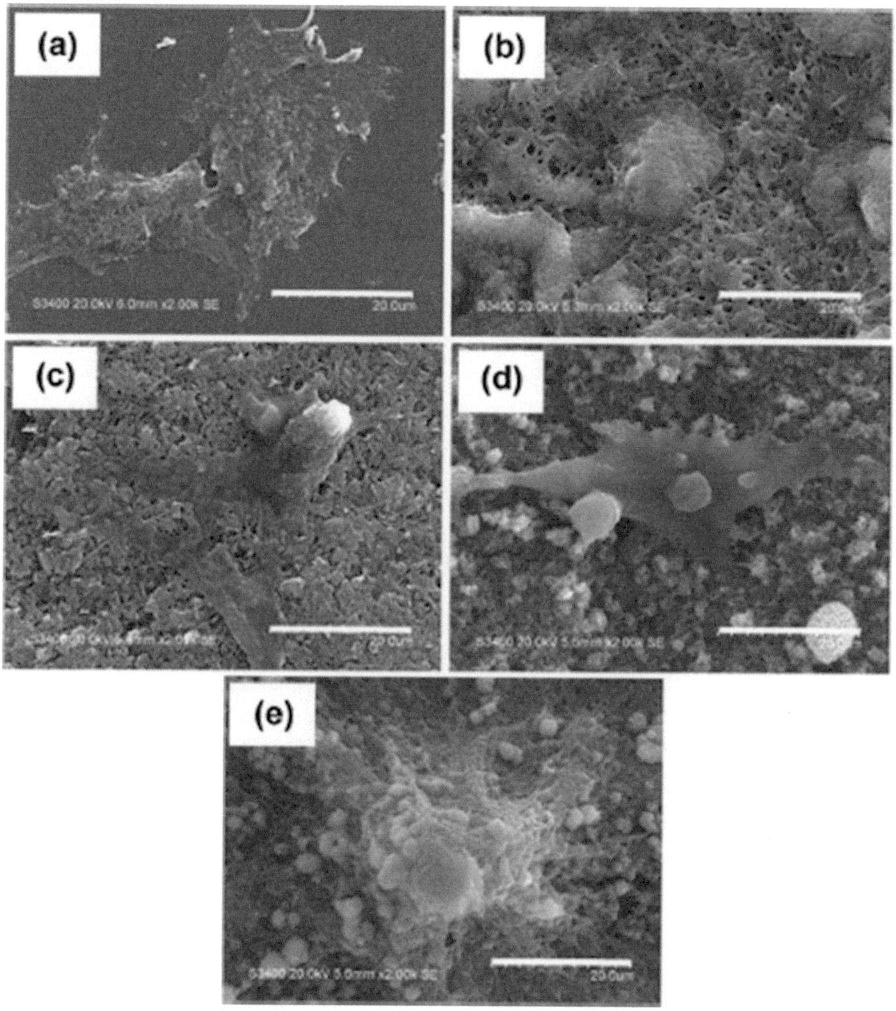

Fig. 7.5 Scanning electron microscope images of the MC3T3-E1 osteoblast cells adheredon (**a**) control, (**b**) GA-crosslinked CS/PVA, (**c**) CS/PVA-HA, (**d**) CS/PVA/CTWK-1-HA, and (**e**) CS/PVA/CTWK-2-HA scaffolds after 7 days of culture, showing that the CTWK combined with HA in bio-nanocomposite can promoteosteoblast cell adhesion (Adapted with permission from Ref. [89]. Copyright 2016 Elsevier Ltd)

change mechanical properties of the bio-nano-composite fibers containing their HA-hybridized forms, while have less impact on their biocompatibility and biodegradability. The increase of CTWK contentled to a improving the Young's modulus of the bio-nano-composite fibers, and particularly, of the nanofibers with HA hybridization. The studies on MTT assessment showed that, firstly, the HA-hybridized bio-nano-composite scaffolds had no toxic effect on osteoblast cells, and secondly, the cell viability and ability to proliferate could be improved by increasing CTWK as reinforcement for chitosan nanofibrous scaffolds. It was found that the CS/PVA/CTWK-2-HA hybrid nanofibers with good mechanical and biological properties containing biomineralization are promising scaffolds for bone tissue engineering applications.

Over the last two decades, biodegradable polymers, such as poly(lactic acid) (PLA), poly(glycolic acid) (PGA) and PCL, have emerged as a class of biomaterials of growing interest for application in surgery, drug delivery, and tissue engineering. They can be used in sutures for wound healing, devices for bone fracture internal fixation, carriers for delivery of bioactive molecules, scaffolds for the regeneration of tissues or organs. Li et al. prepared the porous three-dimensional PLLA scaffolds reinforced by the chitin fibers (PLLA/CF) with and without the link [90], and evaluated the properties *in vitro* [91]. To get an appropriate content of the chitin fibers in the PLLA/CF, they measured pH value of the phosphate buffered saline lixivium of the PLLA/CF with different content of the chitin fibers. Then, they compared the cell proliferation, attachment, alkaline phosphatase level, total protein per unit cell, and osteopontin, osteonectin, and osteocalcin gene expression of human osteoblast-like cells cultured on the PLLA/CF with or without the link, and PLLA scaffold. The results showed that the link treatment had no significant effect on the pH value of the lixivium of the scaffolds. Moreover, 30% volume content might be a suitable content of the chitin fibers in PLLA/CF scaffold. During the lixiviation time of 16 weeks, the PLLA/CF scaffold was significantly better for the attachment, proliferation, differentiation, and mineralization of the osteoblast than PLLA. Similarly, Guarino et al. developed the composite scaffolds through combining two or more types of biodegradable polymers with selected properties, which meet all the mechanical and physiological demands of the host tissue [92]. A new type of composite material for bone regeneration, consisting of a porous PCL matrix enhanced by continuous poly(L-lactic acid) (PLLA) fibers, was processed by the synergistic use of filament winding and phase inversion/salt leaching technique (Fig. 7.6). The chemistry and the degrees of micro- and macro-porosity gained by such techniques decide the structural and biological performances of the composite. Considering the degradation properties, the treatment with NaOH solution (pH 12) aroused the most significant changes over 5 weeks. These just related to the more hydrophilic PLLA component, which experienced significant structural rearrangements, decrease of polymeric chain length and increase of crystallinity. Furthermore, encouraging results are showed in cellular studies in the case of human osteoblasts and marrow stromal cells: the latter were found more active on the composite, due either to a slightly higher number of cells or to a higher activity of the cells compared to the former.

The stiffening effect on mechanical models used to measure the efficient of stiffening in different mineralization methods on nanofiber polymer scaffolds and compared strictly examine the mechanic nanofiber PLGA scaffolds with gradients in the mineral. Lipner et al. determined the local mechanical properties and the mineral volume fractions, and developed a mathematical model to compare the experimental results of the optical bounds [93]. It is found that the two mineralization methods all stiffened the scaffolds, but a varied substantially of magnitudes,

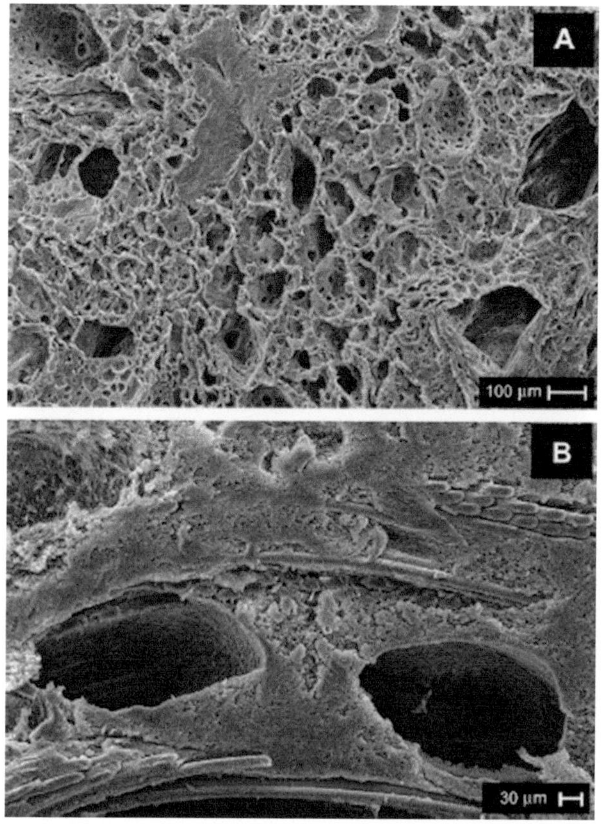

Fig. 7.6 Scanning electron microscope analysis of fiber-reinforced composite scaffolds: (**a**) bimodal pore size distribution and (**b**) organization of PLLA fibers inside porous PCL matrix. The microporosity with small pore sizes ranging 1–10 µm and a macroporosity with 100–400 µm pore size (Adapted with permission from Ref. [92]. Copyright 2008 Elsevier Ltd)

with modified simulated body fluid (m10SBF) showing a more efficient effect than simulated body fluid (10SBF). When compared to the composite bounds we developed, the stiffening by mineralization achieved using 10SBF proved to be weaker than the lowest possible stiffening predicted by homogenization theory, which indicates that mineral linked weakly to the scaffold. In contrast, mineralization-using m10SBF achieved stiffening that was hardly an order of magnitude greater than10SBF. This stiffening was near to that predicted by the Hashin-Shtrikman upper homogenization bound, which suggests that the new method might fit the tendon to bone tissue engineering.

Complexation with other more rigid polymeric constructs, such as electrospun fiber meshes, can be used to provide a template for tissue regeneration and improve structural rigidity [94, 95]. The traditional role of the scaffold as simply a template for tissue formation has evolved and the new generation of scaffolds are increasingly used as delivery vehicles for therapeutic molecules such as drugs, proteins, and genes that initiate biological events leading to the regeneration of tissue [96]. Hoppe et al. found ions can also be classified as therapeutics [97, 98]. One method of delivery is the release of Si and Caions from bioactive glasses, which are defined as inorganic surface-active bioceramics. Moreover, previous studies indicated that

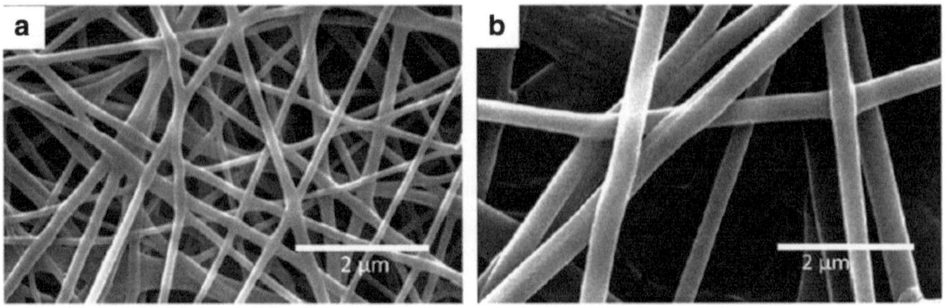

Fig. 7.7 Scanning electron microscopy micrograph of the fibers before heating. (**a**). 45S5 sub-micron BG fibers, (**b**) 45S5 sub-micron BG fibers containing Cu (Adapted with permission from Ref. [101]. Copyright 2016 Elsevier Ltd)

direct mixing of Cu ions with bioactive materials is a feasible way to improve angiogenesis [99, 100]. Sharifi et al. prepared Cu-containing composite scaffolds reinforced by bioglass (BG) fibers and investigate their effects on properties of scaffolds for bone tissue engineering [101]. In that research, three types of scaffolds were fabricated, including gelatin-collagen hydrogel, gelatin-collagen composite containing submicron 45S5 (in wt%: 45% SiO_2, 24.5% Na_2O, 24.4% CaO and 6% P_2O_5) BG fibers, and Cu-doped composites. Sub-micron BG fibers were fabricated by combination of sol-gel and electrospinning processes, the fibers were then mixed with hydrogel matrix containing gelatin and collagen, freeze dried, followed by genipin cross-linking to fabricate final composite scaffolds (Fig. 7.7). Growth and viability of human osteoblast-like cell line SaOS-2 were investigated on these biomimetic scaffolds. Cellular biocompatibility assays illustrated that scaffolds containing Cu ion in the BG structure had more viability and osteoblast growth in comparison with other scaffolds. The scaffolds containing Cu ion in the 45S5 BG network showed better biocompatibility compared to the hydrogel and composite scaffold containing the 45S5 BG. Cu ion increased the growth of osteoblasts on composite scaffold containing sub-micron fibers BG, compared to hydrogel without sub-micron fibers BG. The results showed that the prepared scaffolds, especially Cu-doped BG fibers containing scaffold isnon-cytotoxic and its surface is ideal for osteoblast attachment, growth, viability, and bone regeneration.

Except for those above, ploymer fiber and BG fiber, collagen fiber has been simultaneously studied widely. A series of highly porous scaffolds derived from type I collagen fibers and the polysaccharide have been developed by a controlled freeze-drying process [102–104]. These scaffolds with an optimized composition to facilitate osteogenesis have been shown to promote bone repair *in vivo* in the case of minimally loaded calvarial defects [105]. The emphasis of this study was to incorporate cobalt bioactive glass into CG scaffolds that have been developed and optimized for bone tissue regeneration to improve the mechanical and structural properties of the CG scaffold, and vitally importantly, enhance the initial angiogenic step vital for bone regeneration. Specifically, the aims were to evaluate the effect of the bioactive glass on the pore structure, including porosity, compressive module and biological activity of the resultant composites, by inspecting their ability to induce an angiogenic and osteogenic response from cells. When Combined novel hypoxia-mimicking cobalt bioactive glasses with CG scaffolds, bone were repaired. As the results demonstrated, these scaffolds with collagen fibers could create a microenvironment that stimulates both angiogenesis and vascularization through the release of cobalt, a known hypoxia mimic, as well as supporting osteogenesis, which benefit from the addition of osteoinductive bioactive glass particles. Moreover, Liu et al. produced hydrogel scaffolds reinforced by collagen fibers *in vitro*, in the aspect of improving adhesion and differentiation of bone cells, most hydrogels fail to be independently used in load bearing sites for their less mechanical integrity [106]. Hydrogels, networked and hydrophilic polymers, are typically formed via cross-linking or chain entanglement, absorbing significant amounts of water [107–109]. Therefore, there has been great interest in preparing composites on basis of hydrogels. The first approach is to compound hydrogel materials with other biomaterials, such as hyaluronate composites and collagen fiber. The composites reinforced by fiber are capable of providing the enhanced mechanical properties. Moreover, the composites performed greater osteoconductivity compared to either material independently, indicating the satisfactory synergy of the components.

The anterior cruciate ligament (ACL) is the ligament of the knee injured most commonly as the number of the patients diagnosed with ACL disruptions is more than 200,000 every year [110–112]. The ACL consisting of elastin, water, cells, proteoglycans, and collagens (typesI, III, and V), is a highly organized, dense, cable-liketissue [113]. Moreover, it is vital to normal stability and kinematics as the key intra-articular ligament of the knee [114]. To repair the knee damage created by ACL disruptions, various materials and structures have been studied for their use in tissue engineered ligament replacements. It also has been investigated that polydioxanone (PDS) can be used for scaffold construction as a promising material [115]. In addition, Dunn et al. have also studied the use of type I collagen fibers in ACL scaffolds [116]. Furthermore, compositing some more complex fibers in tissue-engineered structures can resolve section of these possible problems. A recent tissue-engineered structure using twisted silk fibers conforms two different tissue engineering methods to achieve the regeneration of damaged or lost tissue: using isolated cells that have been expanded in vitro and using a three-dimensional, porous matrix. The matrix composited by silk performs hierarchical structure, and is non-toxic in proliferation tests using bone marrow stromal cells. Wounding bundles of silk fibers into strands, which will be wound into cords can form the matrix. The twisted fiber architecture promotes the similarity to ACL in scaffold mechanical properties. The scaffolds also showed the three phase mechanical behavior observed in ligament and tendon, a linear region (high stress per unit strain), and consecutively, a toe region (low stress per unit strain), which is important for the protection of damage deriving from creep and fatigue. Moreover, in other studies, Altman et al. coated the surface with RGD sequences (Arg-Gly-Asp), having improved the ability to elicit new tissue growth and the biocompatibility of this matrix [117]. The presence of the RGD sequences on the silk fiber surfaces also increased the production of extracellular matrix by bone marrow stromal cells creating the possibility of quicker and more complete tissue regeneration.

More recently, Laurencin and his partners have put forward a tissue-engineered method, which is based on a degradable, cell seeded, 3D braided PLLA fibers scaffold [118]. The scaffold used 3D braiding techniques to product a new scaffold with integrated pores, controlled pore size and mechanical properties compared to natural ACL and resistance to wear and rupture. It is very important to control the size and distribution of pores in the scaffold. Because it enhances bone cell tissue and proliferation in growth. The braided scaffold with hierarchical structure is similar to the ligament. The braided scaffold is consists of fibers in the same diameter to collagen fibers. Then the fibers are integrated into bundles and wound through the thickness of the scaffold. The braids include three regions: ligament region, femoral tunnel attachment site, and tibial tunnel attachment site. The intra-articular zone has a lower-angle fiber orientation and the femoral tunnel attachment sites have a high-angle fiber. The different orientations of fiber lead to diversification about pore size between different regions. The three regions of the scaffold comprise pore sizes within these listed ranges to facilitate capillary supply and tissue (ligament and bone) in growth. The smaller pore size (and higher braiding angle) at the insertion sites improve resistance to wear in bone tunnels and enhance the quality of anchorage within bone tunnels via the integration of bone tissue. Pore size as well as overall scaffold porosity is able to change the response to the implant of cells. The pore interconnectivity stretching via an implant expands the overall surface area of cell attachment, which can simultaneously reinforce the regenerative properties of the implant via permitting tissue growth into the internal matrix.

In addition, bone-anterior cruciate ligament-bone (B-ACL-B) grafts were studied by incorporating surfactants lauryl sulfate (SDS), Triton X-100, and/or an organic solvent (tributyl phosphate (TnBP)) [119]. The study showed that the mechanical and biochemical properties of B-ACL-B grafts satisfy the further investigation of the repair B-ACL-B defect (Fig. 7.8). The further study was to prepare the scaffolds reinforced by fibers to repair the B-ACL-B damage.

Besides the fibers, nanotubes are also used extensively for bone repair *in vitro*. Karla et al. compared the osteogenic cell behavior on titania nanotubes surfaces with carbon-coated titania nanotubes surfaces, including osteoblast and mesenchymal stem cell interactions with the different surface chemistries [120]. Their experiments showed that nanotubes had good influence on the alkaline phosphatase activity of osteoblast cells and promoted osteogenic differentiation. The carbon-coated titania nanotubes had higher levels of osteo-differentiation. Similarly, Wan et al. prepared a three-dimensional network-structured silica nanotubes scaffold, and did cell studies using a mouse fibroblast cell line and a human osteoblast-like cell line [121]. The three-dimensional network structured silica nanotubes scaffold benefited for the attachment, spreading, and proliferation of cells and showed excellent biocompatibility. Thus, silica nanotubes are promising used in bone tissue engineering and regeneration.

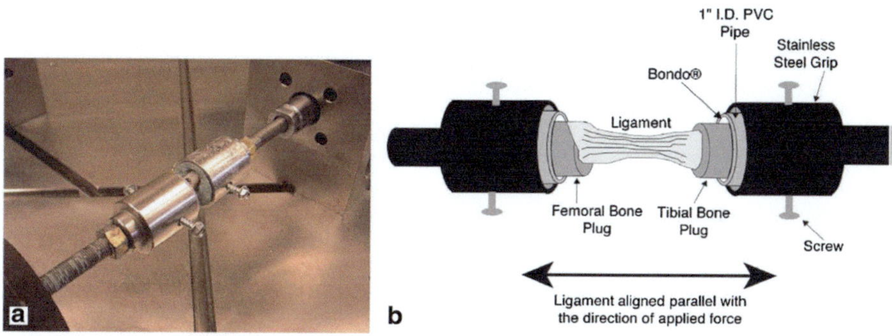

Fig. 7.8 (**a**) Potted B-ACL-B sample mounted onto servo-hydraulic tensile testing apparatus. (**b**) The schematic diagram showing the details of the gripping of the B-ACL-B samples. This set-up was suit for the tensile properties of ACLs loaded along the axis of the ligament independent of the angle of knee flexion and engaging the greatest amount of ligament fibers (Adapted with permission from Ref. [119]. Copyright 2005 Elsevier Ltd)

7.3.2.2 *In Vivo* Studies

Titanium alloys and cobalt chrome have outstanding resistance and biocompatibility with high specific strength. So they are widely applied in dental, orthopaedic and other implants. Titanium fiber greatly improved the rate of repair bone defects. Hirota et al. applied a molecular precursor method, which is a thin hydroxyapatite (HA) film can be opportunely coated onto the titanium fiber web without damaging the interior structure (Fig. 7.9) [122]. The i*n vivo* implantation made in rat cranial bone defects shows that the quantity of newly formed bone is greatly higher in the HA-coated titanium fiber mesh than in the uncoated scaffold. 2 weeks later, the corresponding new bone formation ratio is 35.8% and 7.1%, respectively while the figures increase to 45.1 and 56.2%, demonstrating the earlier osteoinduction of thin HA coatings, which is interrelated to capability of HA to strengthen the formation of apatite crystals brought by human osteoblasts during the culturing process.

Water-soluble phosphorylated chitosan (P-chitosan) fiber is known to increase osteoblastic activity and disodium (1–4)-2-deoxy-2-sulfoamino-β-D- glucopyranuronan(S-chitosan) fiber can inhibit the activities of giant cell tumor of bone [123, 124]. Wang et al. developed a novel bone repair materials incorporating P- and S-chitosan fiber, and tested osteointegration properties of the materials with a standard ulna bone defect model [125]. They prepared composite scaffolds: S-chitosan fiber and P-chitosan fiber mixed separately with PLGA/tricalcium phosphate (TCP) to 1,4-dioxanesolution. A rapid prototyping machine was used to fabricate PLGA/TCP/P-chitosan (P) and PLGA/TCP/S-chitosan (S) scaffolds. Similarly, Chen et al. further studied on the biocompatibility of scaffolds materials, and they found that new bone formation and mineralization took place within the pores of the scaffold and then remodeled together with degradation of the scaffold [126]. There is a significant enhancement of bone regeneration in the groups of P-chitosan fiber and PSP-chitosan fiber compared to the control group at 12 weeks post-operation.

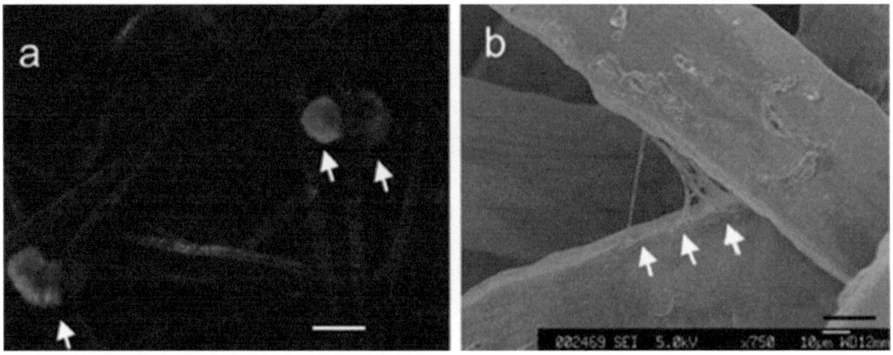

Fig. 7.9 Scanning electron microscope of hydroxyapatite (HA)-coated titanium fiber without damaging the interior structure: (**a**) Stereoscope finding of apatite crystal after 14 days of human osteoblast culture. Apatite crystals are deposited on titanium fiber (*arrows*). Scale bar 200 μm. (**b**) human osteoblast cultured in HA-coated titanium fiber web for 7 days. Human osteoblast extends projections to adhere to titanium fibers, secreting calcification matrix (*arrows*). Scale bar 20 μm (Adapted with permission from Ref. [122]. Copyright 2012 Elsevier Ltd)

In addition, Li et al. prepared nano-hydroxyapatite/collagen/PLA (nHACP) scaffolds reinforced by chitin fibers to repair the goat shank bone defect. They found that the fiber reinforced scaffolds could successfully repair the 40-mm adult goat shank bone defect while the bare nHACP implants could not [127]. They evaluated the performance of bone growth in the experimental groups during 15 weeks by radiograph, mechanical strength, histology and bone mineral density. The results demonstrated that chitin fibers reinforced nHACP implants showed satisfactory repairing defect capacity. Figure 7.10 showed the radiographs in the experimental groups during 15 weeks after surgery. For the defect group, we can see that there was hardly no new bone formation at 15 weeks after surgery. For the PLLA group, there was a little new bone formation at 15 weeks after surgery and the defect was far from the repair. For the the nHACP group, some new bone formed at the both ends of the defects at 15 weeks after surgery, which was significantly more than that for the PLLA group. However, the new bone have not been connected, showing that the defect was not fully repaired because the defect was too large for the nHACP/CF implants. In contrast, for the fiber reinforced group, at only 5 weeks after surgery continuous callus appeared. Interestingly, marrow cavity in continuous new cortical bone could be observed at 10 weeks after surgery. At 15 weeks after surgery, the newly formed bone were successfully connected, and the bone defect was smoothly repaired. In addition, the mean breaking load of the reinforced group was higher than that in both pure PLLA and nHACP groups. Moreover, the mean breaking load of the implants for the reinforced group raised much quicker from 5 to 15 weeks than that for the nHACP group. At 15 weeks after surgery, the mean breaking load of the implants for the fiber reinforced group was about three times as that for PLLA group and about twice as that for nHACP group. Figure 7.11 showed the histological appearances of the defect tissue for all the experimental groups at the three time points. For the defect group, no osteoblasts but only fibrous-connective

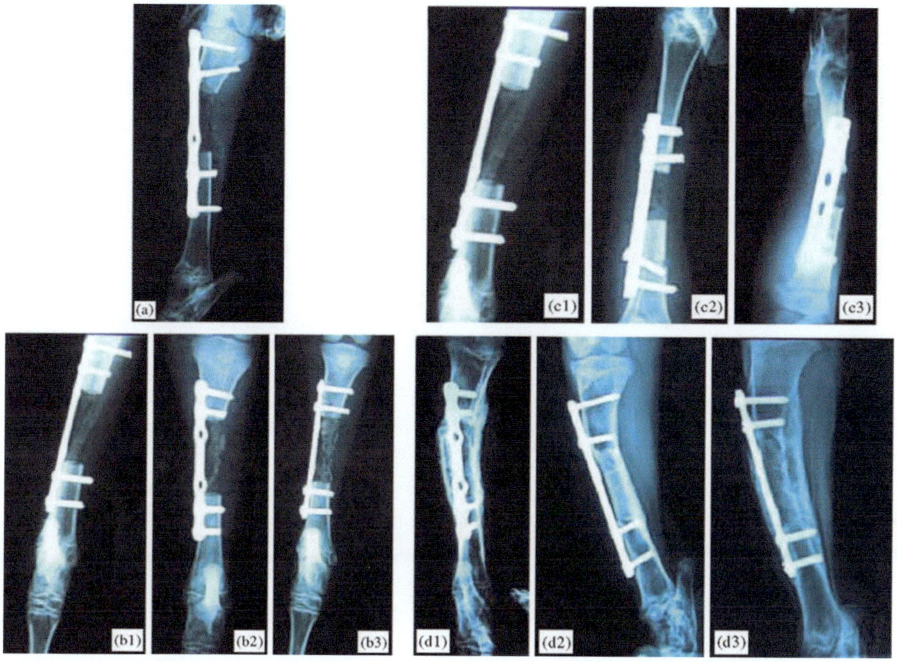

Fig. 7.10 Radiographs in defect group at 15 weeks after surgery (**a**);PLLA group at 5 weeks (**b1**), at 10 weeks (**b2**) and 15 weeks (**b3**); nHACP group at 5 weeks (**c1**), at 10 weeks (**c2**) and at 15 weeks (**c3**); the fiber reinforced nHACP group at 5 weeks (**d1**), at 10 weeks (**d2**) and 15 weeks (**d3**), showing that the fiber reinforced group evidently enhanced the continuous cortex of regenerated bone and a marrow cavity at 10 weeks, and achieved full bone defect repair at 15 weeks (Adapted with permission from Ref. [127]. Copyright 2006 Elsevier Ltd)

tissue can be found at 15 weeks after surgery. For the PLLA group, hardly any osteoblasts but a lot of mononuclear cells (white arrow area), neutrophilic granulocytes (red arrow area) could be observed, indicating that foreign-body feedbacks happened. Even at 15 weeks after surgery, no osseous callus could be found. For the the nHACP group, although not only osteoclasts (white arrow area) but also osteoblasts (red arrow area) were found at 5 weeks after surgery, and fibrous callus (red arrow area) could be observed at 10 weeks after surgery, no mature bone generated at 15 weeks after surgery. In contrast, for the fiber reinforced group, a lot of osseous callus (red arrow area) could be found at only 10 weeks after surgery. Most importantly, bone lacunas (red arrow area) and the bone cells (white arrow area) in the osseous callus could be observed at 15 weeks after surgery, indicating that mature new bone has generated. Furthermore, the analysis results of bone mineral density showed that the value hardly changed during the whole 15 weeks of implantation for the defect group, and although the bone mineral density for PLLA and nHACP increased with the increase of implantation, the value for the fiber reinforced group increased most. At 15 weeks after surgery, the mean bone mineral density for the fiber reinforced group was about twice as that for the nHACP group.

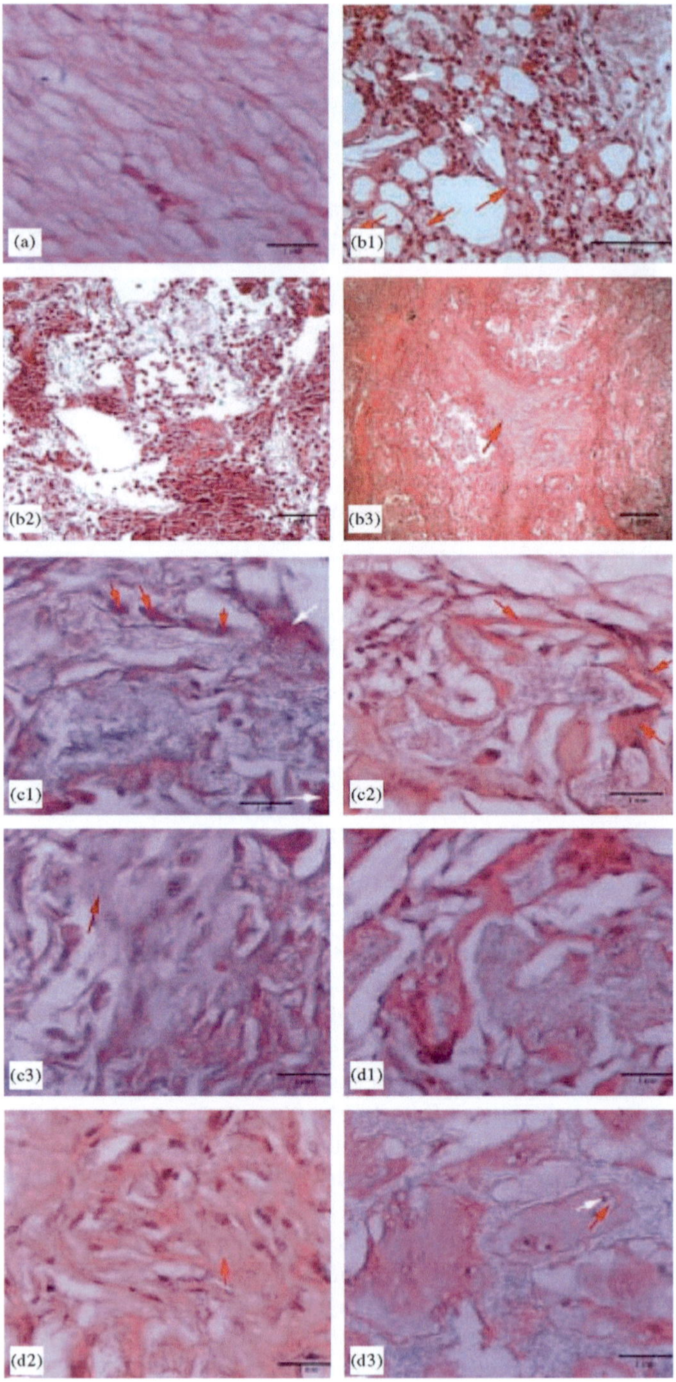

Fig. 7.11 (continued)

In 2016, Li et al. further provided a novel method to study the biocompatibility of nanoscale scaffolds reinforced by fibers or tubes for tissue engineering *in vitro* and *in vivo* [128]. They used the *in vitro* biodegradation products of nanohydroxyapatite/collagen (nHAC), nHACP, and the chitin fiber reinforced nHACP (nHACP/CF) in the D-Hank's Balanced Salt Solution as the testing solution, and found that. The results showed that the survival rate of the neutrophils was almost same in the different three testing solutions. But during the whole testing period the tumor necrosis factor-alpha and lactate dehydrogenase of the cells in the nHACP/CF testing solution were found lowest. In addition, *in vivo* experiments showed that hardly no inflammation reactions could be observed for the nHACP/CF group after the implantation into the subcutaneous dorsum of mice for 2 weeks after surgery while very serious inflammation reactions appeared for nHAC and poly (L-lactic acid) groups (Fig. 7.12). The results indicated that the addition of the reinforcing fibers effectively coordinated the interactions of the components, thereby appropriately controlling the local metal iron concentration and pH value at the implantation sites, which directly improved the biocompatibility of the scaffolds. Their study showed by a novel method that some appropriate fibers or tubes have the potential to not only enhance the mechanical properties of tissue engineering scaffolds but also improve their biocompatibility.

Moreover, Srivastava et al. inserted their composites in linking the big osteoperiosteal gap in rabbits, assessing the role of polyester resin composites strengthened by unidirectional carbon fiber, and filled by bone marrow impregnated tricalcium phosphate-polyvinyl alcohol with radiological, histological and clinical methods [129]. In their study, 20 healthy mature Indian rabbits were operated independently and implanted with the same composites. After the implantation, incision wound was checked every day for 7 days to avoid the gaping, swelling and infection of wound even skin necrosis. The rabbits were observed for 12–32 weeks (Fig. 7.13). The single particle and new bone formation were showed in the histopathology of tricalcium phosphate (purple) neocapillarisation. They implanted polyester resin composites filled with tricalcium phosphate-polyvinyl alcohol and strengthened by carbon fiber into the bone marrow of rabbit to bridge large osteoperiosteal gap. The clinical observations results performs a normal body weight improving. The roentgen graphic and histological observations of implanted points showed that the new bones form on the distal and proximaljunction side. Tricalcium phosphate as a good bone conduction material allowed bone growth, because of which the proximal

Fig. 7.11 (continued) Histological appearances of the tissue in defects for defect group at 15 weeks after surgery, (**a**); PLLA group at 5 weeks (**b1**), at 10 weeks (**b2**) and 15 weeks (**b3**); nHACP group at 5 weeks (**c1**), at 10 weeks (**c2**) and at 15 weeks (**c3**); the fiber reinforced nHACP group at 5 weeks (**d1**), at 10 weeks (**d2**) and 15 weeks (**d3**), showing that only in the fiber reinforced group, could bone lacunas (*red arrow area*) and the bone cells (*white arrow area*) in the osseous callus be observed at 15 weeks after surgery, indicating that mature new bone has generated (Adapted with permission from Ref. [127]. Copyright 2006 Elsevier Ltd)

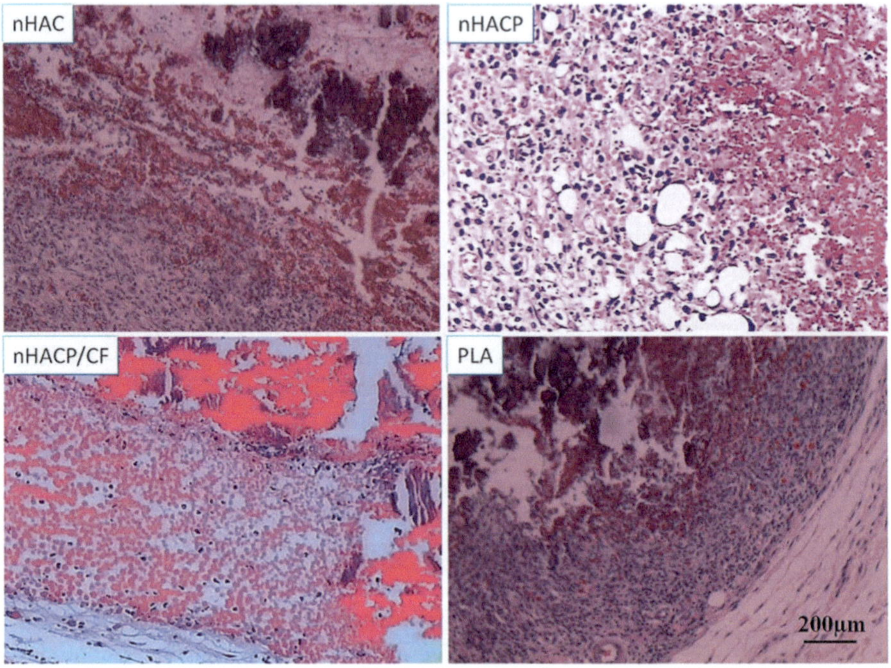

Fig. 7.12 Histological appearances of the tissues of mouse subcutaneous dorsum where the material testing solutions were injected 2 weeks before. A lot of lymphocytes, mononuclear cells, and neutrophilic granulocytes could be observed in the poly (L-lacticacid) (PLA) and nanohydroxyapatite/collagen (nHAC) groups, suggesting that serious foreign-body feedbacks have been caused by the two kinds of implants while the inflammatory cells was less than that in nanohydroxyapatite/collagen/PLA (nHACP) group, and no obvious foreign-body feedback happened in the fiber reinforced nHACP (nHACP/CF) group (Adapted with permission from Ref. [128]. Copyright 2016 John Wiley & Sons)

bone implants represented that the collagen tissue formed between new bone and tricalcium phosphate particles. Therefore, with good biocompatibility and mechanical properties, carbon fiber reinforced tricalcium phosphate-polyvinyl alcohol filled polyester resin composites are capable of being used for space implantation of the large osteoperiosteal gap.

It have recognized that biomimetic nanofibrous scaffolds derived from natural biopolymers for bone tissue engineering applications require good mechanical and biological performances including biomineralization. Moreover, bone tissue engineering scaffolds are usually expected to be resorbable and to guide incoming progenitor and vascular cells until neotissue has formed [130]. Zhou et al. successfully synthesized biodegradable rhBMP-2 loaded zein-based scaffolds with macroporous structures and silica (SBA-15) particles and hydroxypropyltrimethyl ammonium chloride chitosan (HACC) fibers were incorporated into the scaffolds to produce an anti-infective composite scaffold suitable to osteogenic factor delivery in the functional repair of bone defects (Fig. 7.14) [131]. By using the 3D micro-CT technique

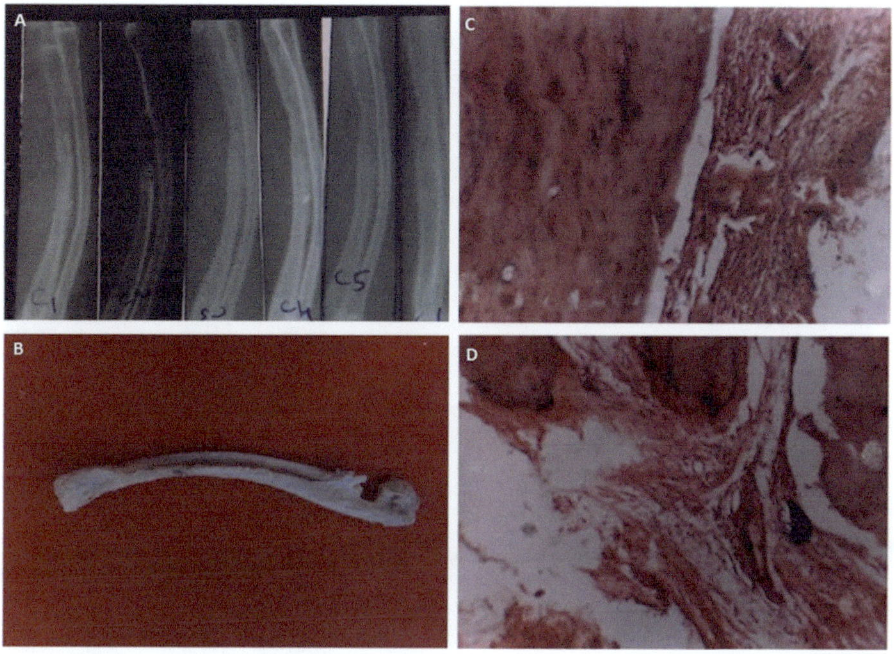

Fig. 7.13 Roentgenographs of implanted bone after a different interval of periods: (**a**) 3rd day, 4th, 10th, 16th, 24th and 32nd weeks. (**b**) implanted bone. (**c**, **d**) Histopathology of tricalcium phosphate (*purple*)neocapillarisation: showing asingle particle of and new bone formation around it (Adapted with permission from Ref. [129]. Copyright 2007 Elsevier Ltd)

Fig. 7.14 SEM of rhBMP-2 loaded silica/hydroxypropyltrimethyl ammonium chloride chitosan (HACC)/zein composite scaffold (Adapted with permission from Ref. [131]. Copyright 2014 Elsevier Ltd)

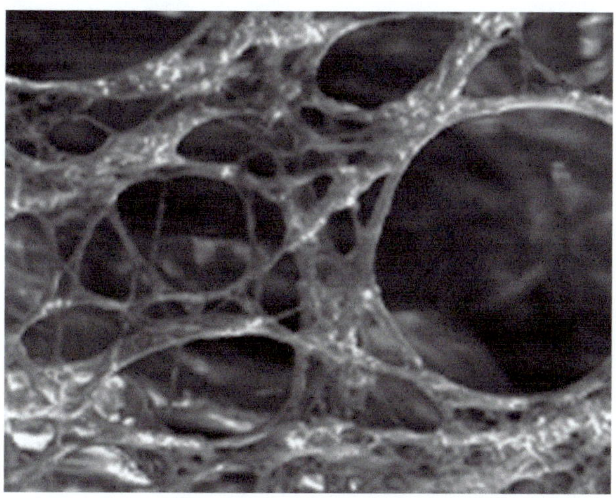

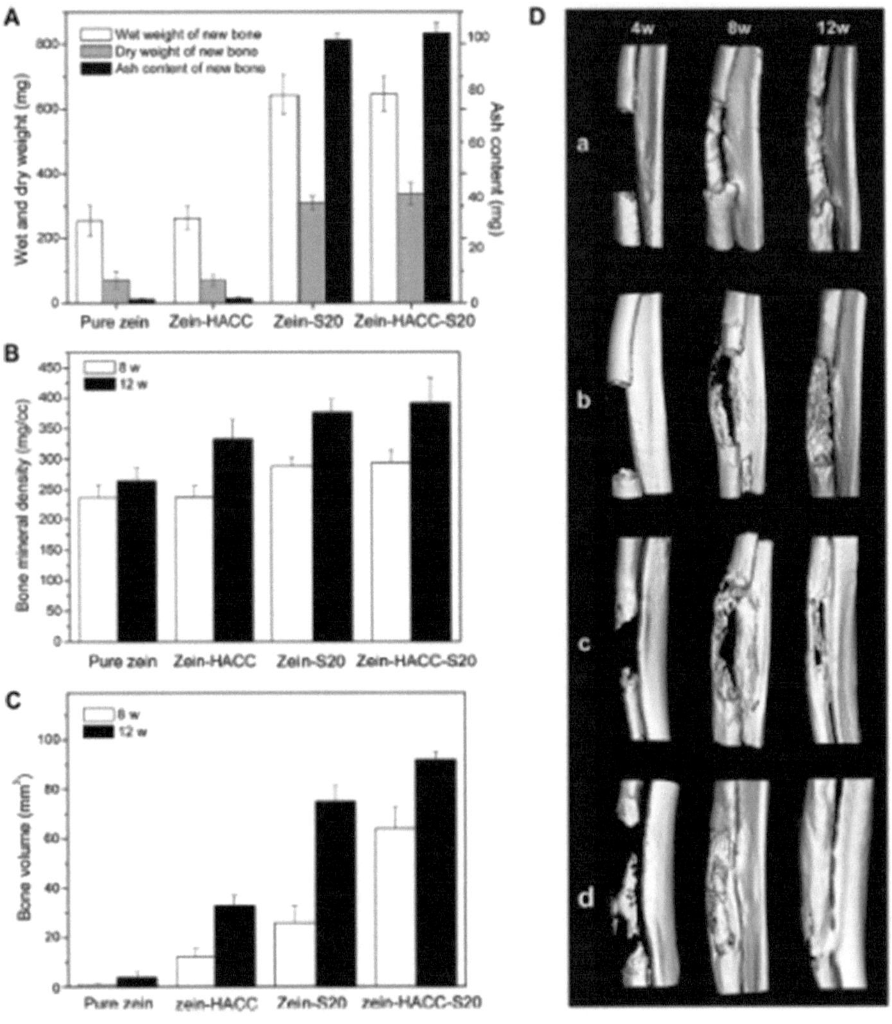

Fig. 7.15 (**a**) Wet, dry, and ash weight of ectopically formed bone induced by different composite caffolds after 4 weeks of implantation. (**b**) bone mineral density at 8 and 12 weeks. (**c**) Regenerated bone volume at both 8 and 12 weeks. (**d**) 3D Micro-CT reconstructed images of rabbit segmental radius at 4, 8, and 12 weeks with different implants: (**a**) pure zein-scaffold, (**b**) zein-hydroxypropyltrimethyl ammonium chloride chitosan (HACC), (**c**) zein-S20, and (**d**) zein-HACC-S20 scaffolds, indicating the zein-HACC-S20 scaffolds a significant improvementon the bone repair (Adapted with permission from Ref. [131]. Copyright 2014 Elsevier Ltd)

and histological analysis, the zein-HACC-S20 (SBA-15, 20 wt%) scaffolds were found to significantly promote the bone repair in a rabbit model of critical size radial defects (Fig. 7.15), which demonstrated considerable promise for bone repair through tissue engineering.

7.4 Conclusion and Perspectives

It could be concluded from this chapter that the use of fibers or tubes reinforced scaffolds have generally become a common method to repair hard tissues. Due to the satisfactory mechanical properties, the fibers or tubes reinforced scaffolds can not only provide more excellent three-dimensional support environment for the ingrowth of the desired cells and tissues, but also stimulate better the related cellular functions, thereby inducing more effectively the osteogenic differentiation of cells. Furthermore, the addition of fibers or tubes can actually increase the specific surface area, which can absorb more proteins, including osteogenesis-related proteins, thereby promoting the osteogenic differentiation and biomineralization of cells. Therefore, the reinforcement using fibers or tubes have been shown to enhance significantly hard tissue repair or regeneration efficacy, and be well recognized as one of promising means to develop new high-performanced hard tissue repair implants.

Although great progress in the research and development of fibers or tubes reinforced scaffolds for hard tissue repair have been made, there are still lots of work to do.

Firstly preparation of novel reinforcing fibers or tubes with both higher mechanical properties and perfect compatibility with the matrix scaffold has been always a worth studying point because mechanical properties of the fibers or tubes reinforced scaffolds are largely dependent on the bonding strength between the fibers or tubes and the matrix. Therefore, more efforts are needed to optimize surface properties of the fibers or tubes, or utilize more effective techniques to crosslink the fibers or tubes and the matrix, thereby achieving more satisfactory bonding strength between the fibers or tubes and the matrix. Also more practical techniques should be developed to study better the interactions between the fibers or tubes and the matrix, thereby inspiring effective improvement measures for their bonding strength.

Secondly, more comprehensive studies should be launched on the biodegradation of the fibers or tubes reinforced scaffolds because the keep of the reinforcement efficiency during the scaffolds biodegradation is crucial, especially for hard tissue repair.

Thirdly, since the interactions between scaffolds and cells or tissue are vital for the final hard tissue repair functions of the scaffolds, more elaborate efforts should be made to investigate into the effects of the fibers or tubes addition on responses of cells or tissues, thereby achieving the optimal choice of the reinforcing fibers or tubes that not only have desired effects on the related cells or tissue but also are able to counteract undesired effects of the matrix on them.

Finally, because the biomechanical environments *in vivo* have particularly important effects on the hard tissue implants, more advanced investigation and evaluation techniques are urgently needed to developed so as to analyze more comprehensively and accurately the effects of the biomechanical stimulations *in vivo* on the efficiency of the fibers or tubes addition, thereby achieving the optimal choice of the reinforcing fibers or tubes that not only can play their roles synergistically with the favorable biomechanical effects, but also have the ability to counteract the unfavorable biomechanical effects.

References

1. Yi JY, Njoroge DM, Sila DN, Kinyanjui PK, Christiaens S, Bi JF, Hendrickx ME. Detailed analysis of seed coat and cotyledon reveals molecular understanding of the hard-to-cook defect of common beans. Food Chem. 2016;210:481–90.
2. Desai MJ, Hutton WC, Jarrett CD. Arthroscopic repair of triangular fibrocartilage tears: a biomechanical comparison of a knotless suture anchor and the traditional outside-in repairs. J Hand Surg [Am]. 2013;38A:2193–7.
3. Koppes RA, Park S, Hood T, Jia XT, Poorheravi NA, Achyuta AH, Fink Y, Anikeeva P. Thermally drawn fibers as nerve guidance scaffolds. Biomaterials. 2016;81:27–35.
4. Sasaki N, Takakuwa J, Yamada H, Mori R. In vitro evaluation of allogeneic bone screws for use in internal fixation of transverse fractures created in proximal sesamoid bones obtained from equine cadavers. Am J Vet Res. 2010;71:483–6.
5. Cole BJ, ElAttrache NS, Anbari A. Arthroscopic rotator cuff repairs: an anatomic and biomechanical rationale for different suture-anchor repair configurations. Arthroscopy. 2007;23:662–9.
6. Farnebo S, Woon CY, Kim M, Pham H, Chang J. Reconstruction of the tendon-bone insertion with decellularized tendon-bone composite grafts: comparison with conventional repair. J Hand Surg [Am]. 2014;39:65–74.
7. Noyes FR, Barber-Westin SD. Arthroscopic repair of meniscus tears extending into the avascular zone with or without anterior cruciate ligament reconstruction in patients 40 years of age and older. Arthroscopy. 2000;16:822–9.
8. Li DQ, Li M, Liu PL, Zhang YK, Lu JX, Li JM. Tissue-engineered bone constructed in a bioreactor for repairing critical-sized bone defects in sheep. Int Orthop. 2014;38:2399–406.
9. Mozdzen LC, Rodgers R, Banks JM, Bailey RC, Harley BAC. Increasing the strength and bioactivity of collagen scaffolds usingcustomizable arrays of 3D-printed polymer fibers. Acta Biomater. 2016;33:25–33.
10. Aya R, Ishiko T, Noda K, Yamawaki S, Sakamoto Y, Tomihata K, Katayama Y, Yoshikawa K, Kubota H, Nakamura T, Naitoh M, Suzuki S. Regeneration of elastic fibers by three-dimensional culture on acollagen scaffold and the addition of latent TGF-b binding protein 4 to improve elastic matrix deposition. Biomaterials. 2015;72:29–37.
11. Hutmacher DW. Scaffolds in tissue engineering bone and cartilage. Biomaterials. 2000;21:2529–43.
12. Agrawal CM, Ray RB. Biodegradable polymeric scaffolds for musculoskeletal tissue engineering. J Biomed Mater Res. 2001;55:141–50.
13. Akin FA, Zreiqat H, Jordan S, Wijesundara MBJ, Hanley L. Hanley preparation and analysis of macroporous TiO2 films on Ti surfaces for bone-tissue implants. J Biomed Mater Res. 2001;57:588–96.
14. van Tienen TG, Heijkants RGJC, Buma P, de Groot JH, Pennings AJ, Veth RPH. Tissue ingrowth and degradation of two biodegradable porous polymers with different porosities and pore sizes. Biomaterials. 2002;23:1731–8.
15. Otsuki B, Takemoto M, Fujibayashi S, Neo M, Kokubo T, Nakamura T. Pore throat size and connectivity determine bone and tissue ingrowth into porous implants: three-dimensional micro-CT based structural analyses of porous bioactive titanium implants. Biomaterials. 2006;27:5892–900.
16. Chen CS, Mrksich M, Huang S, Ingber DE. Geometric control of cell life and death. Science. 1997;276:1425–8.
17. Ma T, Li Y, Yang ST, Kniss DA. Tissue engineering human placenta trophoblast cells in 3-D fibrous matrix: spatial effects on cell proliferation and function. Biotechnol Prog. 1999;15:715–24.
18. Huang L, Nagapudi K, Apkarian RP, Chaikof EL. Engineered collagen-PEO nanofibers and fabrics. J Biomater Sci Polym Ed. 2001;12:979–93.

19. Yoshimoto H, Shin YM, Terai H, Vacanti JP. A biodegradable nanofiber scaffold by electrospinning and its potential for bone tissue engineering. Biomaterials. 2003;24:2077–82.
20. Khan AS, Azam MT, Khan M, Mian SA, Rehman IU. An update on glass fiber dental restorative composites: a systematic review. Mater Sci Eng C Mater Biol Appl. 2015;47:26–39.
21. de Araujo MD, Miranda RBD, Fredericci C, Yoshimura HN, Cesar PF. Effect of fiber addition on slowcrack growth of a dental porcelain. J Mech Behav Biomed Mater. 2015;44:85–95.
22. Wang T, Tsoi JKH, Matinlinna JP. A novelzirconia fibre-reinforced resincomposite for dentaluse. J Mech Behav Biomed Mater. 2016;53:151–60.
23. Fisher MB, Mauck RL. Tissue engineering and regenerative medicine: recent innovations and the transition to translation. Tissue Eng Part B Rev. 2013;19:1–13.
24. Lu MC, Huang YT, Lin JH, Yao CH, Lou CW, Tsai CC, Chen SY. Evaluation of a multilayer microbraided polylactic acid fiber-reinforced conduit for peripheral nerve regeneration. J Mater Sci Mater Med. 2009;20:1175–80.
25. Moutos FT, Guilak F. Functional properties of cell-seeded three-dimensionally woven poly(e-caprolactone) scaffolds for cartilage tissue engineering. Tissue Eng A. 2010;16:1291–301.
26. Li XM, Yang Y, Fan YB, Feng QL, Cui FZ, Watari F. Biocomposites reinforced by fibers or tubes as scaffolds for tissue engineering or regenerative medicine. J Biomed Mater Res A. 2014;102:1580–94.
27. Leeuwenburgh SCG, Jo J, Wang HA, Yamamoto M, Jansen JA, Tabata Y. Mineralization, biodegradation, and drug release behavior of gelatin/apatite composite microspheres for bone regeneration. Biomacromolecules. 2010;11:2653–9.
28. Young S, Wong M, Tabata Y, Mikos AG. Gelatin as a delivery vehicle for the controlled release of bioactive molecules. J Control Release. 2005;109:256–74.
29. Shi YN, Han H, Quan HY, Zang YJ, Wang N, Ren GZ, Xing M, Wu QL. Activated carbon fibers/poly(lactic-co-glycolic) acid composite scaffolds: preparation and characterizations. Mater Sci Eng C. 2014;43:102–8.
30. Wang HA, Boerman OC, Sariibrahimoglu K, Li YB, Jansen JA, Leeuwenburgh SCG. Comparison of micro-vs. nanostructured colloidal gelatin gels for sustained delivery of osteogenic proteins: bone morphogenetic protein-2 and alkaline phosphatase. Biomaterials. 2012;33:8695–703.
31. Sabbagh J, Vreven J, Leloup G. Dynamic and static moduli ofelasticity of resin-based materials. Dent Mater. 2002;18:64–71.
32. Li WJ, Laurencin CT, Caterson EJ, Tuan RS, Ko FK. Electrospun nanofibrous structure: a novel scaffold for tissue engineering. J Mater Res. 2002;60:613–21.
33. Li XM, Feng QL. Dynamic rheological behaviors of the bone scaffold reinforced by chitinfibres. Mater Sci Forum. 2005;475-479:2387–90.
34. Li XM, Feng QL, Cui FZ. In vitro degradation of porous nano-hydroxyapatite/collagen/PLLA scaffold reinforced by chitin fibres. Mater Sci Eng C. 2006;26:716–20.
35. Tuzlakoglu K, Reis RL. Biodegradable polymeric fiber structures in tissue engineering. Tissue Eng Part B Rev. 2009;15:17–27.
36. Park SJ, Lee BK, Na MH, Kim DS. Melt-spun shaped fibers with enhanced surface effects: fiber fabrication, characterization and application to woven scaffolds. Acta Biomater. 2013;9:7719–26.
37. Zein I, Hutmacher DW, Tan KC, Teoh SH. Fused deposition modeling of novel scaffold architectures for tissue engineering applications. Biomaterials. 2002;23:1169–85.
38. Hutmacher DW. Scaffold design and fabrication technologies for engineering tissues – state of the art and future perspectives. J Biomater Sci Polym Ed. 2001;12:107–24.
39. Vozzi G, Previti A, De Rossi D, Ahluwalia A. Microsyringe-based deposition of two-dimensional and three-dimensional polymer scaffolds with a well-defined geometry for application to tissue engineering. Tissue Eng. 2002;8:1089–98.
40. Sherwood JK, Riley SL, Palazzolo R, Brown SC, Monkhouse DC, Coates M, Griffith LG, Landeen LK, Ratcliffe A. A three-dimensional osteochondral composite scaffold for articular cartilage repair. Biomaterials. 2002;23:4739–51.

41. Taboas JM, Maddox RD, Krebsbach PH, Hollister SJ. Indirect solid free form fabrication of local and global porous, biomimetic and composite 3D polymer-ceramic scaffolds. Biomaterials. 2003;24:181–94.
42. Giordano RA, Wu BM, Borland SW, Cima LG, Sachs EM, Cima MJ. Mechanical properties of dense polylactic acid structures fabricated by three dimensional printing. J Biomater Sci Polym Ed. 1996;8:63–75.
43. Landers R, Hubner U, Schmelzeisen R, Mulhaupt R. Rapid prototyping of scaffolds derived from thermoreversible hydrogels and tailored for applications in tissue engineering. Biomaterials. 2002;23:4437–47.
44. Woodfield TBF, Malda J, de Wijn J, Peters F, Riesle J, van Blitterswijk CA. Design of porous scaffolds for cartilage tissue engineering using a three-dimensional fiber-deposition technique. Biomaterials. 2004;25:4149–61.
45. Moroni L, de Wijn JR, van Blitterswijk CA. 3D fiber-deposited scaffolds for tissue engineering: influence of pores geometry and architecture on dynamic mechanical properties. Biomaterials. 2006;27:974–85.
46. Kenawy ER, Layman JM, Watkins JR, Bowlin GL, Matthews JA, Simpson DG, Wnek GE. Electrospinning of poly(ethylene-co-vinyl alcohol) fibers. Biomaterials. 2003;24:907–13.
47. Wintermantel E, Mayer J, Blum KL, Eckert P, Lüscher MM. Tissue engineering scaffolds using superstructures. Biomaterials. 1996;17:83–91.
48. Ge ZG, Yang F, Goh JCH, Ramakrishna S, Lee EH. Biomaterials and scaffolds for ligament tissue engineering. J Mater Res A. 2006;77A:639–52.
49. Lee JH, Park JH, El-Fiqi A, Kim JH, Yun YR, Jang JH, Han CM, Lee EJ, Kim HW. Biointerface control of electrospun fiber scaffolds for bone regeneration: engineered protein link to mineralized surface. Acta Biomater. 2014;10:2750–61.
50. Horner CB, Ico G, Johnson J, Zhao Y, Nam J. Microstructure-dependent mechanical properties of electrospun core-shell scaffolds at multi-scale levels. J Mech Behav Biomed Mater. 2016;59:207–19.
51. Liu J, Cheng F, Grenman H, Spoljaric S, Seppala J, Eriksson JE, Willfor S, Xu CL. Development of nanocellulose scaffolds with tunable structures to support 3D cell culture. Carbohydr Polym. 2016;148:259–71.
52. da Silva LPN, Issa JPM, Bel EAD. Action of nitric oxide on healthy and inflamed human dental pulp tissue. Micron. 2008;39:797–801.
53. Food, Drug Administration (FDA). Dental composites-premarket notification. US Department of Health and Human Services. 1998.
54. Tezvergil A, Lassila LVJ, Vallittu PK. The effect of fiber orientation on the thermal expansion coefficient of fiber reinforced composites. Dent Mater. 2003;19:471–7.
55. Budai Z, Sulyok Z, Vargha V. Glass-fibre reinforced composite materials based on unsaturated polyester resins. J Therm Anal Calorim. 2012;109:1533–44.
56. Zhang M, Matinlinna JP. E-glass fiber reinforced composites in dental applications. Silicon. 2012;4:73–8.
57. Zhu HY, Li DH, Zhang DX, Wu BC, Chen YY. Influence of voids on interlaminar shear strength of carbon/epoxy fabric laminates. Trans Nonferrous Metals Soc China. 2009;19:s470–5.
58. da Silva Pinto CE, Carbajal A, Wypych F, Ramos LP, Satyanarayana KG. Studies of the effect of molding pressure and incorporation of sugarcane bagasse fibers on the structure and properties of poly (hydroxy butyrate). Compos A: Appl Sci Manuf. 2009;40:573–82.
59. Moriwaki T. Mechanical property enhancement of glass fibre reinforced polyamide composite made by direct injection. Composites. 1996;27:379–84.
60. Leenslag JW, Pennings AJ. High-strength poly(L-lactide) fibres by a dry-spinning/hot-drawing process. Polymer. 1987;28:92–4.
61. Nazhat SN, Kellomaki M, Tormala P, Tanner KE, Bonfield W. Dynamic mechanical characterization of biodegradable composites of hydroxyapatite and polylactides. J Biomed Mater Res. 2001;58:335–43.

62. Kulkova J, Moritz N, Huhtinen H, Mattila R, Donati I, Marsich E, et al. Bioactive glass surface for fiber reinforced composite implants via surface etching by Excimer laser. Med Eng Phys. 2016;38:664–70.
63. Abdal-haya A, Sheikh FA, Lim JK. Air jet spinning of hydroxyapatite/poly(lactic acid) hybrid nanocomposite membrane mats for bone tissue engineering. Colloids Surf B: Biointerfaces. 2013;102:635–43.
64. Abdal-Hay A, Hasan A, Yu-Kyoung LMH, Hamdy AS, Khalil KA. Biocorrosion behavior of biodegradable nanocomposite fibers coated layer-by-layer on AM50 magnesium implant. Mater Sci Eng C Mater Biol Appl. 2016;58:1232–41.
65. Erkmen E, Meric G, Kurt A, Tunc Y, Eser A. Biomechanical comparison of implant retained fixed partial dentures with fiber reinforced composite versus conventional metal frameworks: a 3D FEA study. J Mech Behav Biomed Mater. 2011;4:107–16.
66. Boudeau N, Liksonov D, Barriere T, Maslov L, Gelin JC. Composite based on polyetheretherketone reinforced with carbon fibres, an alternative to conventional materials for femoral implant: manufacturing process and resulting structural behaviour. Mater Des. 2012;40:148–56.
67. Rameshbabu AP, Mohanty S, Bankoti K, Ghosh P, Dhara S. Effect of alumina, silk and ceria short fibers in reinforcement of Bis-GMA/TEGDMA dental resin. Compos Part B. 2015;70:238–46.
68. Silva RM, Santos PHN, Souza LB, Dumont VC, Soares JA, Santos MH. Effects of cellulose fibers on the physical and chemical properties of glass ionomer dental restorative materials. Mater Res Bull. 2013;48:118–26.
69. Anderson JB, Marcus W, Estevam AB, Paulo GC. Physicochemical characterization of three fiber-reinforced epoxide-basedcomposites for dental applications. Mater Sci Eng C. 2016;69:905–13.
70. Qidwai M, Sheraz MA, Ahmed S, Alkhuraif AA, Ur Rehman I. Preparation and characterization of bioactive composites and fibers for dental applications. Dent Mater. 2014;30:E253–63.
71. Abdulmunem M, Dabbagh A, Naderi S, Zadeh MT, Halim NFA, Khan S, et al. Evaluation of the effect of dental cements on fracture resistance and fracture mode of teeth restored with various dental posts: a finite element analysis. J Eur Ceram Soc. 2016;36:2213–21.
72. Ereifej N, Silikas N, Watts DC. Initial versus final fracture of metal-free crowns, analyzed via acoustic emission. Dent Mater. 2008;24:1289–95.
73. Chen YC, Li HY, Fok A. In vitro validation of a shape-optimized fiber-reinforced dental bridge. Dent Mater. 2011;27:1229–37.
74. Basaran EG, Ayna E, Vallittu PK, Lassila LVJ. Load bearing capacity of fiber-reinforced and unreinforced composite resin CAD/CAMfabricated fixed dental prostheses. J Prosthet Dent. 2013;109:88–94.
75. Sheikhhassani R, Anvari P, Taei S, Sheikhhassani Y. Potential use of a polycarbonate-urethane matrix reinforced with polyethylene fibers for shock-absorbing dental implants. Med Hypotheses. 2015;85:241–4.
76. Lee YH, Bhattarai G, Park IS, Kim GR, Kim GE, Lee MH, Yi HK. Bone regeneration around N-acetyl cysteine-loaded nanotube titanium dental implant in rat mandible. Biomaterials. 2013;34:10199–208.
77. Lee YH, Lee NH, Bhattarai G, Oh YT, Yu MK, Yoo ID, Jhee EC, Yi HK. Enhancement of osteoblast biocompatibility on titanium surface with terrein treatment. Cell Biochem Funct. 2010;28:678–85.
78. Feng YF, Wang L, Zhang Y, Li X, Ma ZS, Zou JW, Lei W, Zhang ZY. Effect of reactive oxygen species overproduction on osteogenesis of porous titanium implant in the present of diabetes mellitus. Biomaterials. 2013;34:2234–43.
79. Dafara MO, Grol MW, Canhama PB, Dixon J, Rizkalla AS. Reinforcement of flowable dental composites withtitanium dioxide nanotubes. Dent Mater. 2016;32:817–26.
80. Chen Q, Zhao Y, Wu WD, Xu T, Fong H. Fabrication and evaluation of Bis-GMA/TEGDMA dental resins/composites containing halloysite nanotubes. Dent Mater. 2012;28:1071–9.

81. Zhao B, Hu H, Mandal SK, Haddon RC. A bone mimic based on the self-assembly of hydroxyapatite on chemically functionalized single-walled carbon nanotubes. Chem Mater. 2005;17:3235–41.
82. Cheng Q, Rutledge K, Jabbarzadeh E. Carbon nanotube-poly(lactide-co-glycolide) composite scaffolds for bone tissue engineering applications. Ann Biomed Eng. 2013;41:904–16.
83. Mohamed E, Taheri M, Mehrjoo M, Mazaheri M, Zahedi AM, Shokrgozar MA, Golestani-Fard F. In vitro biocompatibility and ageing of 3Y-TZP/CNTs composites. Ceram Int. 2015;41:12773–81.
84. Nejatidanesh F, Momeni G, Savabi O. Flexural strength of interim resin materials for fixed prosthodontics. J Prosthodont. 2009;18:507–11.
85. Barry DH, Jeffrey AH. Fiber-reinforced interim fixed dental prostheses: a clinical protocol. J Prosthet Dent. 2016;116:496–500.
86. Kumbuloglu O, Ozcan M. Clinical survival of indirect, anterior 3-unit surface-retained fibre-reinforced composite fixed dental prosthesis: up to 7.5-years follow-up. J Dent. 2015;43:656–63.
87. Mendes RM, Silva GA, Caliari MV, Silva EE, Ladeira LO, Ferreira AJ. Effects of single wall carbon nanotubes and its functionalization with sodium hyaluronate on bone repair. Life Sci. 2010;87:215–22.
88. Sa MA, Andrade VB, Mendes RM, Caliari MV, Ladeira LO, Silva EE, Silva GAB, Correa JD, Ferreira AJ. Carbon nanotubes functionalized with sodium hyaluronate restore bone repair in diabetic rat sockets. Oral Dis. 2013;19:484–93.
89. Pangon A, Saesoo S, Saengkrit N, Ruktanonchai U, Intasanta V. Hydroxyapatite-hybridized chitosan/chitin whisker bionanocomposite fibers for bone tissue engineering applications. Carbohydr Polym. 2016;144:419–27.
90. Li XM, Feng QL, Wang WJ, Cui FZ. Chemical characteristics and Cytocompatibility of collagen-based scaffold reinforced by chitin fibers for bone tissue engineering. J Biomed Mater Res B Appl Biomater. 2006;77B:219–26.
91. Li XM, Liu XH, Dong W, Feng QL, Cui FZ, Uo M, et al. In vitro evaluation of porous poly(L-lactic acid) scaffold reinforced by chitin fibers. J Biomed Mater Res B Appl Biomater. 2009;90B:503–9.
92. Guarino V, Causa F, Taddei P, di Foggia M, Ciapetti G, Martini D, et al. Polylactic acid fibre-reinforced polycaprolactone scaffolds for bone tissue engineering. Biomaterials. 2008;29:3662–70.
93. Lipner J, Liu W, Liu Y, Boyle J, Genin GM, Xia Y, Thomopoulos S. The mechanics of PLGA nanofiber scaffolds with biomimetic gradients in mineral for tendon-to-bone repair. J Mech Behav Biomed Mater. 2014;40:59–68.
94. Kolambkar YM, Dupont KM, Boerckel JD, Huebsch N, Mooney DJ, Hutmacher DW, et al. An alginate-based hybrid system for growth factor delivery in the functional repair of large bone defects. Biomaterials. 2011;32:65–74.
95. Anselme K, Davidson P, Popa AM, Giazzon M, Liley M, Ploux L. The interaction of cells and bacteria with surfaces structured at the nanometrescale. Acta Biomater. 2010;6:3824–46.
96. Mallick KK, Cox SC. Biomaterial scaffolds for tissue engineering. Front Biosci. 2013;5:341–60.
97. Hoppe A, Güldal NS, Boccaccini AR. A review of the biological response to ionic dissolution products from bioactive glasses and glassceramics. Biomaterials. 2011;32:2757–74.
98. Maeno S, Niki Y, Matsumoto H, Morioka H, Yatabe T, Funayama A, et al. The effect of calcium ion concentration on osteoblast viability, proliferation and differentiation in monolayer and 3D culture. Biomaterials. 2005;26:4847–55.
99. Mouriño V, Cattalini JP, Boccaccini AR. Metallic ions as therapeutic agents in tissue engineering scaffolds: an overview of their biological applications and strategies for new developments. J R Soc Interface. 2011;9:401–19.

100. Wu CT, Zhou YH, Xu MC, Han PP, Chen L, Chang J, et al. Copper-containing mesoporous bioactive glass scaffolds with multifunctional properties of angiogenesis capacity, osteostimulation and antibacterial activity. Biomaterials. 2013;34:422–33.
101. Sharifi E, Azami M, Kajbafzadeh AM, Moztarzadeh F, Majidi Faridi R, Shamousi A, Karimi R, Ai J. Preparation of a biomimetic composite scaffold fromgelatin/collagen and bioactive glass fibers for bone tissue engineering. Mater Sci Eng C: Mater Biol Appl. 2016;59:533–41.
102. O'Brien FJ, Harley BA, Yannas IV, Gibson L. Influence of freezing rate on pore structure in freeze-dried collagen-GAG scaffolds. Biomaterials. 2004;25:1077–86.
103. O'Brien FJ. Biomaterials & scaffolds for tissue engineering. Mater Today. 2011;14(3):88–95.
104. O'Brien FJ, Harley BA, Yannas IV, Gibson LJ. The effect of pore size on cell adhesion in collagen-GAG scaffolds. Biomaterials. 2005;26:433–41.
105. Tierney CM, Jaasma MJ, O'Brien FJ. Osteoblast activity on collagen-GAG scaffolds is affected by collagen and GAG concentrations. J Biomed Mater Res A. 2009;91A:92–101.
106. Liu LS, Thompson AY, Heidaran MA, Poser JW, Spiro RC. An osteoconductive collagen/hyaluronate matrix for bone regeneration. Biomaterials. 1999;20:1097–108.
107. Li ZZ, Wen JH, Jia WJ, Ding S, Xia XH, Zhou CR, Huang YD. Bio-inspired cell membrane ingredient cholesterol-conjugated chitosan as a potential material for bone tissue repair. Chem Res Chin Univ. 2016;32:406–13.
108. Mashak A. In vitro drug release from silicone rubber-polyacrylamide composite. Silicon Chem. 2008;3:295–301.
109. Hutmacher DW. Scaffolds in tissue engineering bone and cartilage. Biomaterials. 2000;21:2529–43.
110. Beynnon BD, Fleming BC. Anterior cruciate ligament strain invivo: a review of previous work. J Biomech. 1998;31:519–25.
111. Pennisi E. Tending tender tendons. Science. 2002;295:1011–11.
112. Albright JC, Carpenter JE, Graf BK, et al. Knee and leg: soft-tissue trauma. In: Beaty JH, editor. Orthopaedic knowledge, update 6. Rosemont: American Academy of Orthopedic Surgery; 1999. p. 78.
113. Makhmalbaf H. ACL deficient knee, an approach to patient selection and reconstruction. Arch Bone Jt Surg. 2013;1:1457–9.
114. Moglo KE, Shirazi-Adl A. Biomechanics of passive knee joint in drawer: load transmission in intact and ACL-deficient joints. Knee. 2003;10:265–76.
115. Hehl G, Strecker W, Richter M, Kiefer H, Wissmeyer T. Clinical experience with PDS II augmentation for operative treatment of acute proximal ACL ruptures-2-year follow-up. Knee Surg Sports Traumatol Arthrosc. 1999;7:102–6.
116. Altman GH, Horan RL, Lu HH, Moreau J, Martin I, Richmond JC, et al. Silk matrix for tissue engineered anterior cruciate ligaments. Biomaterials. 2002;23:4131–441.
117. Majima T, Lo IKY, Randle JA, Marchuk LL, Shrive NG, Frank CB, et al. ACL transection influences mRNA levels for collagen type I and TNF-α in MCL scar. J Orthop Res. 2002;20:520–5.
118. Freeman JW, Woods MD, Cromer DA, Ekwueme EC, Andric T, Atiemo EA, et al. Evaluation of a hydrogel-fiber composite for ACL tissue engineering. J Biomech. 2011;44:694–9.
119. Woods T, Gratzer PF. Effectiveness of three extraction techniques in the development of a decellularized bone-anterior cruciate ligamentbone graft. Biomaterials. 2005;26:7339–49.
120. Brammer KS, Choi C, Frandsen CJ, Oh S, Johnston G, Jin S. Comparative cell behavior on carbon-coated TiO2 nanotube surfaces for osteoblasts vs. osteo-progenitor cells. Acta Biomater. 2011;7:2697–703.
121. Wan YZ, Liu P, Zhang C, Yang ZW, Xiong GY, Zheng XR, Luo HL. Synthesis of a three-dimensional network-structured scaffold built of silica nanotubes for potential bone tissue engineering applications. J Alloys Compd. 2015;647:711–9.
122. Hirota M, Hayakawa T, Yoshinari M, Ametani A, Shima T, Monden Y, et al. Hydroxyapatite coating for titanium fibre mesh scaffold enhances osteoblast activity and bone tissue formation. Int J Oral Maxillofac Surg. 2012;41:1304–9.

123. Sahoo S, Toh SL, Goh JCH. A bFGF-releasing silk/PLGA-based biohybrid scaffold for ligament/tendon tissue engineering using mesenchymal progenitor cells. Biomaterials. 2010;31:2990–8.
124. Zhu Y, Wang XH, Ma J, Feng QL, Cui FZ. In vitro cytocompatibility of phosphorylated chitosan (P-chitosan) with osteoblasts. J Bioact Compat Polym. 2003;18:375–90.
125. Wang XH, Feng QL, Cui FZ, Ma JB. The effects of S-chitosan onthe physical properties of calcium phosphate cements. J Bioact Compat polym. 2003;18:45–57.
126. Chen SH, Tang T, Liu Z, Lau PY, Xie XH, Wang XL, et al. Study on osteointegrative properties of a functional gradient tumor bone repair materials in ulna bone defect model of rabbits. Bone. 2010;47:S432–32.
127. Li XM, Feng QL, Liu XH, Dong W, Cui FZ. Collagen-based implants reinforced by chitin fibres in a goat shank bone defect model. Biomaterials. 2006;27:1917–23.
128. Li XM, Wang Z, Zhao TX, Yu B, Fan YB, Feng QL, Cui FZ, Watari F. A novel method to in vitro evaluate biocompatibility of nanoscaled scaffolds. J Biomed Mater Res A. 2016;109:2117–25.
129. Srivastava VK, Rastogi A, Goel SC, Chukowry SK. Implantation of tricalcium phosphate-polyvinyl alcohol filled carbon fibre reinforced polyester resin composites into bone marrow of rabbits. Mater Sci Eng A. 2007;448:335–9.
130. Liu XH, Ma PX. Polymeric scaffolds for bone tissue engineering. Biomaterials. 2004;32:477–86.
131. Zhou PY, Xia Y, Cheng XS, Wang PF, Xie Y, Xu SG. Enhanced bone tissue regeneration by antibacterial and osteoinductive silica-HACCzein composite scaffolds loaded with rhBMP-2. Biomaterials. 2014;35:10033–45.

Chapter 8
Scaffolds Reinforced by Fibers or Tubes for Soft Tissue Repair

Baoqing Pei, Wei Wang, and Xiaoming Li

Abstract Soft tissue is a dynamic and hierarchical organic structure, and its natural extracellular matrix (ECM) not only provides a microscopic mechanical environment but also regulates a variety of cellular function and exercise through the continuous interaction with the cells, mechanical transmission and the expression of chemical signals. Scaffolds in soft tissue engineering as an artificial extracellular matrix (ECM) should have the appropriate appearance and functions to provide material support for cell proliferation and survival. In nature, many soft tissues are usually mechanically reinforced by fibers to form a three-dimensional structure. Therefore, the application of reinforcing scaffolds has a number of potential advantages that can't be ignored in soft tissue repair and regeneration. In the past few years, novel fibers or tubes reinforced scaffolds with controlled microstructures, mechanical properties and degradation rates have emerged in soft tissue engineering and been considered to be a very effective means for engineering materials development for soft tissue engineering or regenerative medicine. Studies have shown that fiber and tube reinforcements with their unique properties and functions play a key role in improving the biomechanics, biocompatibility, bioactivity, integration and degradation of synthetic scaffolds in soft tissue repair and regeneration. In this chapter, the properties, structures and applications of fiber- or tube- reinforced scaffolds in soft tissue engineering are reviewed in conjunction with recent advances.

Keywords Scaffolds • Reinforce • Fibers • Tubes • Soft tissue

8.1 Introduction

Soft tissue in the anatomy usually contains the blood vessel, skin, subcutaneous shallow fascia, intervertebral disc, muscle, joint capsule, tendon, synovial sac, ligament, nerves, cornea, etc., which are used to connect, support or enclose the body structure and organs rather than hard tissue (E.g., bone). Soft tissue injury often

B. Pei (✉) • W. Wang • X. Li
Key Laboratory for Biomechanics and Mechanobiology of Ministry of Education, School of Biological Science and Medical Engineering, Beihang University, Beijing 100191, China
e-mail: pbq@buaa.edu.cn

X. Li (ed.), *Tissue Repair*, DOI 10.1007/978-981-10-3554-8_8

occurs in daily life, which mainly refers to a variety of acute trauma or chronic stress and its own disease, leading to the human body soft tissue pathological damage. Although scientists have found a variety of methods to repair damaged soft tissue, autologous implantation is still the main treatment of soft tissue repair and regeneration. However, autologous tissue is easily absorbed and quickly lost its volume, and only about 40–60% of the graft tissue can remain active [1–4]. As an emerging field of research, tissue engineering offers a novel approach to generate new or substituted tissues to repair or regenerate damaged soft tissues and organs, overcoming the limitations of current treatments. This technique mainly consists of in vitro culture of cells and the incorporation into temporary scaffolds with tissue morphologies. The scaffold acts as an artificial temporary tissue matrix that directs the formation of new tissue by modulating the interaction of cells-matrix or even cells-cells [5–7]. Therefore, scaffolds used in tissue engineering are usually made into an artificial ECM having a structure, function and ability similar to that of human tissue, and play a vital role in the process of soft tissue regeneration.

Natural ECM is mainly composed of polysaccharides and proteins or proteoglycans to form a complex network structure, support and connect tissue structure, regulate the occurrence of tissue and cell physiological activities. Extracellular matrix as part of animal tissue provides a suitable place for the survival and activity of cells and affects their shape, metabolism, function, migration, proliferation and differentiation through the mechanical forces and signal transduction system. Therefore, in addition to providing physical support for cell growth, soft-tissue engineering scaffolds should meet several requirements, namely, biocompatibility, mechanical stability, and proper morphology as well as chemically modulating signal and mechanical transduction [8–10]. In view of the above-mentioned factors, the scaffold is usually made of biocompatible, biodegradable and resorbable materials and developed into a new biological tissue through advanced manufacturing and culture methods. The current study of soft tissue have found that cell adhesion, proliferation and differentiation have a great relationship with micro-mechanical environment in which the cells are located [11–13]. When soft tissues have high mechanical activity, such as blood vessels, heart valves and corneas, it is particularly important that the scaffold has sufficient mechanical properties to effectively transfer the mechanical stimulus between the cells and scaffold [11, 14, 15]. In fact, the failure of the scaffold grafted in the blood vessel is usually caused by intimal hyperplasia due to compliance mismatch between graft stent and host vessel [16]. Also, studies of bladder tissue have shown that the mechanical properties of the membrane influence on cell growth, that is, the proliferation of cells in the membrane with the elastic modulus closer to the natural bladder tissue will be enhanced [5]. It is critical that the scaffold has good mechanical properties and is able to maintain relatively stable during the tissue growth phase to be successfully applied in soft tissue engineering. However, most of the scaffolds currently utilized in soft tissue engineering are not capable of satisfying the biomechanical properties of the microenvironment in cells and tissues throughout the implantation process.

Human soft tissues and organs have very complex components and microstructures. Elastin and collagen fibers are the most important constituents of most soft tissues. Elastin is the main component of elastic fibers. Elastic fibers are mainly present in ligaments and vessel walls. Elastic fibers and collagen fibers co-exist to provide flexibility and tensile capacity for soft tissues [17]. Collagen is the most important extracellular water-insoluble fibrin, used to form the skeleton of ECM [18]. Collagen in the ECM forms semi-crystalline fibers, provides cells with tension and elasticity and plays a role in the cell migration and development. These semi-crystalline fibrils are arranged into fibers, that is, collagen fibers, which are widely distributed in various organs, especially in load carrying components such as skin, sclera and tendon [19]. Many soft tissue structures are that collagen fibers are encapsulated in the hydrogel-like elastin matrix, such as blood vessels [20], heart valves [21] and myocardium [22], and their biomechanical properties are jointly determined by the arrangement and composition of the two components. The fibers can directly affect the stiffness and strength of tissues under tension and shorten the time response of tissues under compression. Gel phase swelling can support and distribute compression loads through fluid pressurization [23, 24].

Therefore, we can often see in nature that soft tissue is supported by fibers to enhance mechanical properties. Articular cartilage is a typical example, that tough collagen fibrils are distributed in a soft proteoglycan matrix [25]. In addition, the bladder cytoskeleton is also composed of microtubules and intermediate fibers embedded in highly hydrogel-like cytoplasmic matrix [26]. In organisms, the structure of fiber- or tube- reinforced hydrogels is numerous. This concept has been successfully applied in the field of materials engineering, such as glass fiber, carbon fiber and carbon nanotube composite material [27–29]. The concept of scaffold with reinforcements also has a great potential advantage in the field of soft tissue regeneration. In fact, in recent years fiber- or tube- reinforced scaffolds have attracted considerable interest from investigators, and some of these findings have proven that reinforced scaffolds are an engineering material with high performance and utility for soft tissue engineering.

As the reinforcement material has greater controllability and moldability, the reinforced scaffold can better simulate the structure and composition of soft tissues. Fiber or tube-reinforced scaffolds that have been used in soft tissue repair and reconstruction significantly enhance tissue repair or regeneration, for they can provide sufficient mechanical strength for cell attachment, migration, differentiation and proliferation in the whole growth state, which means a reasonable micro-mechanical environment. And in most cases the reinforcement can improve the biocompatibility and biodegradability of the scaffold. Furthermore, recent studies have shown that the mechanical properties of the reinforced scaffold depend largely on the fiber or tube reinforcement' properties, arrangement, volume content and aspect ratio, the combined manner of the matrix and fibers or tubes, and the matrix properties [1, 30–33].

This chapter describes the design and manufacture of novel fiber- or tube- reinforced scaffolds in soft tissue engineering. We discuss currently known fiber- or tube- reinforced scaffolds having been used in the field of soft tissue engineering, and how the fiber/tube influence structural characteristics, mechanical strength and biological activities *in vitro* and *in vivo* in repairing soft tissues. When the design and processing methods are reasonable, the reinforced scaffold will become a promising technology to produce bio-related microenvironments similar to the ECM of soft tissues and organs, and thus promote cell proliferation, migration, resulting in the final formation of new functional tissues.

8.2 Biomedical Applications: Fiber or Tube Reinforced Scaffolds for Soft Tissue

8.2.1 Cartilage Tissue

Cartilage, composed of chondrocytes, fibers and matrix, is a slightly resilient tough tissue with the functions of support and protection in the body. As a typical cartilage tissue, articular cartilage is smooth and elastic and has the function of maximizing absorption and buffer stress. Articular cartilage often gets hurt in daily life. Injured articular cartilage will reduce the absorption of the force, resulting in joint damage and degeneration [34–36]. Due to the lack of blood supply, chondrocytes are not capable of providing adequate metabolism and proliferation, so the repair and regeneration capacity of cartilage after injury is very weak [1]. There is still a formidable task for successful long-term regeneration of cartilage tissues by using current surgical and nonsurgical treatment concepts. Although some new treatments and related products have entered clinical trials stage, and even commercialized, tissue engineering is the preferred target for most products to make up for the lack of traditional surgical treatment [37]. Scaffolds used to repair the cartilage can replace injured tissue with good time, cost efficiency and no donor site complications and lead to new tissue growth. The ideal scaffold should have stiffness and strength close to the surrounding cartilage, so as not to damage the mechanical environment of original tissues [38]. In addition, the improvement in mechanical properties can't reduce the porosity and interconnectivity of scaffolds [39]. However, most of the scaffolds lack reasonable mechanical properties and physical structures that are detrimental to the long-term efficacy of cartilage repair [40].

There are two basic targets for artificial cartilage regeneration based on the scaffold. One is to obtain an interconnected structure with highly porous. The other is to offer the appropriate mechanical environment for cell growth and secretion of the matrix, which can lead to the regeneration of hyaline cartilage tissue. At present, popular materials for cartilage tissue engineering are mainly synthetic materials, such as PGA, PLA, PLGA, PCL, and PEG [41–43]. The stiffness of the scaffold can't be too large nor too small, which may lead to high shear and opposing surface wear of the scaffold or high relative stress concentration in the other adjacent cartilage. The main limitation of the engineering scaffolds is the relatively poor mechanical properties, for an implant without strong mechanical properties may be crushed inside the joint.

Multifarious synthetic scaffolds have been studied to provide adequate support for the cartilage regeneration. Among many efforts on improving or reinforcing mechanical properties of scaffolds for cartilage tissue engineering, fiber- or tube-reinforced scaffolds have made impressive results for the successful implantation. Slivka et al. [40] have prepared a porous, resorbable scaffold reinforced by fibers for the use of repairing articular cartilage. The scaffold was made of PGA fiber as the reinforcements with about 15 lm in diameter and about 2.5 mm in average length and a random copolymer of poly($_{\text{D, L}}$-lactide-co-glycolide) (PLGA 75:25) as the matrix. The method used to prepare scaffolds has been published and validated for several times [44, 45]. First, the PLGA was dissolved in the acetone solution, and the PGA was distributed into the ethanol. Both the PLGA and PGA were then mixed to form a matrix gel. Next, the matrix gel was blended for dispersing and preferring orientation of fibers. Finally, the composite was exposed under a high temperature vacuum, which could allow the gel to foam and remove residual acetone and ethanol. The mechanical properties of the composites was significantly improved by the addition of the PGA fibers. As the fiber content increased from 0 to 20%, the yield strength and the compressive modulus increased from 0.7 to 2.5 MPa and from 10 to 50 MPa, respectively. Based on the relationship between degradation and compression stiffness, PGA fiber reinforced PLGA in smaller amounts with proper porosity appeared to be an optimized articular cartilage repair scaffold. At the same time, PGA fibers with anisotropic optical properties appeared brighter in the orthogonally polarized light.

Agrawal et al. [46] also designed a new type of elastic fiber-reinforced hydrogel composites through 3D rapid prototyping technology. The fiber-reinforced gel formed a cartilage-like structure by impregnating elastic fibers, which had been arranged into crossed “log-piles”, with an epoxy-based hydrogel. Three different log-pile structures were shown in Fig. 8.1. This is a common problem in the traditional fiber-reinforced composite that the modulus and strength of the fibers are higher than that of the matrix, but a lower elongation at break, so the fiber’s fracture strain limits the maximum strain at the interface. Just like a hard glass fiber-reinforced gel has high strength and modulus but low the fracture strain. Therefore, it is not enough that a fiber-reinforced gel has a high strength, and it should have a high breaking strength. Elastic fiber-reinforced hydrogel scaffolds incorporated the mechanical properties of both elastic fibers and gels, which made them similar to the cartilage tissues. The results proved that the strength, modulus and toughness of the composites were improved, while the swell of the composite was limited. The effect of the addition of fibers on the strength was more complex, but the main effect of them was to increase the strain. The new gel composites had a much higher fracture strain, which could be compared with metal reinforced glass system. As for the elastic modulus of the composite, the fiber parallel to the stress axis in the half layer made about twice as much as the unreinforced gels. The new composite had significantly higher fracture toughness than that of natural cartilage, which made it pos-

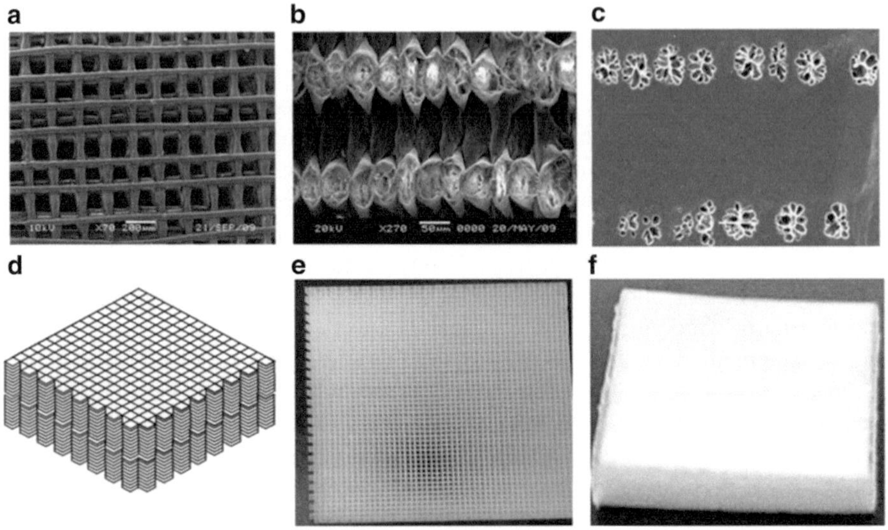

Fig. 8.1 (**a**) SEM image of polyurethane construct (scale = 100 lm). (**b**) Cross-section image of the 200 × 200 construct (scale = 50 lm). (**c**) Cross-section image of reinforced composite gel; the pores correspond to the fiber and the smooth surface corresponds to the gel (scale = 50 lm). (**d**) Processing method sketch. (**e**) Fiber construct image of 1 in. × 1 in. dimensions. (**f**) Fiber reinforced hydrogel of 1 in. × 1 in. dimensions and about 0.25 in. thickness (Adapted with permission from Ref. [46]. Copyright 2013 Elsevier Ltd)

sible to be suitable for cartilage replacement and regeneration. Furthermore, the new type of elastic fiber-reinforced hydrogel composites with the modulus and strength in a MP range would have a potential to be applied in load bearing tissue engineering.

Visser et al. [47] combined the melt-electrospinning writing technology with 3D printing to create reinforced hydrogels with highly organized, high-porosity microfiber networks. Reinforced hydrogel scaffolds with different porosities were prepared from gelatin methacrylamide (GelMA) and PCL fibers. The high-porosity PCL fibers were fabricated by melt electrospinning in a direct writing mode, and then infused and crosslinked with GelMA [48]. By analyzing the stiffness and recovery of the composite in cyclic compression, the study found that the composite had a strong elasticity similar to native articular cartilage and was recovered after physiological axial strain. Compared with the hydrogel or microfiber scaffold alone, the stiffness of the composite was significantly increased by 54-fold, similar to the native articular cartilage. The reinforced GelMA hydrogel also had the stress-strain curve resembling that of native articular cartilage. Thus, the reinforced hydrogel was reinforced in a synergistic manner depending on the porosity of PCL fiber and the degree of crosslinking of the GelMA, and this cooperative effect can be adjusted by varying the porosity of the reinforcing scaffold. As for primary human chondrocytes encapsulated in the reinforced scaffold, they were keeping their circulation pattern and became more susceptible to physiological compressive loading *in vitro* culture, which resulted in a significant change *in vitro* gene expression. These results provide the potential for building cell-culture platforms of tissue engineering to better simulate *in vivo* mechanical environment of natural cartilage. Thus, the new reinforced composite material provides an advantageous environment in the mechanical and biological aspects for cartilage repair or other tissue engineering. And the current method by using melt-electrospinning writing and 3D printing technology to fabricate microfiber reinforced hydrogels also provides the basis for the design and manufacture of new biomechanical organizational structures. In the other study, Boere and Visser et al. [49] also created a reinforced gelMA hydrogel scaffold with 3D-fabricated methacrylate groups of poly(hydroxymethylglycolide-co-ε-caprolactone) (pHMGCL)/PCL as reinforcements to repair the focal articular cartilage defect. The fiber-reinforced composite composed of pMHMGCL/PCL covalently grafting to gelMA was fabricated by using a 3D fiber deposition technology. The reinforced composite constructs had a significantly improved mechanical, especially the resistance capacity to repeated axial and rotational loads. Both *in vitro and in vivo*, chondrocytes embedded in these fiber-reinforced scaffolds produced cartilage-specific matrix components. This means that the fiber-reinforced structure has the ability to guide matrix formation, so this study has the potential to be used to manufacture the substitute of patient-specific cartilage with improved mechanical properties.

Yodmuang et al. [50] reported a kind of silk-based fiber-reinforced hydrogel scaffolds, whose matrix and reinforcements were composed of the same material, silk from Bombyx mori, to imitate the collagen fiber and proteoglycan complex system structure of native cartilage. It was the first time that two forms of silk materials, microfiber and hydrogel, were jointly used for preparing an improved mechanical reinforced scaffold in cartilage tissue engineering. The silk hydrogels reinforced with silk microfibers provided a suited structural and mechanical microenvironment for chondrocyte attachment and growth, thus their mechanical constructs became more robust than those of silk hydrogels alone after 42 days in culture. Furthermore, the silk hydrogel with optimized diffusivity and mechanical properties had an excellent performance in supporting the cartilage matrix deposition. Through studying the role of silk microfibers as reinforcements in silk hydrogel, this work demonstrates that the composite hydrogel provides an excellent scaffold with properties close to native cartilage.

As for the other research of the reinforcing material for the cartilage regeneration, Seifzadeh and cooperators [51] studied the material parameters fiber-reinforced of biphasic poroviscoelastic (BPVE) for cartilage tissue engineering. The cartilage was modeled as a finite element (FE) model and assigned the material property of fiber-reinforced nonlinear biphasic elasticity. The constitutive parameters of articular cartilage materials were obtained from a series of stress relaxation indentation tests. A coupled finite element (FE) optimization scheme was a simulated annealing (SA) algorithm to optimize the fiber-reinforced biphasic poroviscoelastic material parameters. 15 material parameters were optimized and led to the consistency between the relaxation response and the measured stress. The study may provide not only a new fiber-reinforced composite material of scaffolds but also an effective research method for the cartilage tissue engineering.

8.2.2 *Vascular Tissue*

Vascular system throughout the body is biological pipeline for the blood transportation, and is also the main site for the material exchange between blood and tissue. The blood in the vessel carries a variety of nutrients and oxygen to all the tissues and organs of the body and discharges metabolic products. At the beginning of the twenty-first century, about 17 million people died of cardiovascular and cerebrovascular diseases [52, 53], and the number is continuing to rise, expecting to reach 25 million by 2020. With the rise in the incidence of cardiovascular disease (CVD), clinical treatment requires a variety of different diameter vascular substitutes to repair damaged vascular tissue. In recent years, vascular tissue engineering has been recognized as an effective way to provide a variety of methods to create vascular networks for the treatment of cardiovascular diseases [54, 55]. Vascular tissue engineering is the science of preparing, reconstructing, and regenerating vascular replacement materials by using normal cells in the vessel wall and biodegradable materials. With a large number of studies in the field going on, vascular scaffolds have been developed with no immunogenicity, high tissue and cell compatibility and certain mechanical properties, and show the potential to be widely used in clinical vascular replacement to solve the clinical treatment of ischemic diseases caused by vascular stenosis or occlusion [56–58].

Like most regenerative tissue cultures, regenerated vascular designs consist of three components, namely, vascular cells, scaffolds, and nurturing environment [59]. As an artificial ECM of vascular tissue, the scaffolds need to act as an effective temporary structure for supporting and promoting cell adhesion, differentiation and proliferation And then the scaffold should degrade over time when the cells precipitate to form a new regenerative ECM. Extensive materials from natural materials to synthetic materials have been studied as design candidates for tissue engineering scaffolds for vascular tissue engineering. Biomaterials include natural biomaterials, such as collagen, decellularized matrix and fibrin [60–64], synthetic polymeric degradable materials, such as polyglutamic acid, polylactic acid, polyurethanes, and polycaprolactone [16, 65, 66], and their composites or copolymers [67, 68]. Because of the complexity of the structure and function of natural blood vessels, a single material is difficult to meet the requirements of scaffolds. In recent years, fiber- or tube-reinforced scaffolds have been used for vascular tissue and have become the focus of extensive research. Such a reinforcing scaffold can be prepared by the use of two or more materials with different characteristics and mechanical properties through certain methods and manufacturing techniques.

Caves et al. [69] used ECM protein analogues to produce a multi-layered vascular graft consisting of a synthetic collagen microfibers-reinforced elastin-like protein matrix for the repair and regeneration of smaller diameter (<6 mm) grafts. For

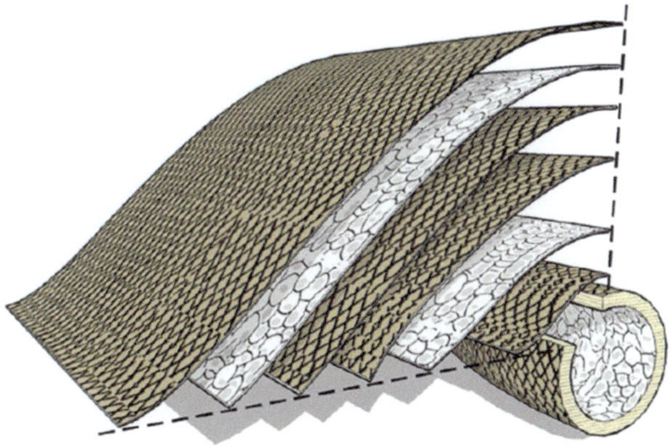

Fig. 8.2 Multilamellar protein polymer microfiber composite vascular graft. The vessel wall has a multilamellar structure consisting of collagen microfiber (thickness 100 mm) embedded in protein polymer, and its orientation and density are established to achieve the mechanical design goals (Adapted with permission from Ref. [69]. Copyright 2010 Elsevier Ltd)

vein graft bypass transplantation, compliance under physiological stress is associated with intimal hyperplasia, which affects vascular patency. Early vascular tissue engineering has demonstrated the feasibility of the concept of ECM protein scaffolds from collagen gel. However, the strength of the scaffold is insufficient due to the deviation of the microstructures from the natural collagen fiber orientation, architecture, and packing density [60]. In order to create a densely packaged oriented collagen scaffold, continuous ultrafine fibers consisting of aligned, D-type periodic collagen fibrils were prepared by using a scalable spinning process [70]. In addition, a composite material has recently been described, consisting of a continuous, aligned, micro-crimped collagen fibers embedded into in an elastin-like protein matrix [71]. As Fig. 8.2 shown, the generation of cell-free arterial substitutes had a multilayer structure formulated from the recombinant elastin-like protein and integrated synthetic collagen microfibers. A simple description of manufacturing small diameter vascular grafts was shown in Fig. 8.3. The manufacturing scheme allowed a series of fiber orientation and volume fraction, which resulted in variable mechanical properties. The composites had a rupture strength of 239–2760 mm Hg, and their compliance and suture retention strength were 2.8–8.4%/100 mm Hg and 35–192 gf, respectively. According to the expected design criteria, the breaking strength, the compliance and the suture retention strength of the most reasonable material structure were 1483 ± 143 mm Hg, 5.1 ± 0.8%/100 mm Hg, and 173 ± 4 gf, respectively. Thus, recombinant elastin-based biomaterials can play an important role in improving mechanical properties of the scaffold by adding collagen microfibers.

In another study, Pooyan and colleagues [59] recognized that microstructural and mechanical properties of vascular stents were important parameters for further cellular activity and new tissue development, and designed a new cellulose nanowhiskers (CNWs) reinforced scaffold. As the most widely distributed and abundant polysaccharide in nature, cellulose is a macromolecular polysaccharide composed of glucose, and it is also the main component of plant cell wall. CNWs are usually extracted from cellulose by intense multistage chemical/mechanical separation methods, typically 1–100 nm in diameter and 0.5–2 μm in length [72, 73]. Figure 8.4

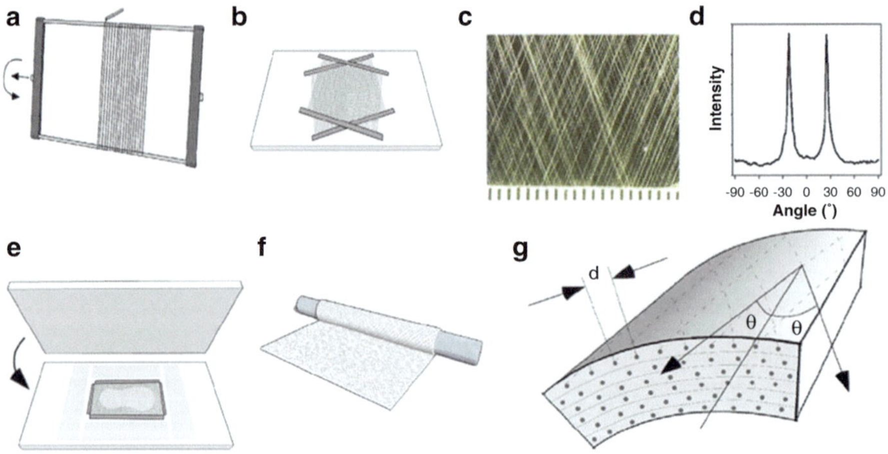

Fig. 8.3 Fabrication of fiber-reinforced small diameter vascular grafts. The graft consists of oriented synthetic collagen microfibers and an elastin protein polymer matrix. (**a**) Parallel arrays of fiber produced by winding about a frame. (**b**) Two arrays were transferred to a glass sheet at the desired angle. (**c**) Characterize the average fiber spacing and orientation with digital photographs. (**d**) Fiber orientation fast fourier transform analysis. (**e**) Fibers were surrounded with precision-thickness shims and a solution of elastin protein polymer was applied into a thin film. (**f**) The gelled film was rolled about a Teflon rod to produce a six-layered tube. (**g**) Schematic diagram of the average fiber spacing (*d*) and angle (*θ*) (Adapted with permission from Ref. [69]. Copyright 2010 Elsevier Ltd)

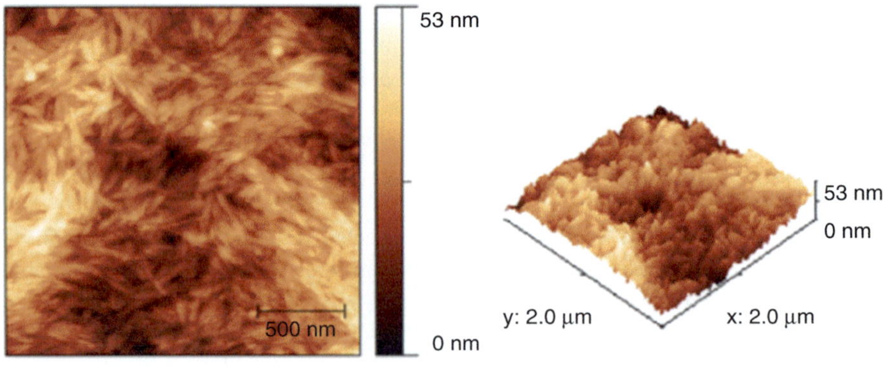

Fig. 8.4 AFM morphology images of cellulose nanowhiskers (Adapted with permission from Ref. [59]. Copyright 2012 Elsevier Ltd)

showed the microstructure and general morphology of the CNW by AFM detection. For CNW has sufficient tensile strength and elastic modulus, the incorporation of CNW into fiber networks may help to get a variety of structures and functions. CNWs have a porous permeability, controllable biodegradability, stable mechanical strength, moldable properties and long-term storage capacity, whose excellent properties make them a revolutionary candidate for the use as a reinforcing material in

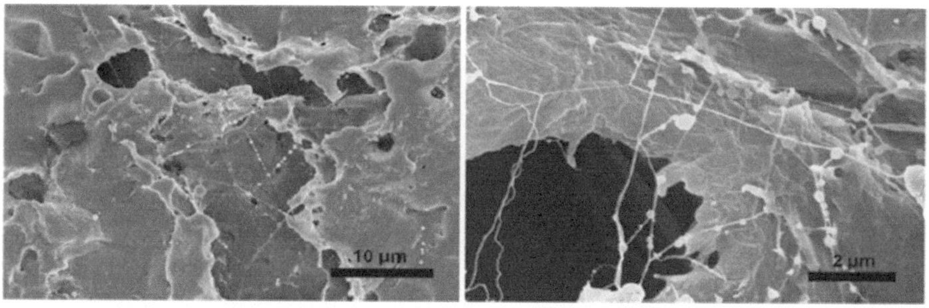

Fig. 8.5 SEM images of the dispersion of CNWs within the CAP medium at 3% volume fraction (Adapted with permission from Ref. [59]. Copyright 2012 Elsevier Ltd)

vascular tissue engineering to provide a biocompatible environment for the cells growth and tissue development [74, 75]. In order to make full use of the properties of cellulose and its derivatives, a biological scaffold consisting of CNWs embedded in the cellulose acetate propionate matrix was designed by Parisa Pooyan et al. As Fig. 8.5 shown, when compared to carbon nanotubes or Kevlar, reinforced scaffolds with 0.2 wt.% CNWs exhibited significant strength and directional rigidity. When adding 3.0 wt.% CNWs, the value was almost double. In order to predict the enhancement effect of CNWs, the mechanical behavior of the nanocomposite was subjected to tensile tests in the non-linear range at a body temperature of 37 ° C. Based on these comparisons, the reason for the excellent mechanical stability of the composite structure at such a low CNWs filler content was that CNWs were connected to each other by strong hydrogen bonds to form a 3D rigid percolation network. Thus, the use of fibrous porous CNW microstructures to reinforce the scaffold under specific operating conditions provides a potential to improve mechanical properties for withstanding the physiological pressure, and simulates the structural features of natural ECMs in human blood vessels. The newly designed composite materials not only greatly expand the use of cellulosic-based materials in tissue engineering, but also provide potential candidates as reinforcements for vascular tissue, especially in small-diameter grafts.

Allen et al. [76] developed an absorbable arteriolar graft that didn't require cell seeding or culture prior to implantation, so the substitute minimized the cost of production and eliminated the waiting time of patient. The composite vascular scaffold was made of microporous tubes of rapidly degradable elastomeric poly (glycerol sebacate) (PGS), whose outer surface was reinforced by PCL nanofibers to improve the long-term survival of acellular scaffold. The scaffold was inserted into the rat abdominal aorta. One year later, the graft regenerated a new "neoarteries" with a general appearance similar to that of the native rat aorta. At 30 days after implantation, the vast majority of composite scaffolds were completely absorbed, except for some PCL residues [77]. Rapid degradation and absorbability of the scaf-

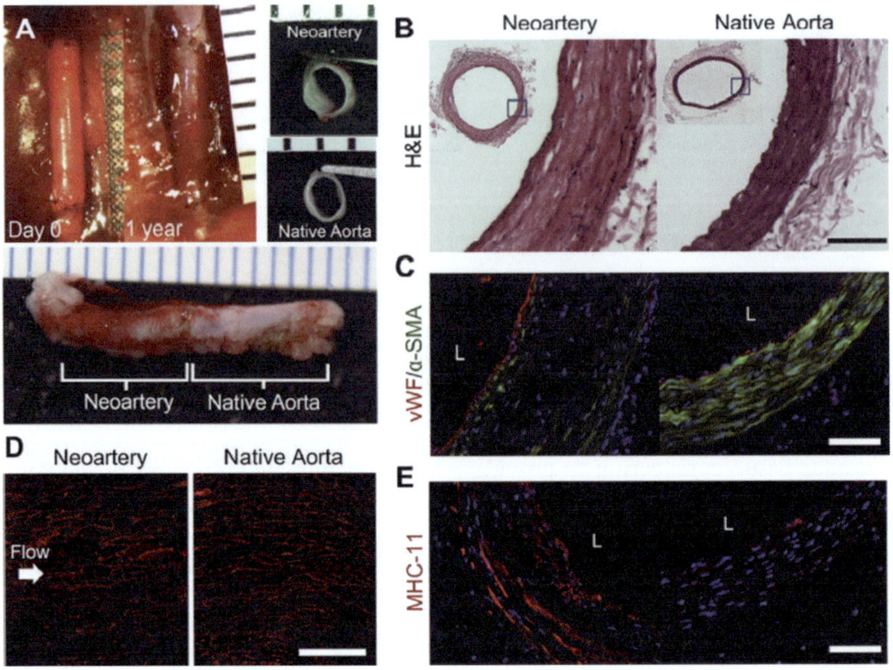

Fig. 8.6 Gross morphology and tissue structure of neoarteries similar to native arteries. (**a**) Gross morphology of neoarteries. *Top left*: transformation of graft into neoartery in situ in 1 year. *Top right*: Transverse view of exploratory neoarteries. *Bottom*: Longitudinal view of exploratory neoarteries. All ruler ticks are 1 mm. (**b**) The H & E staining transverse sections of the middle of neoarteries show similar tissue architecture with native aortas. (**c**) Neoartery sections immunostained for von Willebrand factor (vwf, *red*) and α-smooth muscle actin (α-SMA, *green*). (**d**) En face view of the luminal surface of neoarteries shows complete coverage by CD31 positive cells. (**e**) Immunostaining for smooth muscle myosin heavy chain 11 (MHC-11, *red*). Scale bar 100 mm. *L* indicates vessel lumen. Nuclei stained with DAPI (*blue*) (Adapted with permission from Ref. [76]. Copyright 2013 Elsevier Ltd)

fold material accelerated the ability of host remodeling. The advantage of this scaffold was that it minimized the duration of host response to foreign body, which helped to reduce intimal hyperplasia [78], calcification [79, 80] and tissue degeneration [81] in the late-term. In fact, no signs of tissue degeneration (Fig. 8.6), neointimal hyperplasia (Fig. 8.7a), dilation (Fig. 8.7), or calcification (Fig. 8.7b) were seen at one year after implantation. Moreover, the regenerative arteries contained not only the nerves in perivascular tissues (Fig. 8.8) but also mature elastin in the same amount as native arteries. And mature elastin achieved complete regeneration in native tissue. Although the magnitude of neonatal arteries differed from the native aortas, neonatal arteries responded to vasomotor agents, which were primarily used to modulate vasotonia. There were some differences in matrix tissue, and the regenerative arteries *in vivo* showed a dynamic mechanical compliance similar to that of native arteries. This study shows that cell-free scaffolds with rapid resorbability and

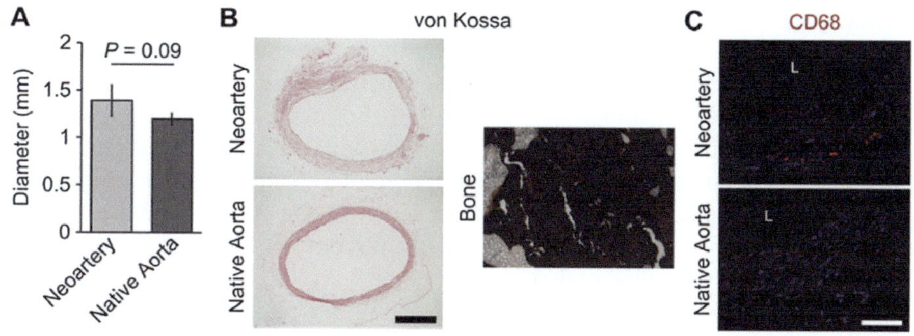

Fig. 8.7 Neoarteries resist common modes of late-term graft failure. (**a**) Statistical diameter of neoartery and native aortas. (**b**) von Kossa staining of Neoartery and Native Aortas. Scale bar 500 mm. (**c**) Neoarteries with cells positive of macrophage marker CD68 in the outermost layer of the neoartery wall. Scale bar 100 mm. DAPI stained nuclei (*blue*) (Adapted with permission from Ref. [76]. Copyright 2013 Elsevier Ltd)

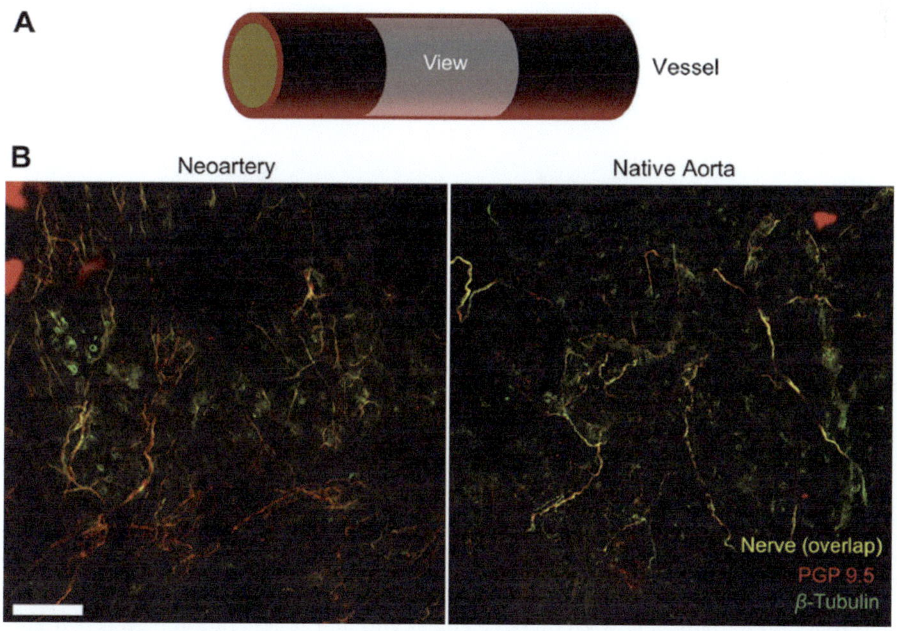

Fig. 8.8 Regeneration of nerves in neoarteries. (**a**) Schematic of en face imaging of the adventitial surface of neoarteries. (**b**) Nerves cover the adventitial surface of neoarteries with similar morphology to those in native aortas. Scale bar 100 mm (Adapted with permission from Ref. [76]. Copyright 2013 Elsevier Ltd)

elasticity can be reconstructed at damaged tissue and result in regenerative arteries with long-term stability. Additionally, this design improves the performance of the cell-free scaffolds in repairing small arteries, which also can be applied to other soft tissue repair and regeneration.

Liu et al. [82] designed a silk-based scaffold as small diameter vessel prostheses, whose matrix and reinforcements were composed of the same material, silk fibroin. The reinforced scaffolds had a bilayer structure, that was an internal silk fiber reinforced silk fibroin tube embedded in the nanofibrous silk layers. The outer nanofibrous silk layers were prepared by lyophilization with a highly porous structure, in which the nanofibers similar to the ECM comprised the macropore walls. The function of the outer layers was to provide an appropriate growth environment for fibroblasts and smooth muscle adhesion and proliferation, while the inner silk fiber reinforcement utilized internal heparin to provide excellent blood compatibility for at least 1 month. Compared with previous silk scaffolds, the bilayer fiber-reinforced ones not only provided a better mechanical strength, burst pressure, and suture retention strength but also achieved a similar compliance to saphenous veins. The silk-based reinforced scaffold also had a good cytocompatibility and hemocompatibility *in vitro*. These results confirm that the silk reinforced bilayer constructs have the potential as small diameter vascular grafts, and the follow-on *in vivo* studies still need to be conducted to reduce the risk of graft failure, due to endometrial hyperplasia, poor suture retention and weak mechanical properties.

8.2.3 Hair Tissue

Hair loss is divided into physiological and pathological. Physiological alopecia refers to the normal shedding of hair. Pathological alopecia refers to abnormal or excessive hair loss. In addition to trauma resulting in partial- or full-thickness skin lesions, increasing pressure on life, environmental degradation, and poor eating habits also make the number of hair loss patients continue to rise. Pathology has a negative impact not only on people's appearance but also on the mental health of patients [83]. Despite the many strategies and methods of treating hair loss, the treatment is still a worldwide problem, because there is no direct and effective therapeutic agent [84–86]. The development of tissue engineering as a new strategy has begun to emerge, but the reorganization of hair tissue with complete organization and function remains a new challenge. The key to hair regeneration is to repair and rebuild the growth environment of the hair, so the scaffold should support epithelial stem cell proliferation and differentiation and induce the formation of new tissue [83]. In numerous studies of hair tissue engineering, the reinforced scaffold is also a promising tool for repairing and reconstructing hair epithelial tissue.

Itoh et al. [87] designed a PGA fiber-reinforced collagen scaffold for reconstructing hair. The process of hair growth is that the epithelial cells grow down into the dermal tissue to form a head, called the dermal papilla [88, 89]. Cell implantation position and diffusion movement are necessary condition for hair follicle reconstruction. Collagen is a commonly used material for repair and reconstruction of skin tissue, but the traditional collagen sponge material is not suitable for the repair of scalp tissue [90–92]. Because of poor mechanical property and stability, the pores of the scaffolds formed by the pure collagen sponge decrease with the shrinkage of the collagen sponge, which hinders the diffusion and deposition of the grafted cells, and even leads to the leakage of the cell suspension. To observe the effect of fiber enhancement, pure collagen scaffolds and fiber-reinforced collagen scaffolds transplanted within hair and hair follicles of athymic mice were implanted into the dorsal skin for 3 weeks. The results showed that the nascent hair with a regular arrangement tilted from beginning to end, and the growth of the sebaceous glands on the fiber-reinforced scaffold was normal. Thus, the formation and reconstruction of the hair did not receive the effects of the implant and lasted for 6 months at least. The average hair densities on the collagen scaffold and fiber-reinforced scaffold were 531 ± 179 and 2808 ± 1237 wool/cm 2, respectively, based on the result of quantitative analysis on both implants. There was no significant difference in the shape and size of the two grafts. Since the scaffold has sufficient mechanical properties to maintain the structural stability of the scaffold *in vivo*, the scaffold-embedded cells can be well preserved and viable. Moreover, the transfer movement of cells could be observed in the reinforced scaffold, which also contributed to the formation of epidermal cysts. In addition, due to the plasticity of the PGA fibers, the addition of them enhanced the ability of collagen sponge scaffolds to be deformed, which was conducive to resist mechanical deformation and increase the possibility of implantation of the reinforced scaffold into an irregularly implanted bed. Therefore, PGA fiber-reinforced collagen scaffolds, as the most typical fiber-reinforced scaffolds, represent the promising to be used in hair tissue repair and regeneration.

8.2.4 Nerve Tissue

Nerve gap, as a common soft tissue defect, affects many people's normal life and work. Researchers have been working to explore a variety of techniques that can restore neural motion and sensory functions in the spinal cord. However, since the injured axons can't regenerate themselves, neural remodeling in the central nervous system has always been a serious challenge [93]. Autologous nerve grafts are the standard method for the clinical treatment of nerve injury, when a severely damaged nerve without tension can't be sutured directly [94]. Although autologous nerve grafts are generally accepted by the medical community as the gold standard, donor tissue availability, surgical trauma, and associated complications can lead to the failure of treatment [95, 96]. Nerve tissue engineering has proposed the use of bioartificial nerve conduits as an alternative of autografts to treat nerve tissue damage, especially for large defects.

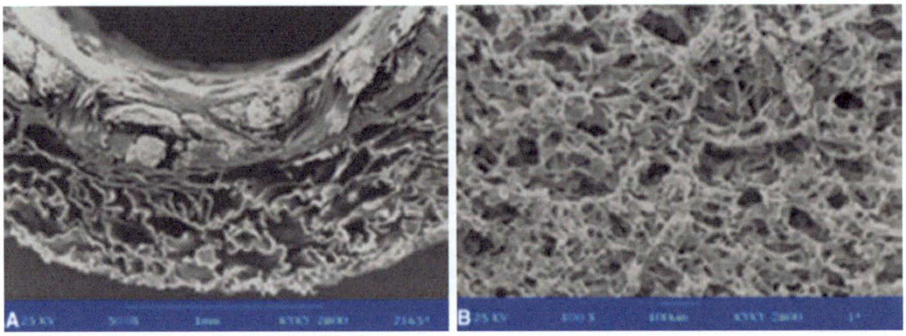

Fig. 8.9 SEM of the fiber-reinforced chitosan scaffold for nerve regeneration (**a**) a well-integrated three-layer structure of the scaffold. The outer surface of the scaffold (**b**) a well-interconnected porous structure. Scale bars 1 mm (Adapted with permission from Ref. [100]. Copyright 2007 Springer)

As a means of guiding the regeneration of axons to promote nerve reconstruction, nerve guidance conduit (NGC) is one of clinical treatments in nerve tissue engineering. A variety of NGCs have been developed as grafts to repair damaged nerves, such as microchannel and microfilaments, hollow fiber membranes, multilayers tubes and fiber bundles [93]. Compared with autografts and allografts, synthetic NGCs have many advantages, such as improved applicability and effectiveness, complex structures and the excellent ability of regenerating axons [97]. Studies have shown that they have excellent performance in inducing axonal regeneration *in vitro*, but there are still many limitations to be overcome *in vivo*. In general, NGCs can't selectively induce sensory and motor axon growth to the target organs. Therefore, it is necessary to use reasonable materials to synthesize NGC, and the ideal biological material should not only have good histocompatibility, biodegradability and bio-activity but also directional guidance ability. During nerve regeneration, the scaffold wall absorbs body fluids and expands, and the scaffold lumen may be obstructed by such swelling, which is detrimental to axon regeneration. Thus, the ideal biomaterial scaffold should provide sufficient longitudinal mechanical support to avoid significant swelling [98, 99]. Fortunately, the addition of fiber or tube seems to be effective in improving mechanical properties of the nerve guide scaffold.

Wang et al. [100] designed a porous chitosan-based fiber-reinforced guide conduit scaffold for nerve regeneration. The chitosan nerve conduits reinforced by braided chitosan fibers were fabricated by a combination of the mold casting and the lyophilization technique. The reinforced guide conduit scaffold with the porous structure had improved physical properties and biocompatibility. The reinforced guide conduit scaffolds had sufficient permeability that allowed molecules penetration ranging from 180 to 66,200 Da in size (Fig. 8.9). Comparing with the chitosan guide conduit scaffold, the compressive load of the reinforced guide conduit scaffold was obviously increased under the same strain. Likewise, the tensile strength value increased significantly from 0.41 ± 0.17 to 3.69 ± 0.64 MPa. Cytotoxicity assays *in vitro* demonstrated that Neuro-2a cells in the scaffold were not cytotoxic. Moreover, the scaffold had no effect on the surrounding tissue *in vivo* in primary implantation testing. The chitosan fibers improve the biocompatibility and mechanical properties of the guide conduit scaffold and make the reinforced scaffold fulfill the requirements of nerve regeneration.

Lu et al. [98] designed a microbraided PLA fiber-reinforced scaffold with a biodegradable multilayer for peripheral nerve regeneration. The scaffold was implanted into the sciatic nerve of rats with 10-nm nerve gap. The PLA fiber-reinforced scaffolds were not significantly swollen or deformed during the 8 week culture period. There was a brown fibrous tissue on the surface of the scaffold, and the new fibrous tissues also could be seen at parts of the nerve gap. After the removal of fibrous tissues and the walls of the conduit, regenerated nerve could be found. As a result, the PLA reinforced scaffold with good biodegradability and effectiveness is successful in repairing a 10 mm nerve gap and provides a promising tool for peripheral nerve regeneration.

Kim et al. [101] prepared an aligned polymer fiber-based NGC for repair the long peripheral nerve gaps. Aligned fibers differing from randomly oriented ones can establish topographical cues. Even if there was no exogenous growth to promote protein, the endogenous neural repair mechanisms could be triggered. The reinforced conduit guided and enhanced neurite outgrowth and axonal regeneration to repair neuromuscular junctions with a 17 mm nerve gap. Significantly, enhanced electrophysiological and behavioral results demonstrated that aligned fibrous structures had a role in promoting sensory and motor nerve regeneration. Additionally, sub-micron scale topographical cues of the aligned oriented fiber reinforced NGC were more conducive to encourage endogenous nerve repair mechanisms, according to the quantitative results *in vivo* of DRG outgrowth and nerve regeneration between randomly oriented and aligned fiber films. Thus, the aligned polymer nanofiber-based NGCs as a new synthetic scaffold have great potential for facilitating the regeneration of peripheral nerves.

Bini et al. [102] initially studied random poly PLGA nanofiber-reinforced scaffolds and formed 11 NGCSs though the electrospinning for peripheral nerve regeneration. Five of these scaffolds had the ability to support nerve regeneration of nerve defects in 10 mm based on chemical and histomorphometric results, but there was no obvious characterization function providing for recovery. Five PLA-fiber-reinforced scaffolds with single, double and triple layer structures all had the ability to induce nerve tissue growth to complete nerve repair and regeneration. In addition, the regenerative nerve had a structure similar to that of thick fibrous tissue, including cells and vascularized endomembranes. The clusters schwann cells also could be observed under the condition of nerve cuff bridging the bridging, which is a unit of regeneration around the unmyelinated axons, a common tissue structure [103]. As a result, the PLGA nanofiber-reinforced NGCSs can be used as nerve substitutes to promote the nerve regeneration.

Gao et al. [104] deisigned a 3D new-typed fiber-reinforced scaffold encapsulated Schwann cells and vascular endothelial cells *in vivo* to support the vascularization and induce the nerve regeneration for the repair of long-distance sciatic nerve defects in rabbit. The fiber-reinforced scaffolds were composed of a composite of nano-hydroxyapatite, collagen and poly($_L$-lactic acid) (PLLA) (nHACP) reinforced by chitin fibers, and prepared by using a previously method published by Li et al. [105]. In order to explore the role of vascularized grafts in the early recovery of neurological function, two groups of the scaffolds, nonvascularized and vascularized nerve scaffolds, were implanted into the rabbit sciatic nerve. The experiment group was implanted the prevascularized nerve scaffold, while the control group only cultured with Schwann cells. According to the results of neurological functional recovery at 4, 8, and 16 weeks after surgery, the experimental group had more significant growth in the recovery of ulcer healing, nerve conduction velocity, nerve fiber number, and ultrastructure of myelin sheath and regenerating nerve, as compared to the control group. Thus, the fiber-reinforced nHACP scaffolds cultured with Schwann cells and vascular endothelial cells can induce vascular sciatic nerve formation. The prevascularization of the scaffold can effectively stimulate the recovery of nerve function in sciatic nerve regeneration. Therefore, the study demonstrates that the fiber-reinforced nHACP scaffolds cultured with Schwann cells and vascular endothelial cells can induce vascularized sciatic nerve formation. Prevascularized scaffolds can effectively promote the restoration of neurological function in sciatic nerve defects regeneration. This study provides not only a new idea for finding a method to promote vascularization of tissue engineered nerve but also an ideal nerve substitutes for the regeneration of vessel and nerve.

In the other work, carbon nanotubes (CNTs) as a new kind of materials have attracted enormous attention as the most commonly used tubes reinforced materials for unique applications in the nerve tissue or other soft tissues repair [106]. CNTs have excellent structures and properties that are unmatched by other materials, such as high aspect ratio [107, 108], high conductivity [109, 110], strength [111], and chemical stability. It is interesting to note that aligned CNTs within the scaffold provide a key advantage of directional electrical conductivity, which helps to promote effective and rapid healing of damage nerve tissue [112–114]. Therefore, the researchers should spare no efforts to develop more extensive research and study on CNTs in the future.

In recent years, many studies have been conducted on the resulting properties of carbon nanotubes as scaffold reinforcement materials in never tissue engineering. Based on previous studies on the use of CNTs for nerve repair, Hu et al. [115] used carbon nanotubes as substrates for successful cultured hippocampal neuronal. Mattson et al. [116] successfully cultured embryonic rat brain neurons on the nanotubes, and the neurons survived and continued to grow for 8 days. Shin et al. [114] designed a CNT-reinforced gelatin methacrylate (GelMA) hybrid to create a 3D constructs with biocompatibility and cellular responses. This study also presented a simple and direct method to fabricate a porous CNT-reinforced scaffold. Through the use of GelMA hydrophobic interaction between the polypeptide chain and the sidewall of carbon nanotubes, a thin layer of GelMA were coated to CNTs, and their complete physical properties were still maintained. The addition of CNTs successfully enhanced the performance of GelMA hydrogel without the reduction of the

Fig. 8.10 Digital images of the fabricated scaffolds (A colour version of this figure can be viewed online) (Adapted with permission from Ref. [107]. Copyright 2015 Elsevier Ltd)

porosity and inhibition of cell growth. NIH-3T3 cells encapsulated in the CNT/GelMA scaffold kept their viability in 48 h of the culture, and readily spread and proliferated in the cytotoxicity test. As for the results of the mesenchymal stem cells (hMSCs) embedded into the CNT/GelMA scaffold, they appeared good biocompatibility. The nanofiber CNTs with a web-like structure enhanced mechanical properties of the GelMA hydrogel without interfering with their beneficial bioactive properties. Interestingly, the mechanical properties of the CNT/GelMA composite can be adjusted via controlling the content ratio of CNTs. CNT-reinforced GelMA composites provide the hope to generate 3D scaffolds to guide cell deposition into nervous tissue in a manner.

In the other study, Gupta et al. [107] fabricated macroscale aligned carbon nanotube (MWCNT) reinforced chitosan scaffolds by using electric field alignment technique for neural tissue regeneration (Fig. 8.10). This study provides an efficient platform for adjusting the mechanical and electrical properties of CNT-reinforced scaffolds by changing the alignment of CNTs. Compared with the random CNT-chitosan and pure chitosan scaffolds, the yield strength, elastic modulus and ultimate tensile strength of MWCNT-chitosan scaffolds were increased by 21.9%, 12.7% and 11.2%, respectively. This was because carbon nanotubes in the latter had a uniform distribution, good interface and aligned network. The MWCNT-chitosan scaffolds exhibited an anisotropic conductivity of 100,000 times higher in the alignment direction than that of the transverse direction, and had the ability to guide the direction of neurites growth and cell migration. The fabricated scaffolds were evaluated by incubating HT-22 mouse hippocampal neurons *in vitro* (Fig. 8.11). HT-22 hippocampal neurons on the MWCNT-chitosan had a significantly increased viability, which proved the scaffolds with a good biocompatiblity. Moreover, 50–60% of neurons were arranged in the alignment direction of the scaffold, which was also evidenced that the addition of aligned carbon CNTs supplied mechanical and electrical clues in favor of regular regeneration of the nerves.

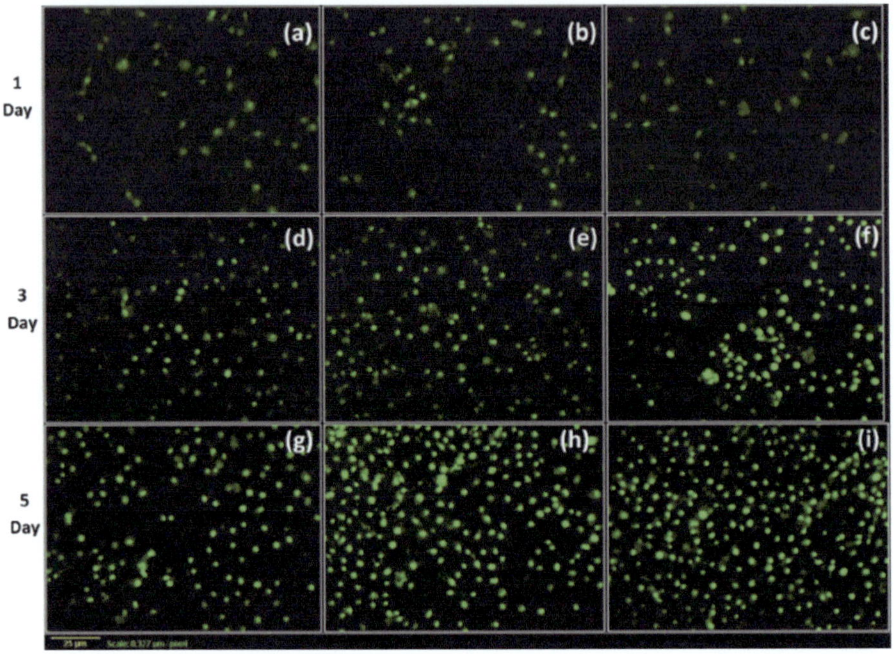

Fig. 8.11 Acridine orange-stained fluorescent images of HT-22 rat hippocampal neuron on chitosan (**a**, **d**, **g**), hybrid chitosan-random MWCNT film (**b**, **e**, **h**), and hybrid chitosan-aligned MWCNT film (**c**, **f**, **i**), respectively, after cultured for 1, 3 and 5 days (A colour version of this figure can be viewed online) (Adapted with permission from Ref. [107]. Copyright 2015 Elsevier Ltd)

8.2.5 Connective Tissue: Tendon and Ligament

Connective tissue is one of the basic organizations of people and higher animals and consists of cells, fibers, and ECM. Connective tissue is widely dispersed with diverse forms, including tendons, ligaments, fascia and dermis, and has multiple functions of support, connection, nutrition, protection, etc. Connective tissue injury is very common in the clinical and surgical care [117].

Ligaments and tendons are dense connective tissues, composed of polysaccharide phases, proteoglycans, and protein phases, collagen and elastin, whose function is mainly to maintain the stability and normal movement of joints in the musculoskeletal system. Each year more than 800,000 people require medical care due to these structures damage, which can lead to significant joint instability and the development of other tissue injuries or degenerative joint disease [118–120]. Surgical repair is a commonly used clinical procedure, but it does not maintain normal activity and has a certain failure rate [121, 122]. Autologous and allografts have also become increasingly common for tendon and ligament repair [1, 123]. The structural properties of grafts should mimic native tendons and ligaments with sufficient mechanical properties and reasonable structure morphology [118, 120, 124, 125]. Although some scaffolds derived from natural and synthetic polymers have been investigated for repairing tendons and ligaments, the success of these materials as substitutes is limited due to the lack of mechanical properties, such as stiffness, strength, creep, ductility, axial splitting, elasticity and stress-shielding. There are many problems that most biomechanical properties of the scaffolds can't simulate or sustainably maintain the biomechanical properties of the original ligaments and tendons [120, 125, 126]. Centered on these design goals of overcoming the limitations, a new generation of the synthetic composites in ligament and tendon tissue engineering tends to be heterogeneous from two or more materials through one or more manufacturing techniques. Many studies have found that the use of fiber- or tube- reinforcement is a viable method to obtain the ideal scaffold for the tendon and ligament repair.

Numerous groups have reinforced scaffolds with fibers to develop biomechanically functional tissue constructs for tendon and ligaments repair. As far as more methods are available for producing reinforced scaffolds for the repair of tendons and ligaments, Shepherd et al. [127] designed lyophilized structures of collagen-reinforced collagen-chondroitin-6-sulfate (C6S) material to produce bioactivities and biomechanical scaffolds for repairing and regenerating soft tissues such as tendons, ligaments as well as cartilage. The pore structures of the collagen-based composite were produced by axially aligned collagen fibers derived from acidic swollen gel type I collagen, and generally considered to be conducive to cell migration, proliferation and attachment. Although collagen-glycosaminoglycan has biological activities and porous structures favorable for cell infiltration and attachment, the poor mechanical properties reduce their range of applications in tissue engineering, particularly for the applications requiring tensile strength. Considering the carbodiimide-based cross-linking method, the collagen fiber bundles were extruded and then crosslinked in the 1-ethyl-3-(3-dimethylaminopropyl) carbodiimide and N-hydroxysuccinimide solution. Next, they were mixed with the collagen C6S matrix slurry to be lyophilized for two times. The collagen C6S matrix reinforced by crosslinked fibers remained a pore structure and good interaction between fiber and matrix, and they bear the maximum load of about 60N before failure. Compared to the matrix that was not crosslinked, the collagen C6S matrix reinforced by cross-linked fibers had higher mechanical properties and more pore channels, which enhanced cell migration and provided the potential to guide aligned tissue regeneration, like tendons and ligaments.

De Santis et al. [19] designed a continuous fiber-reinforced polymer scaffold with mechanical properties like the native connective tissue that could be used as substitutes of tendons and ligaments. Many soft tissue stress-strain curves are similar in shape, and the Fig. 8.12 shows this nonlinear relationship. HydroThane™, Hydrophilic Thermoplastic Polyurethane by Cardio-Tech International, has excellent mechanical properties, but native connective tissues have higher the mechanical properties, as Fig. 8.12 shown. The polyethylenterephtalate (PET) fibers were used to reinforce HydroThane™ polyurethane matrix by filament winding to simulate the mechanical properties of ligaments and tendons (Fig. 8.13). The mechanical behavior of the composite could be seen as a function of the winding angle, so the value of the elastic modulus can be easily controlled by the angle of the different

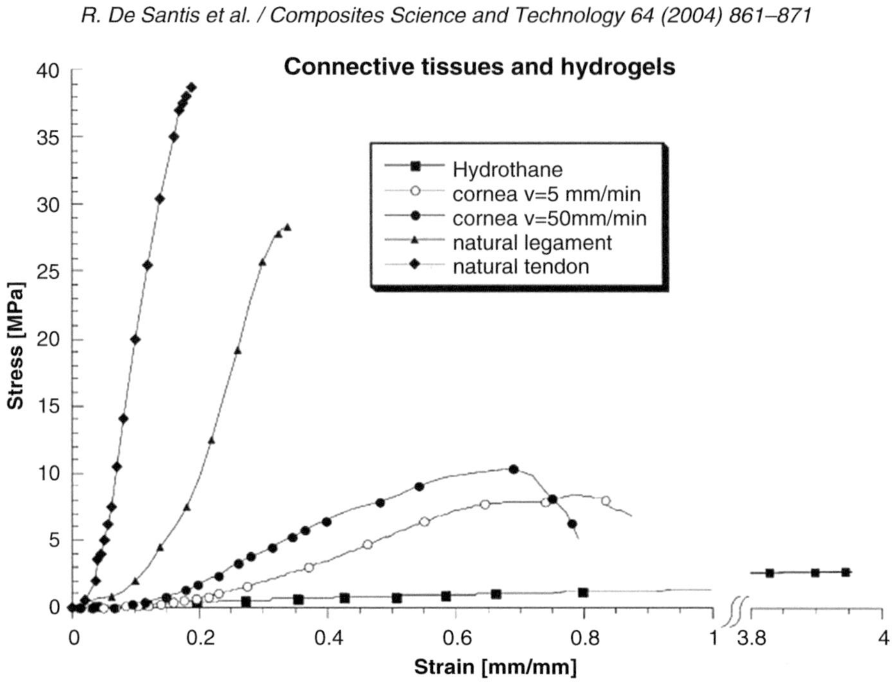

Fig. 8.12 Stress–strain behavior of connective tissues and Hydrotane. The mechanical properties of natural connective tissue are higher than that of hydrogels, thus a composite structure must be designed to match the mechanical properties of tendons and ligaments (Adapted with permission from Ref. [19]. Copyright 2004 Elsevier Ltd)

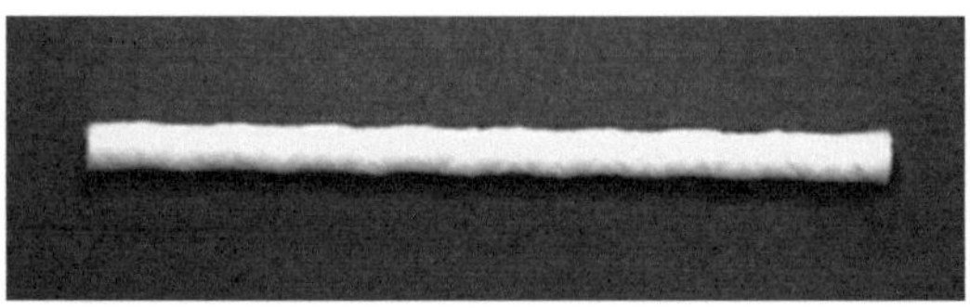

Fig. 8.13 HydroThaneTM reinforced PET as ligament prostheses. The composite structures was achieved by using the filament winding technology in order to simulate the morphology and mechanical properties of natural ligaments (Adapted with permission from Ref. [19]. Copyright 2004 Elsevier Ltd)

fiber winding. In fact, the reinforcing effect of PET fibers in the HydroThane™ matrix with 35 °C winding angle was a better than that of 20 °C and 15 °C. The wide range of the fibers aligned along the load direction increased the modulus and strength of the scaffolds. The structure of the reinforced composite had a static stress-strain behavior close to the native ligament.

Webb et al. [128] investigated a poly(3-hydroxybutyrate-co-3 -hydroxyhexanoate) (PHBHHx) scaffold with a PHBHHx fiber-reinforced collagen core used in a rat Achilles tendon repair mode, which had a good delayed biodegradability and

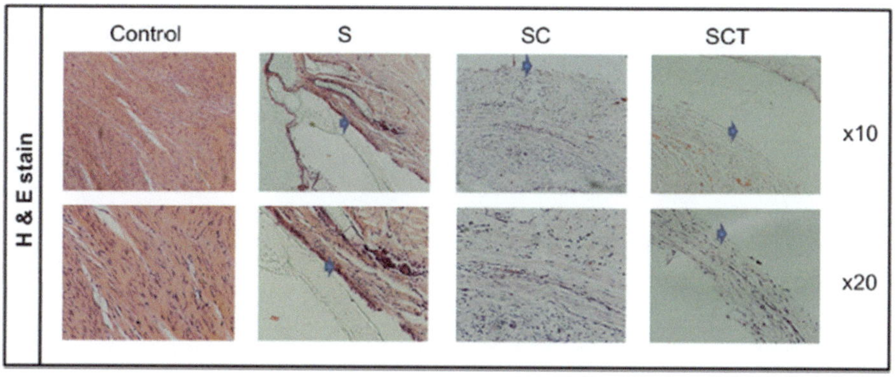

Fig. 8.14 Histological staining of explanted scaffolds. A microscopic assessment of the damage to the tendon site using Haematoxylin and Eosin confirmed evidence of cellular infiltration and partial structure regeneration along the fiber-tissue interface (indicated by the *arrows*). Different levels of sinusoidal morphology are observed at the interface with control (*C*), PHBHHx scaffold (*S*), PHBHHx scaffold-collagen (*SC*) and PHBHHx scaffold-collagen-tenocyte (*SCT*), where the last two present the greater degree of sinusoidal morphology (Adapted with permission from Ref. [128]. Copyright 2013 Elsevier Ltd)

biocompatibility. The mechanical properties PHBHHx scaffolds with maximal loads at 23.73 ± 1.08 N in the mechanical testing could be compared with health rat tendons with maximal loads at 17.35 ± 1.76 N. Therefore, PHBHHx fiber-reinforced collagen gels were implanted into the lumen of the PHBHHx tube to produe a PHBHHx-based composite scaffold. The composite scaffolds encapsulated with tenocytes were implanted in rat model to verify that PHBHHx could be used to promote the repair of tendon tissue damage without inducing immune response or prolonging inflammation. During 40 days of culture *in vivo*, prolonged immunological or inflammatory responses did not occur in the rat model, and the functional recovery of the Achilles tendon was accelerated. The tenocytes not only migrated to the PHBHHx-collagen core but also formed new tendon tissue along the fiber surface of the core. Thus, the core promoted a regenerative cellular response based on the contribution of the mechanical properties of the scaffold and facilitated subsequent tissue reconstruction (Fig. 8.14). The PHBHHx-based scaffold with PHBHHx fiber-reinforced collagen core has a 3D structure morphology, which contributes to the successful applications in an *in vivo* tendon repair model.

Yang et al. [129] introduced a new aligned nanoyarn-reinforced nanofibers (NRS) scaffold by electrospinning with improved mechanical strength and cell infiltration for tendon tissue regeneration. Three different types of scaffolds consisting of the hybridization of poly (L-lactide-co-caprolactone) (P (LLA-CL)) and silk fibroin were prepared, that were aligned nanofibers scaffold, random nanofibers scaffold and NRS scaffold. The aligned and randomly NRS scaffolds had relatively high porosity and large pore sizes, which contributed to cell adhesion and migration, oxygen and nutrient delivery and metabolite discharge. The results of the three scaffolds cultured with marrow-derived mesenchymal stem cells (MSCs) in the bio-

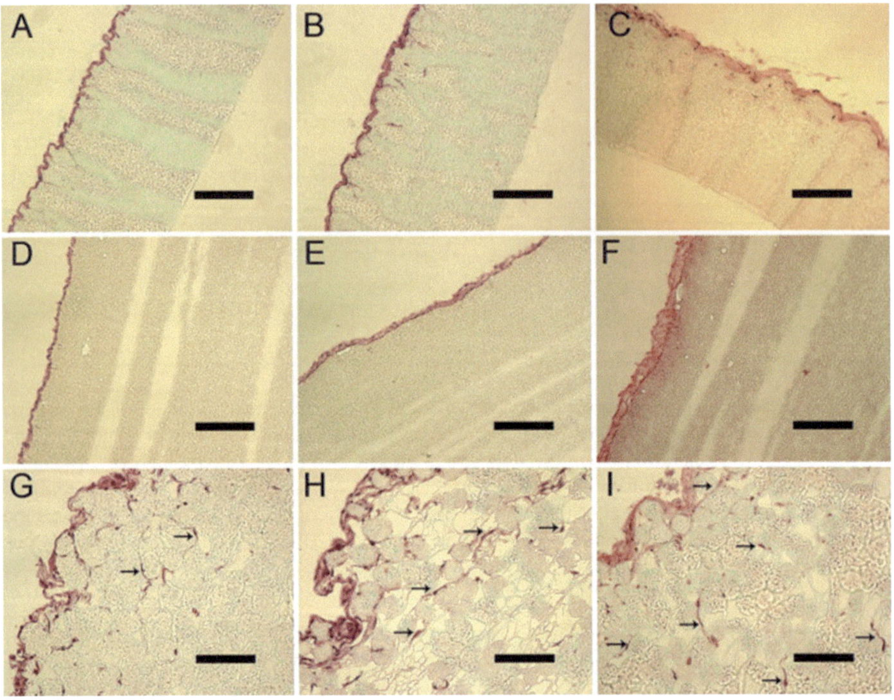

Fig. 8.15 HE-stained histology images of MSCs interactions with the random nanofibrous scaffolds (**a–c**), aligned nanofibrous scaffolds (**d–f**) and NRS (**g–i**) on day 7 (**a**, **d** and **g**), 14 (**b**, **e** and **h**) and 28 (**c**, **f** and **i**). The MSCs (*black arrow*) are infiltrated from the surface of NRS at a depth of about 150 μm on day 7 (**g**), approximately 300 μm on day 14 (**h**) and approximately 500 μm on day 28 (**i**). Scale bars 100 μm for all images (Adapted with permission from Ref. [129]. Copyright 2014 Elsevier Ltd)

compatibility analysis demonstrated that the reinforced nanoyarn nanofibers had the ability to enable cell infiltration and improve cell proliferation (Fig. 8.15). The MSCs on the NRS scaffolds showed an elongated and aligned growth pattern along the nanoyarn, but their distribution was not uniform in the yarns (Fig. 8.16). Furthermore, the tensile properties of the composite were enhanced in parallel to the direction of nanoyarns, which was required for the repair of tendons.

Hakimi et al. [130] developed a layered surgical scaffold by electrospinning and braiding for the repair of endogenous tendon. Electrospinning submicron materials with excellent cellular and tissue responses have promising results to be used as a scaffold material for tendon repair, however, whose mechanical properties are too poor to be surgically implanted or support the damage tendons during the period of recovery. In the published literature on tendon repair, several technologies only

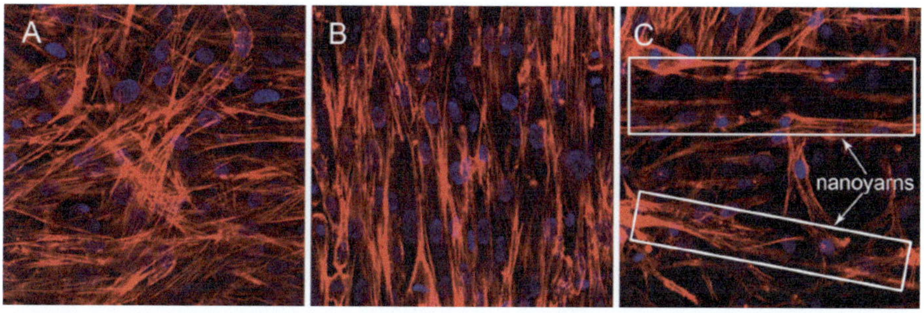

Fig. 8.16 Confocal microscopy fluorescence images of the actin filaments (*red*) and nuclei (*blue*) of MSCs on the random nanofibrous scaffold (**a**), aligned nanofibous scaffold (**b**) and NRS (**c**) after 7-day culture. Magnifications of all images are 400× (Adapted with permission from Ref. [129]. Copyright 2014 Elsevier Ltd)

improve marginal strength of the scaffold, which does not have close mechanical properties to tendon tissue. In this study, a bonding technology with biocompatible and non-destructive was developed to create a first prototype of the scaffold, where a multi-layered woven fabric was used to enhance an aligned electrospun sheet (Fig. 8.17). Both layers were made up of polydioxanone (PDO) with excellent compatibility, and they could be detected by ultrasonography, which could be used to locally inject cells or growth factors postoperatively [131]. As expected, the weaving and electrospinning layer resulted in the scaffold strength of at least 20 times stronger than that of the electrospun layer in the tensile tests. The scaffold with the twill woven fabrics and aligned electrospun sheets exhibited a maximum suture trength at 167 N similar to that of human tendon. The multi-layered scaffold implanted in an *in vivo* rat model was gradually more embedded in dense tissue, and PDO fiber pattern was less visible by weeks 6 and 12 (Fig. 8.18). Moreover, the scaffold with a multi-layered reinforced structure maintained integrity during the duration of culture, and there was no separation or delamination of layers. For the manufacturing process is non-destructive to the aligned electrospun layers, the porosity and thickness of the electrospinning layer can be controlled, which provides the potential to design various scaffolds with reinforced electrospun sheets for more medical applications. The use of woven medical fabrics to enhance aligned electrospun layers has resulted in a promising implantable scaffold, which has the potential to improve the endogenous repair of the rotator cuff tendon.

8.2.6 The Intervertebral Disc

Four fifths of adults in their lives have an experience of back or neck pains, which are often associated with degeneration and injury of the intervertebral disc [132], and about $ 90 billion are needed to treat the intervertebral disc diseases in the United States alone every year [133]. The intervertebral disc (IVD) is a

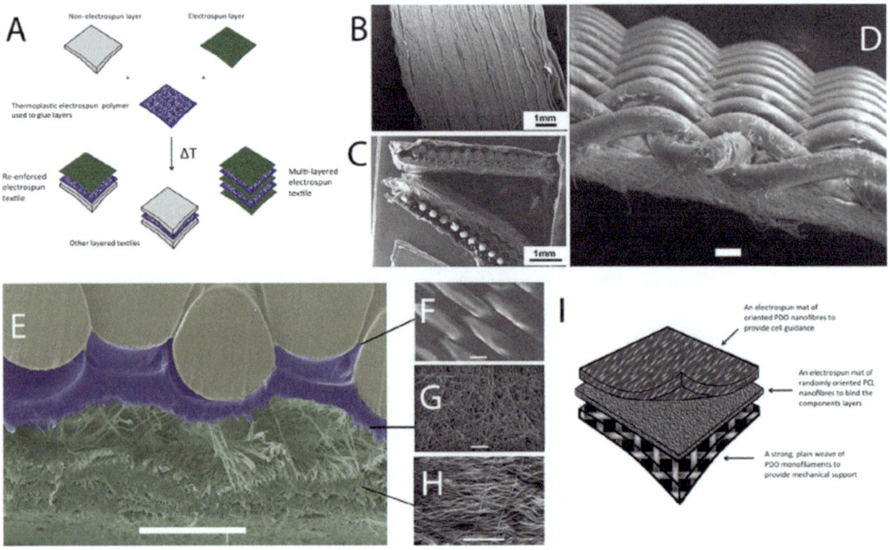

Fig. 8.17 Multi-layered scaffolds from electrospun and non-electrospun mats and prototype layered electrospun/woven scaffold architecture. (**a**) The technique of thermoplastic mats and the incorporation of electrospinning and non-electrospinning. (**b**) Multi-layered electrospun sheets. (**c**) Woven polydioxanone textile sandwiched between two electrospun mats. (**d**) Schematic diagram of the prototype scaffold designed and tested. (**e**) A cross-sectional view of the layered scaffold. (**f**) The woven layer. (**g**) The random electrospun PCL thermoplastic adhesive layer. (**h**) The aligned PDO electrospun mats. (**i**) A schematic description of the design rational behind the layered scaffold. Unless specified scale bars are 100 lm (Adapted with permission from Ref. [130]. Copyright 2015 Elsevier Ltd)

cartilaginous tissue that has the abilities in body of separating the spinal vertebrae, permitting a limited range of motion and protecting the spine from impact loading [134]. Degenerative disc disease (DDD) has been associated with aging and is thought to be caused by many factors, including excessive mechanical load, smoking, malnutrition and cell metabolism [135]. Until now, the current treatment is limited to physiotherapy, analgesic therapy, or highly invasive surgical procedures such as spinal fusion, discectomy, or partial or complete removal of damaged IDVs [136, 137]. Surgical procedures are the primary means of addressing severe disc diseases, but their severe invasiveness often leads to tends to limit the flexibility of the spine and accelerate the degeneration of adjacent vertebrae. Tissue-engineered therapy has attracted increasing interest as a solution to traditional strategies, and in particular the use of scaffolds having similar structures to IVDs is expected to regenerate and remodel the IVD tissues. IVD is nonvascular and makes up with two distinct regions, the outer annulus fibrosus (AF) and the inner nucleus pulposus (NP). AF has a complex structure and consists of aligned collagen fibers, whose function is to provide mechanical stability and integrity under bending and torsion loads [138–140], while NP has a gel-like structure and consists of disperse collagen fibers and glucosaminoglycans, both of which are contribute to compressive and shear loads [141]. Thus, the IVD with a structure of proteoglycan hydrogels

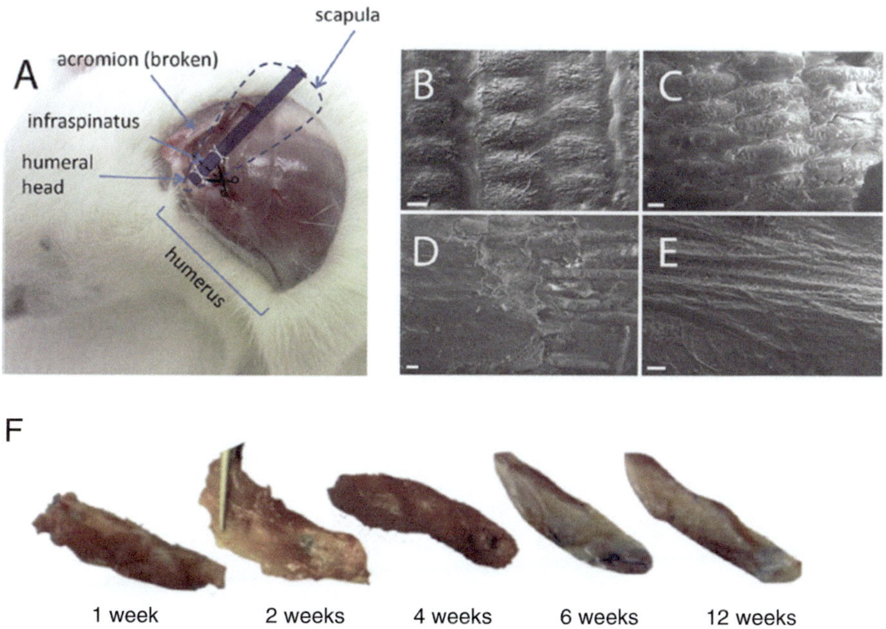

Fig. 8.18 The layered scaffold integrated in an *in vivo* rat model. (**a**) An illustration of the scaffold placement in the rat shoulder. (**b–e**) Scanning electron micrographs show tissue ingrowth into the layered scaffold after 2 weeks (**b**), 4 weeks (**c**), 6 weeks (**d**) and 12 weeks (**e**). (**f**) The capsule size increased up to 2 weeks *in vivo*, after which it was reduced and better-defined over 12 weeks (Adapted with permission from Ref. [130]. Copyright 2015 Elsevier Ltd)

encapsulated into collagen fibers is much like fiber-reinforced composites. For the IVD plays a key role in the spine mechanics of the body, the idea scaffolds used to repair IVD must have sufficient biomechanical properties and can simulate the characteristics of the two regions.

Ambrosio et al. [142] proposed a PET fiber-reinforced hydrogel Poly(2-hydroxyethylmethacrylate)(PHEMA)-PCL composite with high biocompatibility, permeability and hydrophilicity, which has the potential to be used as an artificial disc. Based on the previous studies, due to PCL crystalline microdomains can act as physical crosslinks, the incorporating hydrophobic PCL into PHEMA hydrogels has a good effect in enhancing the compression properties to overcome the lack of mechanical properties of the PHEMA hydrogels in the hydrated state. However, the values of maximum stress and modulus of the PHEMA -PCL composite were significantly lower than those of a canine discs, ranging from 32 to 115 MPa and from 8 to 19 MPa, respectively. When the PHEMA-PCL with semi-interpenetrating networks (s-IPNs) was reinforced by incorporating about 40 vol% PET fiber with winding angles from 45° to 65°. The maximum stress and modulus increased from 12 to 17 MPa and from 30 to 73 MPa, respectively. The trend of compressive stress-strain curve of the reinforced composite system was similar in quality to that of the native disc. When PHEMA/PCL was enhanced with 50 vol% PET fiber with the same winding, the maximum stress and modulus increased up to 20 MPa and

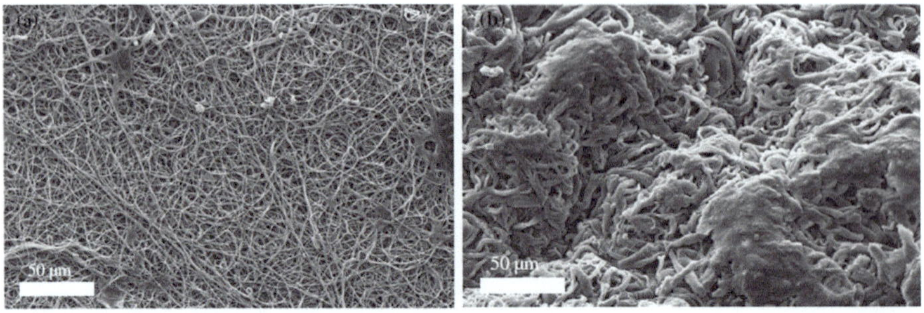

Fig. 8.19 SEM micrograph of (**a**) a PCL electrospun fiber structure and (**b**) a PCL–5%Alg composite (1,0009 magnification) (Adapted with permission from Ref. [134]. Copyright 2014 Springer)

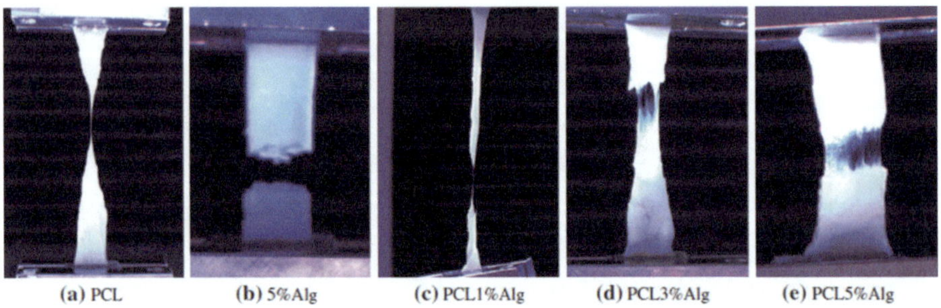

Fig. 8.20 Images of single phase materials and composites undergoing tensile testing, at the point of failure (Adapted with permission from Ref. [134]. Copyright 2014 Springer)

129 MPa respectively, which were typical values of a canine lumbar IVD. The PET and PCL fiber-reinforced PHEMA composite have the compression properties close to that of canine IVD in different position of the spine. Through controlling the hydrogel matrix and the quantity and winding angle of PET fibers, the hydrophilicity and mechanical properties of IVD scaffolds can be adjusted. The composite with semi-IPNs structure can allow for the inward growth of native tissue with the degradation of PCL to leave voids in the network. Although the composites have the mechanical properties similar to canine IVD, they also give good indications for the use of these systems to repair the human IVD.

Strange et al. [134] demonstrated the creation of large 3D electrospun fiber-reinforced hydrogel composites, which could use to achieve a functional engineered IVD scaffold. The thick electrospun PCL fibers with a 3D structure infiltrated into the alginate (Alg) hydrogel to create a PCL-Alg composite, as shown in Fig. 8.19. The mechanical properties of the PCL-Alg composite were strongly affected by their structure based on the dominating responses of individual structural component in indentation and tensile testing. The tensile properties of the PCL-Alg composite were found to be mainly determined by the presence of the electrospun PCL fibers (Fig. 8.20), while the compressive properties were determined by the hydrogel modulus. But it was not clear whether this behavior was caused by the material tensile compression nonlinearity or the structure transverse isotropy. The PCL fibers provided stiffness and strength in tension for the composite, and moderated the time-dependent response in compression at the same time. The gel-phase supported and distributed compressive loads through swelling to tuning the fluid pressurization. The PCL fibers did show a tensile-compressive nonlinearity with varying mechanical properties in the planar and through thickness direction, although they were randomly distributed in the plane. By adjusting the PLC fibers concentration, the nonlinearity of the composite has the potential to match the non-linearity of AF, which consistes of aligned type 1 collagen fibrils [143, 144]. The thick PCL-Alg composite also provides the possibility of preparing a scaffold with comparable thickness to the NP by adjusting the ratio of hydrogel and fibers. This study provides a methodology of designing the tensile and compressive mechanical properties of fiber-reinforced composites by modulating the allocation of fibers and matrix phases to create a tissue-engineered scaffold that is closer to the natural structure of IVD and other tissues.

Thorvaldsson et al. [145] developed a method to combine electrospun PCL nanofibers and aerosol sprayed gelatin gellan gum solution to create a nanofiber-reinforced gellan gum gel scaffold with improved mechanical properties for NP regeneration. When it comes to NP repair, hydrogels have attracted much attention due to their excellent properties. The mechanical stability of hydrogels is usually poor in the case of stretching and compression [146]. To produce electrospun nanofiber-reinforced hydrogels with improve structural integrity and mechanical properties, the gellan gum solution was sprayed onto a mandrel rotating in a calcium chloride bath to produce a hydrogel, while, at the same time, the PCL nanofibers were electrospinning to the same collector to enhance the hydrogel. The incorporation of gellan gum concentration, nanofibers and crosslinker concentration could be applied to adjust the mechanical properties of the material based on optical and rheological evaluation results. Furthermore, the composite was enhanced by adding only a small amount of nanofibers. And their reinforcing effects could be changed by tuning the amount of fibers alone. The incorporation of a small amount of nanofibers allowed the hydrogel scaffold to have rheological properties similar to those of human NP, and the resulting modulus were within the range of previously reported natural human NPs. The study shows the flexibility and potential of the nanofiber-reinforced gel with suitable properties used for NP tissue engineering.

8.2.7 Skin Tissue

The skin refers to the tissue on the surface of body wrapped in the outside of muscles, and it is the largest organ in the body with the functions of protecting the body, perspirating and experiencing temperature and pressure. When people are injured, the skin will be quickly closed to protect the wound from infection and dehydration. Skin damage is common in everyday life, but large areas of skin damage are a serious hazard to human health, such as severe burns, trauma and skin diseases. So far, autologous skin grafts are still the ideal way to replace irreversible damaged skin. However, limitations of autologous grafts, such as suffering from pain, infection and scar formation at the donor site, have led to the development of tissue-engineered skin as substitutes for repair skin tissue [147–150]. Recently, tissue-engineered skin substitutes have made significant advances in the field of wound healing, particularly in the case of burns, where the main limiting factor is the availability of autologous skin. A series of skin substitutes with interconnected porous structures, good biocompatibility and biodegradation have been developed for skin tissue engineering. Recent studies have found that there are more stringent mechanical requirements for scaffolds used in load-bearing soft tissue with low stiffness and high strength in tension [151–153]. In fact, an ideal skin scaffold should have desired mechanical properties to withstand suturing tension during implantation and maintain the stable structures and functions after implantation [154, 155]. However, obtaining the scaffold with appropriate stiffness and strength remains a challenge for a single natural or synthetic material. With the development of the reinforced scaffolds with fibers or tubes, it provides new prospects of scaffolds as skin substitutes for tissue repair and regeneration. In the past years, scaffolds reinforced by fibers or tubes have used for the skin tissue and made some progress.

Han et al. [156] designed a silk nanofibrous scaffold reinforced with woven silk fibers to make up for the lack of mechanical properties, especially suture retention, aimed at dermal tissue engineering. In order to manufacture a silk-based scaffold system with feasible mechanical properties, the silk nanofibrous scaffolds were prepared by lyophilization and then incorporated with degummed woven silk fabric. The suture retention strength, elongation at break and tensile strength were achieved in the fabric-reinforced scaffolds without sacrificing morphology and porous structure. These suitable mechanical properties were important for the clinical application of the skin. The stiffness of the silk-based reinforced scaffold was mainly determined by fibroin porous scaffold, and the degradation rate could be controlled by appropriate post-treatment (e.g., water annealing and methanol impregnation) to make the scaffold more suitable for dermal tissue regeneration. These silk-based reinforced scaffolds maintained their excellent cell compatibility of pure regenerated silk scaffolds after reinforced by woven silk fabrics (Fig. 8.21). The addition of fibers silk fabric fibers can be used to improve and adjust the mechanical properties of the silk nanofibrous scaffold, and at the same remain their biocompatibility and porous structure. The reinforced composite can be served as matrices for repair skin tissue, which extends the future applications of silk-based materials in soft tissue engineering.

Cai et al. [157] designed a unidirectionally aligned halloysite nanotube (HNT) reinforced electrospun PLLA nanofiber scaffold to toughen the mechanical properties, for high porosity and random alignment of PLLA fibers resulted in the poor mechanical properties, which limits their applications, especially as tissue engineering scaffolds. Although various types of nano-fillers have been added into PLLA nanofibers to solve the problem [158], these nano-fills result in unsatisfactory biocompatibility and thus have no effect in expanding their potential applications in the

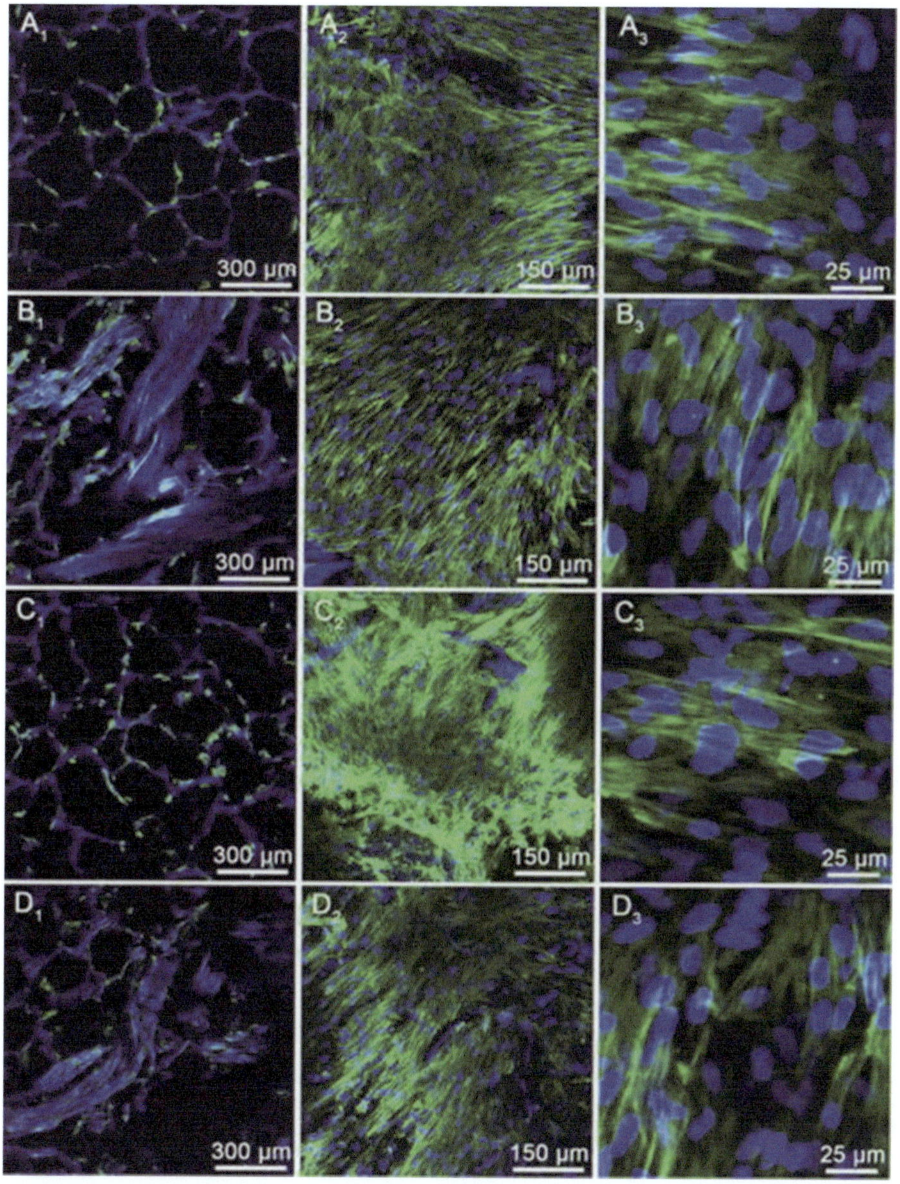

Fig. 8.21 Confocal microscopy of amniotic fluid-derived stem cells cultured on different scaffolds at day 1 and day 12: (*A1*) water-annealed silk scaffolds (WA-S), day 1; (*A2*, *A3*) water-annealed silk scaffolds (WA-S), day 12; (*B1*) water-annealed fabric-reinforced silk scaffolds (WA-C), day 1; (*B2*, *B3*) water-annealed fabric-reinforced silk scaffolds (WA-C), day 12; (*C1*) methanol-annealed silk scaffolds (MA-S), day 1; (*C2*, *C3*) methanol-annealed silk scaffolds (MA-S), day 12; (*D1*) methanol-annealed fabric-reinforced silk scaffolds (MA-C), day 1; (*D2*, *D3*) methanol-annealed fabric-reinforced silk scaffolds (MA-C), day 12. The *blue* DAPI represents silk scaffolds and cell nucleus, while the *green* FITC labeled phalloidin represents actin cytoskeleton (Adapted with permission from Ref. [156]. Copyright 2013 Elsevier Ltd)

future biomedical engineering [159]. As an attractive nanoreinformcent candidate, halloysite clay nanotubes (HNTs) [Al2Si2O5(OH)4·nH_2O] have good biocompatibility and low cost. This study used HNTs to reinforce PLLA nanofibers scaffold to obtain substantial improvement of mechanical properties. The PLLA nanofiber scaffold with adding 4 wt% HNT had the optimum mechanical performance, whose the Young's modulus, tensile strength, fracture energy and elongation at break were increased by 100%, 61%, 181% and 49%, respectively. HNTs can be effectively dispersed and incorporated in the PLLA fibers matrix based on the results of SEM and TEM. Due to the strong interaction between the HNTs and the PLLA fibers, the applied load could be efficiently transferred from the substrate to the HNT, which explained why the HNTs nanofillers had the enhanced effect. The HNT/PLLA nanofiber membranes also had an improvement in crystallinity and thermal stability, and their protein adsorption was higher than that of pure PLLA nafibers, which provides the potential for future applications in tissue engineering. This study provides a HNTs reinforced PLLA nanofiber scaffold, whose mechanical properties and biological characteristics provide a promising to be used in skin tissue engineering.

8.2.8 Corneal Tissue

Corneal disease is the second cause of blindness with about ten million people suffering from corneal blindness every year in the world [160, 161]. Furthermore, the number of sick people increases with the growth of the elderly population, while corneal transplantation is only about 60,000 times a year in the world. At present, the only available treatment of corneal blindness is the human donor corneal transplantation [162, 163]. Apart from technical difficulties in surgery, the shortage of supply of health donor corneas is the international public health difficulties now. There is a need to develop alternatives of donor tissues to meet the growing demand for corneal transplantation [164, 165]. Over the past few years, the development of corneal substitutes has made some progress to fabricate a cell-scaffold constructs as a temporary corneas in corneal tissue engineering [166]. The success of a tissue-engineered cornea is significantly dependent on the scaffold's performance similar to the microstructure and properties of the corneal ECM, whose major components are collagen fibrils and proteoglycans [167]. The highly-organized collagen fibrils with an aligned structure in the corneal matrix provide the physical strength, stable shape and transparency of the cornea [168–170]. The uniformity of collagen fibril diameter and the regularity of interfibrillar spacing are key factors affecting corneal transparency [171, 172]. The direction and orientation of collagen fibrils are directly related to the tensile strength of the cornea, which is resistant to external pressure and maintains corneal shape and curvature [168, 170]. Thus, the highly organized fibrous structure is a major factor that must be considered in the design of corneal tissue substitutes, which is directly related to whether the scaffolds have the desired mechanical and optical properties and spatial guidance for the corneal cell growth

[173]. The use of fiber reinforcements allows the biomaterial to have biological properties similar to those of the cornea, such as sufficient mechanical properties, morphological integrity and transparency of corneal tissue repair [174–176]. Considering these key features in the design, scientists have focused on designing appropriate methods for preparing fiber-reinforced scaffolds for corneal tissue regeneration.

Long et al. [177] designed a toughness silk fibroin reinforced collagen-based (CS) membrane as corneal grafts to improve the mechanical properties. It is well known that collagen having excellent biocompatibility is a promising candidate in corneal repair and regeneration, but their mechanical properties can't meet the requirements of corneal scaffolds, especially the suture retention strength. In this study, collagen-based membranes reinforced by silk fibroin with different ratios of 5 wt%, 10 wt% and wt% were prepared by cross-linking with 1-ethyl-3 -(3-dimethyl (EDC) and N-hydroxysuccinimide (NHS) [127, 178]. By measuring the water content and mechanical properties of the CS membrane, it was found that their water content was depended by the weight percentage of silk fibroin, and the CS10 membrane with silk fibroin at 10 wt% showed the optimal mechanical properties. CS10 also had high suture retention strength, which can be sutured integrally in rabbit eyes according to the results of suturing experiments. The CS membranes had a better optical property in the optical test, with higher transmittance, for their construction and components were close to the native cornea with a multi-layer structure of highly-organized, anisotropic collagen fibrillar lamellae [179]. As for the biocompatibility of the membrane, the CS10 membrane supported the proliferation of human corneal epithelial cells (HCEC) and induced corneal repair *in vitro* animal experiments. The epithelial cells encapsulated in the CS10 membrane were completely epithelialized from 30 to 40 days, and their transparency was rapidly restored during the first month without conical cornea, neovascularization and corneal rejection. These properties make CS membrane a potential candidate for future corneal alternatives. In addition, the composite membranes may provide a promising approach for designing the corneal scaffolds in the field of corneal tissue engineering.

As opposed to the current attempt to regenerate corneal tissue using expensive materials in the design of corneal scaffolds with a similar corneal structure, Tonsomboon et al. [180] developed transparent nanofiber reinforced hydrogels from gelatin and alginate, both of which are cheap and non-immunogenic natural polymers, to mimic the cornea's microstructure. Gelatin nanofibers were produced by the electrospinning and then infiltrated into alginate hydrogel to obtain a transparent fiber reinforced hydrogel. Due to the hydrophilicity of the gelatin, gelatin fibers attracted the accumulation of water molecules in the surrounding when the electrospun gelatin pad contacted the alginate solution. Then, with the water molecules bound to the alginate chain, the gelatin fibers were surrounded by the alginate chains. Under the help of Ca^{2+} ions, the alginate chains were cross-linked together to form a gel, in which the gelatin fibers were wrapped. The gelatin fiber reinforced alginate hydrogel had an elasticity modulus from 0.45 to 0.5 MPa very close to that of the natural cornea from 0.579 to 4.9 MPa [6, 17, 18]. The composite also had a desired optical transparency, which might be due to the fact that the diameter of the gelatin fibers was uniform and relatively small at 67 ± 7 nm as compared to the wavelengths of visible light at 400–700 nm so that the amount of light scattered from the material was minimized. Moreover, the orientation of the fibers also affected the performances of gelatin-alginate hydrogels, so the resulting hydrogel reinforced with the oriented fibers was slightly stiffer and stronger than that reinforced with the randomly oriented fibers, when the applied load was parallel to the fibers. However, the stiffness of the oriented gelatin fibers reinforced hydrogel was only along the fiber axis, and in the transverse direction the composite was very compliant. Thus, the next-generation reinforced composite should be provided with biaxial mechanical properties, i.e., reinforced with biaxially aligned fibers, to better simulate the biaxially rigid structure of the natural cornea. The adding of electrospun nanofibers gelatin into alginate hydrogels increased the mechanical properties of the hydrogels by about one order of magnitude. Thus, this study demonstrates that transparent hydrogels with excellent mechanical properties can be made simply from two inexpensive natural polymers: gelatin and alginate, and these novel nanofiber-reinforced composites have a wide range of prospects as scaffolds for corneal transplantation when the donor tissue is not available.

8.3 Conclusion and Future Directions

As a common seen phenomenon in nature, soft tissues and organs have complex and diverse microstructures and constituents and thus they are usually supported by fibers to form a framework with mechanical properties. Reinforced scaffolds with fibers or tubes can make better use of the structure and properties of the material to simulate the natural soft tissue structure and biomechanical properties. Due to their extremely high aspect ratio, fibers or tubes are particularly suitable as reinforcements that can be used as controllable elements to improve the mechanical and physical properties of engineered biomaterials in soft tissue engineering. Reinforced scaffolds with fibers or tubes are typically made of the same material with different forms or various materials, including natural and synthetic materials, which provide means of combining the advantages of each material. Due to the synergistic interaction between the different materials, the scaffolds can have a wider design space to have suitable morphology and geometry, which may contribute to overcome the limitations of each single-component scaffold. Over the past few years, fiber- or tube- reinforced scaffolds have attracted considerable attention and been suggested to be effective ways to develop engineering implants for soft tissue repair and reconstruction. The fiber and tube reinforcements with their unique properties and functions play a key role in improving the biomechanics, biocompatibility, bioactivity, integration and degradation of artificial composite scaffolds in soft tissue repair and regeneration. In fact, fiber-based or tube-based biomaterials with a wide range of morphological and mechanical properties have been widely used as reinforcing additives in soft tissue engineering to produce the desirable scaffold similar to ECM components. Apart from traditional manufacturing technologies, such as weaving, knitting and braiding, many new methods and strategies based on new technologies, such as electrospinning, wet spinning, melt spinning, interfacial compositing, direct writing and 3D printing, have been used to manufacture fiber or tube reinforced composites for obtaining excellent mechanical properties, porosity, conductivity, chemical stability and anisotropy in soft tissue repair and regeneration. Despite the remarkable achievements of current research, there is still a need to develop advanced theories and satisfactory intensive processing techniques to achieve the fiber- and tube- reinforced scaffolds with uniform structure, performance and composition similar to native ECMs.

The future direction of developing the next generation in soft tissue engineering is that the new scaffold should be made into a space structure with favorable morphology, optical properties, electrical properties, biocompatibility, biochemical and biological properties, especially adequate mechanical properties. More intelligent and diverse materials need to be further studied and will contribute to better soft tissue repair and regeneration with developmental biology disciplines. In addition, all the impact factors dictating properties of the reinforced scaffolds should be made a systematic study in depth, such as the volume fraction, aspect ratio, direction and arrangement of the fiber or tube.

The design of new composites depends to a large extent on its manufacturing technology and experimental conditions. It is also worth mentioning that the processing techniques play a leading role in the development of reinforcing scaffolds. It is necessary to develop new manufacturing techniques and experimental conditions.Reasonable experimental methods and manufacturing strategies used to prepare scaffolds to better simulate natural soft tissue will continue to attract a wide range of attention in the future.

Consideration should be given to the correlation between the manufacturing process and the mechanical properties and degradation kinetics of the reinforced scaffolds during the manufacturing process. It is important not only to choose fibers or tubes, but also to develop efficient, high-quality processing technology to achieve a uniform structure and composition throughout the reinforcing scaffold.

As described in the previous study, good adhesion strength and compatibility between the fiber or tube and the matrix may improve the mechanical properties of the scaffold, increase cell proliferation and diffusion, and accelerate tissue recovery. Therefore, the processing technology should also cause the fiber or tube to be oriented in the main path of the matrix, leading to forming the desired structural morphology.

Although the various materials reinforced by fibers or tubes have been used for a long time to repair or remold soft tissue, from simple use to more complex applications, studies of biocomposite materials with active enzymes, cells and antibodies still need to further develop and improve in the next generation of engineered scaffolds for soft tissue engineering.

Over the past few years, a large number of applications have been explored in fiber or tube- reinforced scaffolds for soft tissue repair and reconstruction, and these findings have opened up a new direction for the reinforced scaffold design. However, there are still more investigation space and rewarding opportunities in the future. Further interdisciplinary investigations and the use of fiber- or tube- reinforced technologies can open up a wider range of biotherapeutic applications, especially in the field of regenerative medicine.

References

1. Li X, Yang Y, Fan Y, Feng Q, Cui FZ, Watari F. Biocomposites reinforced by fibers or tubes as scaffolds for tissue engineering or regenerative medicine. J Biomed Mater Res A. 2014;102(5):1580–94.
2. Bauer-Kreisel P, Goepferich A, Blunk T. Cell-delivery therapeutics for adipose tissue regeneration. Adv Drug Deliv Rev. 2010;62(7):798–813.
3. Xu H, Cai S, Sellers A, Yang Y. Intrinsically water-stable electrospun three-dimensional ultrafine fibrous soy protein scaffolds for soft tissue engineering using adipose derived mesenchymal stem cells. RSC Adv. 2014;4(30):15451–7.
4. Xu H, Cai S, Sellers A, Yang Y. Electrospun ultrafine fibrous wheat glutenin scaffolds with three-dimensionally random organization and water stability for soft tissue engineering. J Biotechnol. 2014;184:179–86.
5. Baker SC, Rohman G, Southgate J, Cameron NR. The relationship between the mechanical properties and cell behaviour on PLGA and PCL scaffolds for bladder tissue engineering. Biomaterials. 2009;30(7):1321–8.
6. Bawolin NK, Zhang WJ, Chen XB. A brief review of the modelling of the time dependent mechanical properties of tissue engineering scaffolds. J Biomimetics Biomater Tissue Eng. 2010;6:19–33. Trans Tech Publications.
7. Dutta RC, Dutta AK. Cell-interactive 3D-scaffold; advances and applications. Biotechnol Adv. 2009;27(4):334–9.
8. Parrag IC, Woodhouse KA. Development of biodegradable polyurethane scaffolds using amino acid and dipeptide-based chain extenders for soft tissue engineering. J Biomater Sci Polym Ed. 2010;21(6–7):843–62.
9. Heydarkhan-Hagvall S, Schenke-Layland K, Dhanasopon AP, Rofail F, Smith H, Wu BM, … MacLellan WR. Three-dimensional electrospun ECM-based hybrid scaffolds for cardiovascular tissue engineering. Biomaterials. 2008;29(19):2907–14.
10. Guan J, Fujimoto KL, Sacks MS, Wagner WR. Preparation and characterization of highly porous, biodegradable polyurethane scaffolds for soft tissue applications. Biomaterials. 2005;26(18):3961–71.
11. Courtney T, Sacks MS, Stankus J, Guan J, Wagner WR. Design and analysis of tissue engineering scaffolds that mimic soft tissue mechanical anisotropy. Biomaterials. 2006;27(19):3631–8.
12. Zhu Y, Dong Z, Wejinya UC, Jin S, Ye K. Determination of mechanical properties of soft tissue scaffolds by atomic force microscopy nanoindentation. J Biomech. 2011;44(13):2356–61.
13. Grover CN, Cameron RE, Best SM. Investigating the morphological, mechanical and degradation properties of scaffolds comprising collagen, gelatin and elastin for use in soft tissue engineering. J Mech Behav Biomed Mater. 2012;10:62–74.
14. He C, Xu X, Zhang F, Cao L, Feng W, Wang H, Mo X. Fabrication of fibrinogen/P (LLA–CL) hybrid nanofibrous scaffold for potential soft tissue engineering applications. J Biomed Mater Res A. 2011;97(3):339–47.
15. Guan J, Fujimoto KL, Wagner WR. Elastase-sensitive elastomeric scaffolds with variable anisotropy for soft tissue engineering. Pharm Res. 2008;25(10):2400–12.
16. Pektok E, Nottelet B, Tille JC, Gurny R, Kalangos A, Moeller M, Walpoth BH. Degradation and healing characteristics of small-diameter poly (ε-Caprolactone) vascular grafts in the rat systemic arterial circulation CLINICAL PERSPECTIVE. Circulation. 2008;118(24):2563–70.
17. Fung YC. Biomechanics: mechanical properties of living tissues. New York: Springer Science & Business Media; 2013.
18. Zhu J, Li J, Wang B, Zhang WJ, Zhou G, Cao Y, Liu W. The regulation of phenotype of cultured tenocytes by microgrooved surface structure. Biomaterials. 2010;31(27):6952–8.
19. De Santis R, Sarracino F, Mollica F, Netti PA, Ambrosio L, Nicolais L. Continuous fibre reinforced polymers as connective tissue replacement. Compos Sci Technol. 2004;64(6):861–71.

20. Shau YW, Wang CL, Shieh JY, Hsu TC. Noninvasive assessment of the viscoelasticity of peripheral arteries. Ultrasound Med Biol. 1999;25(9):1377–88.
21. Vesely I. The role of elastin in aortic valve mechanics. J Biomech. 1997;31(2):115–23.
22. Silver FH, Freeman JW, DeVore D. Viscoelastic properties of human skin and processed dermis. Skin Res Technol. 2001;7(1):18–23.
23. Kumar VA, Caves JM, Haller CA, Dai E, Liu L, Grainger S, Chaikof EL. Collagen-based substrates with tunable strength for soft tissue engineering. Biomater Sci. 2013;1(11):1193–202.
24. Billiet T, Vandenhaute M, Schelfhout J, Van Vlierberghe S, Dubruel P. A review of trends and limitations in hydrogel-rapid prototyping for tissue engineering. Biomaterials. 2012;33(26):6020–41.
25. Jeon JE, Vaquette C, Klein TJ, Hutmacher DW. Perspectives in multiphasic osteochondral tissue engineering. Anat Rec. 2014;297(1):26–35.
26. Darnell JE, Lodish H, Baltimore D. Molecular cell biology, vol. 2. New York: Scientific American Books; 1990.
27. Chand S. Review carbon fibers for composites. J Mater Sci. 2000;35(6):1303–13.
28. Wambua P, Ivens J, Verpoest I. Natural fibres: can they replace glass in fibre reinforced plastics? Compos Sci Technol. 2003;63(9):1259–64.
29. Sathishkumar TP, Satheeshkumar S, Naveen J. Glass fiber-reinforced polymer composites–a review. J Reinf Plast Compos. 2014;33(13):1258–75.
30. Li X, Feng Q, Jiao Y, Cui F. Collagen–based scaffolds reinforced by chitosan fibres for bone tissue engineering. Polym Int. 2005;54(7):1034–40.
31. Wang X, Yan Y, Yost MJ, Fann SA, Dong S, Li X. Nanomechanical characterization of micro/nanofiber reinforced type I collagens. J Biomed Mater Res A. 2007;83(1):130–5.
32. Li X, Feng Q. Porous poly-L-lactic acid scaffold reinforced by chitin fibers. Polym Bull. 2005;54(1):47–55.
33. Moffat KL, Kwei ASP, Spalazzi JP, Doty SB, Levine WN, Lu HH. Novel nanofiber-based scaffold for rotator cuff repair and augmentation. Tissue Eng A. 2008;15(1):115–26.
34. McGann ME, Vahdati A, Wagner DR. Methods to assess in vitro wear of articular cartilage. Proc Inst Mech Eng H J Eng Med. 2012;226(8):612–22.
35. Li L, Patil S, Steklov N, Bae W, Temple-Wong M, D'Lima DD, … Fregly BJ. Computational wear simulation of patellofemoral articular cartilage during in vitro testing. J Biomech. 2011;44(8):1507–13.
36. Mosher TJ, Dardzinski BJ, Smith MB. Human articular cartilage: influence of aging and early symptomatic degeneration on the spatial variation of T2--preliminary findings at 3 T. Radiology. 2000;214(1):259–66.
37. Burdick JA, Mauck RL, editors. Biomaterials for tissue engineering applications: a review of the past and future trends. Wien: Springer; 2010.
38. Brown TD, Pope DF, Hale JE, Buckwalter JA, Brand RA. Effects of osteochondral defect size on cartilage contact stress. J Orthop Res. 1991;9(4):559–67.
39. Li XM, Feng QL. Dynamic rheological behaviors of the bone scaffold reinforced by chitin fibres. Mat Sci Forum. 2005;475:2387–90.
40. Slivka MA, Leatherbury NC, Kieswetter K, Niederauer GG. Porous, resorbable, fiber-reinforced scaffolds tailored for articular cartilage repair. Tissue Eng. 2001;7(6):767–80.
41. Middleton JC, Tipton AJ. Synthetic biodegradable polymers as orthopedic devices. Biomaterials. 2000;21(23):2335–46.
42. Sakata R, Iwakura T, Reddi AH. Regeneration of articular cartilage surface: morphogens, cells, and extracellular matrix scaffolds. Tissue Eng B Rev. 2015;21(5):461–73.
43. Lee J, Choi YJ, Kim CH. Articular cartilage repair with tissue-engineered hyaline cartilage reconstructed in a chitosan-hyaluronic acid scaffold by costal chondrocytes. Tissue Eng Regen Med. 2011;8(5):446–53.
44. Singhal AR, Agrawal CM, Athanasiou KA. Salient degradation features of a 50: 50 PLA/PGA scaffold for tissue engineering. Tissue Eng. 1996;2(3):197–207.

45. Thompson DE, Agrawal CM, Athanasiou K. The effects of dynamic compressive loading on biodegradable implants of 50–50% polylactic acid–polyglycolic acid. Tissue Eng. 1996;2(1):61–74.
46. Agrawal A, Rahbar N, Calvert PD. Strong fiber-reinforced hydrogel. Acta Biomater. 2013;9(2):5313–8.
47. Visser J, Melchels FP, Jeon JE, Van Bussel EM, Kimpton LS, Byrne HM, … Malda J. Reinforcement of hydrogels using three-dimensionally printed microfibres. Nat Commun. 2015;6:6933.
48. Brown TD, Dalton PD, Hutmacher DW. Direct writing by way of melt electrospinning. Adv Mater. 2011;23(47):5651–7.
49. Boere KW, Visser J, Seyednejad H, Rahimian S, Gawlitta D, Van Steenbergen MJ, … Malda J. Covalent attachment of a three-dimensionally printed thermoplast to a gelatin hydrogel for mechanically enhanced cartilage constructs. Acta Biomater. 2014;10(6):2602–11.
50. Yodmuang S, McNamara SL, Nover AB, Mandal BB, Agarwal M, Kelly TAN, … Vunjak-Novakovic G. Silk microfiber-reinforced silk hydrogel composites for functional cartilage tissue repair. Acta Biomater. 2015;11:27–36.
51. Seifzadeh A, Oguamanam DCD, Trutiak N, Hurtig M, Papini M. Determination of nonlinear fibre-reinforced biphasic poroviscoelastic constitutive parameters of articular cartilage using stress relaxation indentation testing and an optimizing finite element analysis. Comput Methods Prog Biomed. 2012;107(2):315–26.
52. de Pascual-Teresa S, Moreno DA, García-Viguera C. Flavanols and anthocyanins in cardiovascular health: a review of current evidence. Int J Mol Sci. 2010;11(4):1679–703.
53. Szabo SM, Levy AR, Rao SR. Increased risk of cardiovascular and cerebrovascular diseases in individuals with ankylosing spondylitis: a population-based study. Arthritis Rheum. 2011;63(11):3294–304.
54. Blanco PJ, Queiroz RA, Feijoo RA. A computational approach to generate concurrent arterial networks in vascular territories. Int J Num Method Biomed Eng. 2013;29(5):601–14.
55. Jurczuk K, Kretowski M, Eliat PA, Saint-Jalmes H, Bezy-Wendling J. In silico modeling of magnetic resonance flow imaging in complex vascular networks. IEEE Trans Med Imaging. 2014;33(11):2191–209.
56. Walles T, Giere B, Hofmann M, Schanz J, Hofmann F, Mertsching H, Macchiarini P. Experimental generation of a tissue-engineered functional and vascularized trachea. J Thorac Cardiovasc Surg. 2004;128(6):900–6.
57. Oh HH, Lu H, Kawazoe N, Chen G. Spatially guided angiogenesis by three-dimensional collagen scaffolds micropatterned with vascular endothelial growth factor. J Biomater Sci Polym Ed. 2012;23(17):2185–95.
58. Zhang B, Montgomery M, Chamberlain MD, et al. Biodegradable scaffold with built-in vasculature for organ-on-a-chip engineering and direct surgical anastomosis. Nat Mater. 2016;15(6):669–78.
59. Pooyan P, Tannenbaum R, Garmestani H. Mechanical behavior of a cellulose-reinforced scaffold in vascular tissue engineering. J Mech Behav Biomed Mater. 2012;7:50–9.
60. Weinberg CB, Bell E. A blood vessel model constructed from collagen and cultured vascular cells. Science. 1986;231:397–401.
61. Ercolani E, Del Gaudio C, Bianco A. Vascular tissue engineering of small-diameter blood vessels: reviewing the electrospinning approach. J Tissue Eng Regen Med. 2015;9(8):861–88.
62. Yao L, Liu J, Andreadis ST. Composite fibrin scaffolds increase mechanical strength and preserve contractility of tissue engineered blood vessels. Pharm Res. 2008;25(5):1212–21.
63. Liu JY, Swartz DD, Peng HF, Gugino SF, Russell JA, Andreadis ST. Functional tissue-engineered blood vessels from bone marrow progenitor cells. Cardiovasc Res. 2007;75(3):618–28.
64. Soletti L, Hong Y, Guan J, Stankus JJ, El-Kurdi MS, Wagner WR, Vorp DA. A bilayered elastomeric scaffold for tissue engineering of small diameter vascular grafts. Acta Biomater. 2010;6(1):110–22.

65. Stekelenburg M, Rutten MC, Snoeckx LH, Baaijens FP. Dynamic straining combined with fibrin gel cell seeding improves strength of tissue-engineered small-diameter vascular grafts. Tissue Eng Part A. 2008;15(5):1081–9.
66. Zhang L, Zhou J, Lu Q, Wei Y, Hu S. A novel small–diameter vascular graft: in vivo behavior of biodegradable three–layered tubular scaffolds. Biotechnol Bioeng. 2008;99(4):1007–15.
67. He W, Ma Z, Teo WE, Dong YX, Robless PA, Lim TC, Ramakrishna S. Tubular nanofiber scaffolds for tissue engineered small–diameter vascular grafts. J Biomed Mater Res A. 2009;90(1):205–16.
68. Gui L, Muto A, Chan SA, Breuer CK, Niklason LE. Development of decellularized human umbilical arteries as small-diameter vascular grafts. Tissue Eng A. 2009;15(9):2665–76.
69. Caves JM, Kumar VA, Martinez AW, Kim J, Ripberger CM, Haller CA, Chaikof EL. The use of microfiber composites of elastin-like protein matrix reinforced with synthetic collagen in the design of vascular grafts. Biomaterials. 2010;31(27):7175–82.
70. Caves JM, Kumar VA, Wen J, Cui W, Martinez A, Apkarian R, … Chaikof EL. Fibrillogenesis in continuously spun synthetic collagen fiber. J Biomed Mater Res B Appl Biomater. 2010;93(1):24–38.
71. Caves JM, Kumar VA, Xu W, Naik N, Allen MG, Chaikof EL. Microcrimped collagen fiber-elastin composites. Adv Mater. 2010;22(18):2041–4.
72. Hubbe MA, Rojas OJ, Lucia LA, Sain M. Cellulosic nanocomposites: a review. Bioresources. 2008;3(3):929–80.
73. Azizi Samir MAS, Alloin F, Dufresne A. Review of recent research into cellulosic whiskers, their properties and their application in nanocomposite field. Biomacromolecules. 2005;6(2):612–26.
74. Czaja W, Krystynowicz A, Bielecki S, Brown RM. Microbial cellulose—the natural power to heal wounds. Biomaterials. 2006;27(2):145–51.
75. Klemm D, Schumann D, Udhardt U, Marsch S. Bacterial synthesized cellulose—artificial blood vessels for microsurgery. Prog Polym Sci. 2001;26(9):1561–603.
76. Allen RA, Wu W, Yao M, Dutta D, Duan X, Bachman TN, … Isenberg JS. Nerve regeneration and elastin formation within poly (glycerol sebacate)-based synthetic arterial grafts one-year post-implantation in a rat model. Biomaterials. 2014;35(1):165–73.
77. Wu W, Allen RA, Wang Y. Fast-degrading elastomer enables rapid remodeling of a cell-free synthetic graft into a neoartery. Nat Med. 2012;18(7):1148–53.
78. Li L, Terry CM, Shiu YTE, Cheung AK. Neointimal hyperplasia associated with synthetic hemodialysis grafts. Kidney Int. 2008;74(10):1247–61.
79. De VS, Tille JC, Mugnai D, Mrowczynski W, Gurny R, Möller M, et al. Long term performance of polycaprolactone vascular grafts in a rat abdominal aorta replacement model. Biomaterials. 2012;33(1):38–47.
80. Mendelson K, Schoen FJ. Heart valve tissue engineering: concepts, approaches, progress, and challenges. Ann Biomed Eng. 2006;34(12):1799–819.
81. He X, Xiao Q, Lu C, Wang Y, Zhang X, Zhao J, … Deng Y. Uniaxially aligned electrospun all-cellulose nanocomposite nanofibers reinforced with cellulose nanocrystals: scaffold for tissue engineering. Biomacromolecules. 2014;15(2):618–27.
82. Liu Y, Zahedmanesh H, Lally C, Cahill PA, McGuinness GB. Compliance properties of a composite electrospun fibre–hydrogel blood vessel scaffold. Mater Lett. 2016;178:296–9.
83. Balañá ME, Charreau HE, Leirós GJ. Epidermal stem cells and skin tissue engineering in hair follicle regeneration. World J Stem Cells. 2015;7(4):711-27.
84. Ohyama M, Zheng Y, Paus R, Stenn KS. The mesenchymal component of hair follicle neogenesis: background, methods and molecular characterization. Exp Dermatol. 2010;19(2):89–99.
85. Shevchenko RV, James SL, James SE. A review of tissue-engineered skin bioconstructs available for skin reconstruction. J R Soc Interface. 2009;7(43):229–58.
86. Festa E, Fretz J, Berry R, Schmidt B, Rodeheffer M, Horowitz M, Horsley V. Adipocyte lineage cells contribute to the skin stem cell niche to drive hair cycling. Cell. 2011;146(5):761–71.

87. Itoh M, Hiraoka Y, Kataoka K, Huh NH, Tabata Y, Okochi H. Novel collagen sponge reinforced with polyglycolic acid fiber produces robust, normal hair in murine hair reconstitution model. Tissue Eng. 2004;10(5–6):818–24.
88. Stenn KS, Paus R. Controls of hair follicle cycling. Physiol Rev. 2001;81(1):449–94.
89. Millar SE. Molecular mechanisms regulating hair follicle development. J Investig Dermatol. 2002;118(2):216–25.
90. Yannas IV, Burke JF. Design of an artificial skin. I. Basic design principles. J Biomed Mater Res. 1980;14(1):65–81.
91. Bell E, Ehrlich HP, Buttle DJ, Nakatsuji T. Living tissue formed in vitro and accepted as skin-equivalent tissue of full thickness. Science. 1981;211(4486):1052–4.
92. Maruguchi T, Maruguchi Y, Suzuki S, Matsuda K, Toda KI, Isshiki N. A new skin equivalent: keratinocytes proliferated and differentiated on collagen sponge containing fibroblasts. Plast Reconstr Surg. 1994;93(3):537–44.
93. Lundborg G. A 25-year perspective of peripheral nerve surgery: evolving neuroscientific concepts and clinical significance. J Hand Surg. 2000;25(3):391–414.
94. Cao J, Xiao Z, Jin W, Chen B, Meng D, Ding W, … Wang J. Induction of rat facial nerve regeneration by functional collagen scaffolds. Biomaterials. 2013;34(4):1302–10.
95. Hung V, Dellon AL. Reconstruction of a 4-cm human median nerve gap by including an autogenous nerve slice in a bioabsorbable nerve conduit: case report. J Hand Surg. 2008;33(3):313–5.
96. Battiston B, Tos P, Conforti LG, Geuna S. Alternative techniques for peripheral nerve repair: conduits and end-to-side neurorrhaphy. Acta Neurochir Suppl. 2007;100:43–50.
97. Wen X, Tresco PA. Fabrication and characterization of permeable degradable poly (DL-lactide-co-glycolide)(PLGA) hollow fiber phase inversion membranes for use as nerve tract guidance channels. Biomaterials. 2006;27(20):3800–9.
98. Lu MC, Huang YT, Lin JH, Yao CH, Lou CW, Tsai CC, Chen YS. Evaluation of a multi-layer microbraided polylactic acid fiber-reinforced conduit for peripheral nerve regeneration. J Mater Sci Mater Med. 2009;20(5):1175–80.
99. Ciardelli G, Chiono V. Materials for peripheral nerve regeneration. Macromol Biosci. 2006;6(1):13–26.
100. Wang A, Ao Q, Wei Y, Gong K, Liu X, Zhao N, … Zhang X. Physical properties and biocompatibility of a porous chitosan-based fiber-reinforced conduit for nerve regeneration. Biotechnol Lett. 2007;29(11):1697–702.
101. Kim YT, Haftel VK, Kumar S, Bellamkonda RV. The role of aligned polymer fiber-based constructs in the bridging of long peripheral nerve gaps. Biomaterials. 2008;29(21):3117–27.
102. Bini TB, Gao S, Wang S, Lim A, Hai LB, Ramakrishna S. Electrospun poly (L-lactide-co-glycolide) biodegradable polymer nanofibre tubes for peripheral nerve regeneration. Nanotechnology. 2004;15(11):1459–64.
103. Cui FZ, Li Y, Ge J. Self-assembly of mineralized collagen composites. Mater Sci Eng R Rep. 2007;57(1):1–27.
104. Gao H, Yang Y, Zhang G, Sha Z, Shen Y. The use of fiber-reinforced scaffolds cocultured with Schwann cells and vascular endothelial cells to repair rabbit sciatic nerve defect with vascularization. Biomed Res Int. 2013;2013(4):362918.
105. Li X, Feng Q, Liu X, Dong W, Cui F. Collagen-based implants reinforced by chitin fibres in a goat shank bone defect model. Biomaterials. 2006;27(9):1917–23.
106. Kang M, Chen P, Jin HJ. Preparation of multiwalled carbon nanotubes incorporated silk fibroin nanofibers by electrospinning. Curr Appl Phys. 2009;9(1):S95–7.
107. Gupta P, Sharan S, Roy P, Lahiri D. Aligned carbon nanotube reinforced polymeric scaffolds with electrical cues for neural tissue regeneration. Carbon. 2015;95:715–24.
108. Wang X, Li Q, Xie J, Jin Z, Wang J, Li Y, … Fan S. Fabrication of ultralong and electrically uniform single-walled carbon nanotubes on clean substrates. Nano Lett. 2009;9(9):3137–41.
109. Saito R, Dresselhaus GDS. Physical properties of carbon nanotubes. London: Imperial College Press; 2003. p. 623–30.

110. Demczyk BG, Wang YM, Cumings J, Hetman M, Han W, Zettl A, Ritchie RO. Direct mechanical measurement of the tensile strength and elastic modulus of multiwalled carbon nanotubes. Mater Sci Eng A. 2002;334(1):173–8.
111. Boccaccini AR, Kaya C, Shaffer MSP. Electrophoretic deposition of carbon nanotubes (CNTs) and CNT/nanoparticle composites. In: Electrophoretic deposition of nanomaterials; 2012. p. 157–79.
112. Thostenson ET, Ren Z, Chou TW. Advances in the science and technology of carbon nanotubes and their composites: a review. Compos Sci Technol. 2001;61(13):1899–912.
113. Allaoui A, Bai S, Cheng HM, Bai JB. Mechanical and electrical properties of a MWNT/epoxy composite. Compos Sci Technol. 2002;62(15):1993–8.
114. Shin SR, Bae H, Cha JM, Mun JY, Chen YC, Tekin H, … Khademhosseini A. Carbon nanotube reinforced hybrid microgels as scaffold materials for cell encapsulation. ACS Nano. 2011;6(1):362–72.
115. Hu H, Ni Y, Mandal SK, Montana V, Zhao B, Haddon RC, Parpura V. Polyethyleneimine functionalized single-walled carbon nanotubes as a substrate for neuronal growth. J Phys Chem B. 2005;109(10):4285–9.
116. Mattson MP, Haddon RC, Rao AM. Molecular functionalization of carbon nanotubes and use as substrates for neuronal growth. J Mol Neurosci. 2000;14(3):175–82.
117. Butler DL, Dressler M, Awad H. Functional tissue engineering: assessment of function in tendon and ligament repair. In: Functional tissue engineering. New York: Springer; 2003. p. 1071–5.
118. Woo SLY, Abramowitch SD, Kilger R, Liang R. Biomechanics of knee ligaments: injury, healing, and repair. J Biomech. 2006;39(1):1–20.
119. Pedowitz RA, O'Connor JJ, Akeson WH, editors. Daniel's knee injuries: ligament and cartilage structure, function, injury, and repair. Philadelphia, PA: Lippincott Williams & Wilkins; 2003. p. 185–201.
120. Laurencin CT, Freeman JW. Ligament tissue engineering: an evolutionary materials science approach. Biomaterials. 2005;26(36):7530–6.
121. Maffulli N, Ajis A, Longo UG, Denaro V. Chronic rupture of tendo Achillis. Foot Ankle Clin. 2007;12(4):583–96.
122. Maffulli N, Longo UG, Gougoulias N, Denaro V. Ipsilateral free semitendinosus tendon graft transfer for reconstruction of chronic tears of the Achilles tendon. BMC Musculoskelet Disord. 2008;9(1):100.
123. Chen J, Xu J, Wang A, Zheng M. Scaffolds for tendon and ligament repair: review of the efficacy of commercial products. Exp Rev Med Devices. 2009;6(1):61–73.
124. Wang JHC. Mechanobiology of tendon. J Biomech. 2006;39(9):1563–82.
125. Vunjak-Novakovic G, Altman G, Horan R, Kaplan DL. Tissue engineering of ligaments. Annu Rev Biomed Eng. 2004;6:131–56.
126. Petrigliano FA, McAllister DR, Wu BM. Tissue engineering for anterior cruciate ligament reconstruction: a review of current strategies. Arthroscopy J Arthrosc Relat Surg. 2006;22(4):441–51.
127. Shepherd JH, Ghose S, Kew SJ, Moavenian A, Best SM, Cameron RE. Effect of fiber crosslinking on collagen–fiber reinforced collagen–chondroitin-6-sulfate materials for regenerating load–bearing soft tissues. J Biomed Mat Res Part A. 2013;101(1):176–84.
128. Webb WR, Dale TP, Lomas AJ, Zeng G, Wimpenny I, El Haj AJ, … Chen GQ. The application of poly (3-hydroxybutyrate-co-3-hydroxyhexanoate) scaffolds for tendon repair in the rat model. Biomaterials. 2013;34(28):6683–94.
129. Yang C, Deng G, Chen W, Ye X, Mo X. A novel electrospun-aligned nanoyarn-reinforced nanofibrous scaffold for tendon tissue engineering. Colloids Surf B: Biointerfaces. 2014;122:270–6.
130. Hakimi O, Mouthuy PA, Zargar N, Lostis E, Morrey M, Carr A. A layered electrospun and woven surgical scaffold to enhance endogenous tendon repair. Acta Biomater. 2015;26:124–35.

131. Hakimi O, Murphy R, Stachewicz U, Hislop S, Carr AJ. An electrospun polydioxanone patch for the localisation of biological therapies during tendon repair. Eur Cell Mater. 2012;24:344–57.
132. Martin BI, Deyo RA, Mirza SK, Turner JA, Comstock BA, Hollingworth W, Sullivan SD. Expenditures and health status among adults with back and neck problems. JAMA. 2008;299(6):656–64.
133. Raj PP. Intervertebral disc: anatomy–physiology–pathophysiology–treatment. Pain Pract. 2008;8(1):18–44.
134. Strange DG, Tonsomboon K, Oyen ML. Mechanical behaviour of electrospun fibre-reinforced hydrogels. J Mater Sci Mater Med. 2014;25(3):681–90.
135. Hudson KD, Alimi M, Grunert P, Härtl R, Bonassar LJ. Recent advances in biological therapies for disc degeneration: tissue engineering of the annulus fibrosus, nucleus pulposus and whole intervertebral discs. Curr Opin Biotechnol. 2013;24(5):872–9.
136. Kandel R, Roberts S, Urban JP. Tissue engineering and the intervertebral disc: the challenges. Eur Spine J. 2008;17(4):480–91.
137. O'Halloran DM, Pandit AS. Tissue-engineering approach to regenerating the intervertebral disc. Tissue Eng. 2007;13(8):1927–54.
138. Koepsell L, Zhang L, Neufeld D, Hao F, Deng Y. Electrospun nanofibrous polycaprolactone scaffolds for tissue engineering of annulus fibrosus. Macromol Biosci. 2011;11(3):391–9.
139. Koepsell L, Zhang L, Neufeld D, Fong H, Deng Y. Electrospun nanofibrous polycaprolactone scaffolds for tissue engineering of annulus fibrosus. Macromol Biosci. 2011;11(3):391–9.
140. Nathan AS, Baker BM, Nerurkar NL, Mauck RL. Mechano-topographic modulation of stem cell nuclear shape on nanofibrous scaffolds. Acta Biomater. 2011;7(1):57–66.
141. Kretlow JD, Klouda L, Mikos AG. Injectable matrices and scaffolds for drug delivery in tissue engineering. Adv Drug Deliv Rev. 2007;59(4):263–73.
142. Ambrosio L, Netti PA, Iannace S, Huang SJ, Nicolais L. Composite hydrogels for intervertebral disc prostheses. J Mater Sci Mater Med. 1996;7(5):251–4.
143. Nerurkar NL, Baker BM, Sen S, Wible EE, Elliott DM, Mauck RL. Nanofibrous biologic laminates replicate the form and function of the annulus fibrosus. Nat Mater. 2009;8(12):986–92.
144. Nerurkar NL, Sen S, Huang AH, Elliott DM, Mauck RL. Engineered disc-like angle-ply structures for intervertebral disc replacement. Spine. 2010;35(8):867–73.
145. Thorvaldsson A, Silva-Correia J, Oliveira JM, Reis RL, Gatenholm P, Walkenström P. Development of nanofiber–reinforced hydrogel scaffolds for nucleus pulposus regeneration by a combination of electrospinning and spraying technique. J Appl Polym Sci. 2013;128(2):1158–63.
146. Xu W, Ma J, Jabbari E. Material properties and osteogenic differentiation of marrow stromal cells on fiber-reinforced laminated hydrogel nanocomposites. Acta Biomater. 2010;6(6):1992–2002.
147. Kahn S A, Beers R J, Lentz C W. Use of acellular dermal replacement in reconstruction of nonhealing lower extremity wounds. Journal of Burn Care & Research Official Publication of the American Burn Association. 2011;32(1):124-8.
148. Priya SG, Jungvid H, Kumar A. Skin tissue engineering for tissue repair and regeneration. Tissue Eng B Rev. 2008;14(1):105–18.
149. Ruszczak Z. Effect of collagen matrices on dermal wound healing. Adv Drug Deliv Rev. 2003;55(12):1595–611.
150. Rnjak-Kovacina J, Wise SG, Li Z, Maitz PK, Young CJ, Wang Y, Weiss AS. Electrospun synthetic human elastin: collagen composite scaffolds for dermal tissue engineering. Acta Biomater. 2012;8(10):3714–22.
151. Dikovsky D, Bianco-Peled H, Seliktar D. Defining the role of matrix compliance and proteolysis in three-dimensional cell spreading and remodeling. Biophys J. 2008;94(7):2914–25.
152. Engler AJ, Sen S, Sweeney HL, Discher DE. Matrix elasticity directs stem cell lineage specification. Cell. 2006;126(4):677–89.

153. Albanna MZ, Bou-Akl TH, Walters HL, Matthew HW. Improving the mechanical properties of chitosan-based heart valve scaffolds using chitosan fibers. J Mech Behav Biomed Mater. 2012;5(1):171–80.
154. McCullen SD, Haslauer CM, Loboa EG. Fiber-reinforced scaffolds for tissue engineering and regenerative medicine: use of traditional textile substrates to nanofibrous arrays. J Mater Chem. 2010;20(40):8776–88.
155. Shepherd JH, Ghose S, Kew SJ, Moavenian A, Best SM, Cameron RE. Effect of fiber cross-linking on collagen–fiber reinforced collagen–chondroitin–6–sulfate materials for regenerating load–bearing soft tissues. J Biomed Mater Res A. 2013;101(1):176–84.
156. Han F, Liu S, Liu X, Pei Y, Bai S, Zhao H, Zhu H. Woven silk fabric-reinforced silk nanofibrous scaffolds for regenerating load-bearing soft tissues. Acta Biomater. 2014;10(2):921–30.
157. Cai N, Dai Q, Wang Z, Luo X, Xue Y, Yu F. Toughening of electrospun poly (L-lactic acid) nanofiber scaffolds with unidirectionally aligned halloysite nanotubes. J Mater Sci. 2015;50(3):1435–45.
158. Ionita M, Pandele MA, Iovu H. Sodium alginate/graphene oxide composite films with enhanced thermal and mechanical properties. Carbohydr Polym. 2013;94(1):339–44.
159. Liu M, Zhang Y, Wu C, Xiong S, Zhou C. Chitosan/halloysite nanotubes bionanocomposites: structure, mechanical properties and biocompatibility. Int J Biol Macromol. 2012;51(4):566–75.
160. Panda A, Vanathi M, Kumar A, Dash Y, Priya S. Corneal graft rejection. Surv Ophthalmol. 2007;52(4):375–96.
161. Polisetti N, Islam MM, Griffith M. The artificial Cornea.Corneal Regenerative Medicine. Humana Press. 2013:45-52.
162. Feder RS, Kshettry P. Noninflammatory ectatic disorders. Cornea. 2005;1:955–74.
163. Snibson GR. Collagen cross-linking: a new treatment paradigm in corneal disease - a review. Clin Experiment Ophthalmol. 2010;38(2):141–53.
164. Huang YX, Li QH. An active artificial cornea with the function of inducing new corneal tissue generation in vivo-a new approach to corneal tissue engineering. Biomed Mater. 2007;2(3):S121–5.
165. Chandra SR. Global blindness: VISION 2020: the right to sight. Arch Ophthalmol. 2008;126(10):1457.
166. Ionescu AM, Alaminos M, de la Cruz Cardona J, Durán JDDGL, González-Andrades M, Ghinea R, del Mar Pérez M. Investigating a novel nanostructured fibrin–agarose biomaterial for human cornea tissue engineering: Rheological properties. J Mech Behav Biomed Mater. 2011;4(8):1963–73.
167. Ruberti JW, Zieske JD. Prelude to corneal tissue engineering–gaining control of collagen organization. Prog Retin Eye Res. 2008;27(5):549–77.
168. Liu Z, Ji J, Zhang J, Huang C, Meng Z, Qiu W, … Wang W. Corneal reinforcement using an acellular dermal matrix for an analysis of biocompatibility, mechanical properties, and transparency. Acta Biomater. 2012;8(9):3326–32.
169. Duan X, Sheardown H. Dendrimer crosslinked collagen as a corneal tissue engineering scaffold: mechanical properties and corneal epithelial cell interactions. Biomaterials. 2006;27(26):4608–17.
170. Meek KM, Boote C. The organization of collagen in the corneal stroma. Exp Eye Res. 2004;78(3):503–12.
171. Connon CJ, Meek KM, Kinoshita S, Quantock AJ. Spatial and temporal alterations in the collagen fibrillar array during the onset of transparency in the avian cornea. Exp Eye Res. 2004;78(5):909–15.
172. Hassell JR, Birk DE. The molecular basis of corneal transparency. Exp Eye Res. 2010;91(3):326–35.
173. Wu J, Du Y, Watkins SC, Funderburgh JL, Wagner WR. The engineering of organized human corneal tissue through the spatial guidance of corneal stromal stem cells. Biomaterials. 2012;33(5):1343–52.

174. Wilson SL, Wimpenny I, Ahearne M, Rauz S, El Haj AJ, Yang Y. Chemical and topographical effects on cell differentiation and matrix elasticity in a corneal stromal layer model. Adv Funct Mater. 2012;22(17):3641–9.
175. Cartwright NEK, Tyrer JR, Marshall J. Age-related differences in the elasticity of the human cornea. Invest Ophthalmol Vis Sci. 2011;52(7):4324–9.
176. Elsheikh A, Alhasso D, Rama P. Biomechanical properties of human and porcine corneas. Exp Eye Res. 2008;86(5):783–90.
177. Long K, Liu Y, Li W, Wang L, Liu S, Wang Y, Ren L. Improving the mechanical properties of collagen-based membranes using silk fibroin for corneal tissue engineering. J Biomed Mater Res A. 2015;103(3):1159–68.
178. Ahn JI, Kuffova L, Merrett K, Mitra D, Forrester JV, Li F, Griffith M. Crosslinked collagen hydrogels as corneal implants: effects of sterically bulky vs. non-bulky carbodiimides as crosslinkers. Acta Biomater. 2013;9(8):7796–805.
179. Saeidi N, Karmelek KP, Paten JA, Zareian R, DiMasi E, Ruberti JW. Molecular crowding of collagen: a pathway to produce highly-organized collagenous structures. Biomaterials. 2012;33(30):7366–74.
180. Tonsomboon K, Strange DGT, Oyen ML. Gelatin nanofiber-reinforced alginate gel scaffolds for corneal tissue engineering. In: Engineering in medicine and biology society (EMBC). IEEE; 2013. p. 6671–4.